Power Rules

JOHN S. DUFFIELD

Power Rules

THE EVOLUTION OF
NATO'S CONVENTIONAL
FORCE POSTURE

STANFORD UNIVERSITY PRESS

STANFORD, CALIFORNIA

Stanford University Press
Stanford, California
© 1995 by the Board of Trustees of the
Leland Stanford Junior University
Printed in the United States of America

CIP data appear at the end of the book

Stanford University Press publications are
distributed exclusively by Stanford
University Press within the United States,
Canada, and Mexico; they are distributed
exclusively by Cambridge University Press
throughout the rest of the world.

To Cheryl

Acknowledgments

Numerous individuals and institutions have contributed to the realization of this project. I owe a substantial intellectual debt to professors Richard H. Ullman and Stephen M. Walt, whose wise guidance and assistance were indispensable during the project's crucial early stages. They took considerable time to read drafts of each chapter and to write lengthy comments, providing incisive criticism, useful suggestions, and encouragement in equal measure. Throughout a long and often demanding process, they have remained unstinting in their support.

I would also like to thank the other teachers, friends, and colleagues who read all or parts of the manuscript at various stages and offered valuable comments. These include Aaron Friedberg, Frank von Hippel, Richard Immerman, Lawrence Kaplan, Richard Kugler, Melvyn Leffler, David Petraeus, and Edward Rhodes. Their help greatly improved the final product.

As any young scholar knows well, the only resource more scarce than money is time. In this regard, I gratefully acknowledge the generous financial support provided by the Woodrow Wilson School and the Center of International Studies at Princeton University, the U.S. Arms Control and Disarmament Agency, the Center for International Studies at the University of Southern California, and the Center for Science and International Affairs at Harvard University, which enabled me to devote full time to the project for several years. I also received vital administrative and research support from the Brookings Institution and the University of Virginia as well as the aforementioned universities and members of their staffs.

Writing this book involved discovering and then digesting large amounts of primary source material. I am particularly indebted to Jane E. Stromseth and Robert A. Wampler for making available their own carefully researched manuscripts, for sharing documents, and for directing me toward rich veins of additional material. I was also greatly aided in this task by archivists at the National Archives, the John F. Kennedy Library, the Lyndon B. Johnson Library, the Dwight D. Eisenhower Library, the Seely G. Mudd Library at Princeton University, the National Security Archive in Washington, D.C., and the British Public Record Office. Both the Kennedy and Johnson library foundations provided grants that defrayed the costs of research trips to those institutions. A number of active and retired government officials generously consented to be interviewed.

Throughout this project I have been blessed by caring and supportive friends and family members, especially my parents, Richard and Mary Rose Duffield, and my parents-in-law, Joseph and Mary Ann Eschbach. Most important of all has been the companionship of my wife, Cheryl Eschbach. Through the example of her own scholarship, she demonstrated that the many challenges I encountered along the way could be overcome. And by broadening my horizons, she helped me keep my work in perspective. Her love and support made it possible for me to write this book; her presence made the effort worthwhile.

J.S.D.

Contents

x   *Contents*

Tables

Power Rules

Introduction

Since 1989, the European security environment has undergone a profound transformation. The overthrow of communist regimes throughout Eastern Europe, the resulting dissolution of the Warsaw Pact, and the breakup of the Soviet Union have relegated the long-standing East-West military confrontation on the continent to the history books. Intense political and ideological competition has been replaced by unprecedented levels of cooperation as the fledgling democracies of the former Soviet bloc seek to join the West. At the same time, new security risks have emerged in the East in the form of domestic political and economic instability, violent ethnic conflict, and nuclear proliferation.

Throughout these momentous developments, the North Atlantic Treaty Organization (NATO) has remained a prominent fixture on the European security landscape. Although born of the Cold War, the alliance has been adapted to address many of the new security challenges faced by its member states. Moreover, many of NATO's established institutional features, such as its elaborate consultative organs and its integrated military structure, have been retained. In particular, the maintenance of an effective, multinational conventional capability has been regarded as critical for the achievement of many of the alliance's post–Cold War security goals.

This book provides a detailed description of the evolution of NATO's conventional force posture from the beginning of the alliance through the dramatic events of the early 1990s. It answers the following questions: What roles has NATO military strategy assigned to conventional forces? How large have the alliance's conventional capabilities actually been, especially in the Central Region of Europe? When and in what ways has NATO's conventional force posture changed over the years? A second and related purpose of the book is to explain this history. How did NATO's conventional force posture come to be what it is? What factors have been most influential in shaping alliance strategy and conventional force structure?

NATO's conventional force posture in the Central Region, the area between the North Sea and the Alps, has constituted one of the most important aspects of the alliance's military arrangements. Although the North Atlantic Treaty commits all parties to assist one another in the event of an attack without regard to geographical location, alliance political and military leaders have considered the maintenance of the political independence of the countries in and immediately adjacent to the region, especially Germany, France, and Britain, especially vital to the well-being of the alliance as a whole. At the same time, this area has been highly exposed to possible Soviet political intimidation, military threats, and even invasion.

Not surprisingly, the Central Region has hosted the highest concentration of NATO forces. Each of the four leading allied powers—the United States, Britain, France, and Germany—has made its primary conventional force contribution to the alliance there, and allied military preparations have been most interdependent in the region. More than half a dozen countries have typically stationed forces in West Germany at any given time.

Even with the end of the Cold War, NATO's conventional force posture remains highly relevant. The alliance's institutional arrangements survived the collapse of Soviet power largely intact, and they are not likely to disappear soon. Although the threat that was primarily responsible for bringing NATO into existence has largely subsided, other purposes served by the alliance have not. NATO has even acquired new political-military roles, such as helping to stabilize the countries of Central and Eastern Europe. And though the competing concepts of a West European defense identity and a European collective security system have received considerable attention, the events of the early 1990s suggest that progress toward their realization will be slow. Thus NATO is likely to remain the premier security organization on the continent for the foreseeable future, even as the details of its military posture continue to evolve, and the alliance's conventional forces will assume greater relative importance as the need for nuclear deterrence recedes.

Historical Contributions

Despite its importance, NATO's conventional force posture has been largely neglected by scholars. No comprehensive account of the subject has been written, notwithstanding the existence of a considerable body of literature on NATO. General surveys of the alliance have dealt primarily with political relations among its members and have paid relatively little attention to its military details.[1] Many of the works devoted to NATO military issues have focused on the alliance's nuclear posture.[2] And those studies that have concentrated on NATO's conventional force posture have either covered only a limited time period, usually during the early years of the alliance,[3] or addressed only a narrow aspect of the subject.[4] Finally, much of the literature that appeared in the 1980s consisted of contemporary assessments of the conventional balance in Europe and prescriptions for correcting perceived NATO deficiencies.[5]

One reason for this neglect is the complexity of the topic. NATO's conventional force posture has comprehended the activities of numerous countries and a variety of forces. In addition, a veil of secrecy has obscured many aspects of the subject for most of its history. Since the late 1970s, however, a large number of previously classified official documents concerning NATO's conventional force posture have become available to the public. Beyond providing additional details, the new material repudiates much of the conventional wisdom that has come to surround the alliance.

Based largely on these recently declassified documents, this book revises the standard account of NATO history in several ways. First, it attaches less importance to the adoption of the strategy of flexible response in 1967. It establishes that the shift away from massive retaliation and the incorporation of a significant degree of flexibility into NATO strategy began a decade before, much earlier than is suggested by most accounts. More generally, the book shows that the role of conventional forces in NATO strategy has varied more often than is commonly thought. Even before the end of the Cold War in Europe, alliance military strategy had passed through at least four distinct phases. The frequency with which NATO strategy was revised declined over time, however, and from the late 1950s on, the alliance's formal strategic concept was ambiguous.

This book also reveals that there has been considerable continuity in the alliance's conventional force requirements and actual force levels. The requirement for ready forces in the Central Region underwent almost no modification from the early 1950s until the 1960s, when formal force requirements were last determined, despite several changes in strategy. The Eisenhower administration's "New Look," which sought to place greater emphasis on nuclear weapons, had little direct impact on U.S. and allied conventional capabilities in Europe. And although conventional force levels in the Central Region varied dramatically in the 1950s for other reasons, they exhibited remarkable stability during the following three decades.

Theoretical Contributions

Beyond amplifying and, where necessary, correcting the historical record, this book also seeks to explain these patterns in terms of international relations theory and thus to help fill a significant void in the theoretical literature on alliances. Numerous works have addressed the formation and disintegration of alliances,[6] and noteworthy progress has been made in identifying the forces that determine when alliances form and fall apart.[7] In contrast, the behavior of alliances, especially the military postures they are likely to adopt, has received comparatively little theoretical attention.[8]

In addition to explaining the particular case of NATO's conventional force posture, this application of international relations theory serves two more general purposes. First, it offers an analytical framework that may be useful for understanding the behavior of other alliances as well. Although NATO has achieved a unique level of security cooperation, it is not the only military alliance to feature a significant degree of coordinated military planning and preparation in peacetime or in periods of crisis short of hostilities.[9] Second, it provides an opportunity to evaluate the utility of existing theories of international relations, thereby contributing to the process of theory development.[10]

To explain the evolution of NATO's conventional force posture, especially the considerable continuity that characterized alliance strategy and force levels from the late 1950s through the 1980s, this book draws upon three leading perspectives in the international relations literature, each of which promises to be useful for understanding alliance behavior. The balance-of-power perspective concerns how the alliance as a whole responds to changes in its international environment. It emphasizes such factors as the capabilities and intentions of the adversary, the state of military technology, and overall alliance capacity. The intra-alliance perspective focuses on differences among alliance members—in military power and potential, geography, extraregional interests and commitments, and historical experience—and asks how these differences affect individual and collective behavior. The institutional perspective considers the influence of the alliance's institutions and the rules they embody on the actions of member countries.

Previous attempts to understand alliance behavior have drawn largely upon the first two perspectives in the form of balance-of-power theory and public goods theory, respectively.[11] In contrast, the institutional perspective has rarely been used in the alliance context, having been developed primarily in the realm of international political economy.[12] Thus the study also seeks to determine whether the institutional perspective can help to explain alliance behavior and international security affairs more generally.

The Argument

Indeed, this book argues that all three perspectives are necessary to provide a satisfactory overall account of the evolution of NATO's conventional force posture. It finds, however, that the role of conventional forces in NATO strategy on the one hand and the alliance's actual conventional capabilities on the other lend themselves to distinct explanations, which draw upon the three perspectives in different ways.

Initially, NATO strategy varied quite rapidly in response to advances in military technology and changing Western estimates of Soviet capabilities and intentions, as the balance-of-power perspective would suggest. By the same token, the declining frequency with which the strategy was subsequently revised was due in part to the increasing stability of these determinants over time. More important, however, was the interaction of Soviet nuclear advances with the allies' uneven exposure to Soviet military power, which caused allied strategic preferences to diverge beginning in the late 1950s, as the intra-alliance perspective would emphasize. This divergence made agreement on a common alliance strategy increasingly difficult to achieve, requiring the allies to resort to ever more ambiguous formulations to paper over their disagreements.

Not surprisingly, disparities in the relative size of the conventional force contributions made by NATO members are best explained in terms of the differences emphasized by the intra-alliance perspective. Early variations in NATO's overall conventional capabilities, moreover, like those in strategy, were strongly influenced by balance-of-power considerations, especially evolving perceptions of Soviet intentions. In contrast, the stability that characterized alliance force levels in later years, especially the absence of significant force reductions in response to domestic pressures and favorable international conditions, can best be understood in terms of the institutional perspective, which posits that the provision of conventional forces by NATO countries has been governed by an international regime. By increasing the costs and reducing the benefits of unilateral force reductions, the regime strongly influenced national force contributions. Over time, moreover, the rules of the regime acquired strong domestic and cognitive roots, which further reinforced the tendency of member states to comply with its behavioral injunctions.

The events of the late 1980s and early 1990s remind us, however, that there are limits to the autonomy of regimes. The sharp decline of the Soviet threat precipitated a profound transformation of NATO's conventional force posture, confirming that the range of possible institutional forms is ultimately bounded by the causal factors emphasized by the balance-of-power and intra-alliance perspectives. Nevertheless, NATO's military strategy was revised and alliance force levels in the Central Region decreased only after the traditional military threat to Western Europe had virtually disappeared, demonstrating that such variables may have to change substantially before institutional resistance can be overcome. Moreover, NATO's retention of a common strategy and its integrated military structure in the post–Cold War era reveals the resilience of institutional arrangements, even in the face of considerable variation in the external security environment.

Chapter Outline

Chapter 1 defines NATO's conventional force posture and identifies those aspects of alliance strategy and force structure that are the focus of this study. It then presents the three analytical perspectives and, where possible, derives explicit, testable hypotheses about alliance behavior that are evaluated in subsequent chapters. Readers who are principally interested in the historical account may skip this chapter without loss of comprehension.

The history of NATO's conventional force posture is divided into five periods, which are described and analyzed in Chapters 2 through 6. Each period represents a distinct phase in the evolution of the alliance's military strategy, its conventional force structure, or both. The time spans covered by the chapters are of unequal length because until the early 1990s, all formal revisions of NATO strategy and most significant variations in the alliance's conventional capabilities took place during NATO's first two decades of existence. Thus the organizational structure underscores the fact that relatively little change occurred in NATO's conventional force posture for more than twenty years prior to the end of the Cold War in Europe.

Chapter 2 reviews NATO's conventional force posture from the formation of the alliance in 1949 until the end of the Truman administration. During this period, conventional forces played a central role in NATO strategy. In the event of war, the alliance would have relied primarily on them to prevent Western Europe from being overrun while the U.S. atomic counteroffensive against the Soviet Union took effect. Initially, however, there was little concern about overt Soviet aggression, and NATO members took few extra military precautions. Only after the outbreak of the Korean War, when the possibility of an attack was viewed much more seriously, did the alliance undertake a concerted effort to erect an effective defense of the region. And although NATO's conventional capabilities increased substantially, they never reached the levels considered necessary to implement the strategy.

The early years of the Eisenhower administration are the subject of Chapter 3. During this period, NATO adopted a new military strategy, labeled MC 48, that placed less emphasis on conventional forces. Henceforth, the alliance was to rely increasingly on tactical nuclear weapons to deter aggression and to defend Western Europe, if necessary. Consequently, NATO's conventional force requirements were lowered somewhat, although by far the greatest reductions occurred in the goals for reserve rather than ready forces. The alliance's actual conventional capabilities nevertheless continued to fall short of even the new requirements and, toward the end of this period, actually declined.

Chapter 4 covers the mid- to late 1950s, when NATO's formal military strategy was modified once again. In contrast to the previous revision, however, the new strategy, which was designated MC 14/2, offered conventional forces a potentially greater role. MC 14/2 recognized, for the first time in NATO history, the possibility of limited forms of aggression in Europe, and it created a place for limited and even purely conventional responses. Nevertheless, the renewed emphasis on conventional forces in NATO strategy had little impact on the alliance's overall non-nuclear capabilities, which showed little or no net growth.

Chapter 5 discusses the evolution of NATO's conventional force posture in the 1960s, culminating in the adoption of the strategy of flexible response. Although highly ambiguous, this strategy placed even greater stress on conventional forces than did its predecessor. Henceforth, the alliance's initial response to non-nuclear aggression would be limited to conventional forces, although the point at which it would choose to escalate across the nuclear threshold was left unclear. Contrary to the conventional wisdom, however, the United States did not seriously consider revising NATO's existing strategy until the middle of the decade. During the early years of this period, the Kennedy administration simply focused on securing increases in NATO's conventional capabilities in the Central Region, although apart from the completion of the German military buildup, this initiative was largely unsuccessful. The eventual adoption of flexible response, moreover, owed as much to other developments as to U.S. efforts, and it was accompanied, ironically, by reductions in the number of U.S. and British forces on the continent.

The period from the late 1960s until the late 1980s is recounted in Chapter 6. This period witnessed relatively little change in NATO's conventional force posture. No attempt was made to modify the strategy of flexible response, although the alliance did undertake a series of initiatives to improve its conventional forces. Apart from a modest increase in the size of the German forces and the restoration of some of the U.S. and British cuts that had occurred in the late 1960s, however, NATO conventional force levels in the Central Region remained virtually constant.

Chapter 7 integrates the findings of the historical chapters. Drawing upon the three analytical perspectives presented in Chapter 1, it elaborates comprehensive explanations of the role of conventional forces in NATO strategy and the alliance's actual conventional capabilities, respectively, during NATO's first forty years. It indicates the contributions of each perspective and suggests how they can be combined to provide a satisfactory overall account of the evolution of NATO's conventional force posture. The Epilogue analyzes the changes that have occurred in the alliance as a result of the end of the Cold War.

CHAPTER I

Conventional Force Posture
and Alliance Behavior

Defining NATO's Conventional Force Posture

Two key aspects of NATO's conventional force posture are explored in this book: the role of conventional forces as defined in NATO military strategy and the alliance's actual conventional capabilities in the Central Region of Europe.[1]

THE ROLE OF CONVENTIONAL FORCES IN NATO MILITARY STRATEGY

NATO's military strategy has been embodied in a series of documents, which have been revised sporadically over the years. These documents have always received high-level political attention, usually being approved by the NATO foreign ministers. They should be distinguished from the alliance's actual defense plans, which have been regularly updated by the NATO military commanders, and from the operational doctrines of its forces. NATO strategy also differs from the strategic doctrines and declaratory policies of individual alliance members, which have not always been compatible with it.

Military strategy is but one element of grand strategy. Grand strategy concerns the use of the entire panoply of instruments available to a state—or an alliance—to achieve its overall foreign policy objectives, beginning with security. Military strategy is that element which deals most explicitly with the application of military means in support of foreign policy.[2] Thus while NATO's "grand strategy" has involved the use of diplomatic, economic, psychological, and other tools to attain alliance goals, the following analysis primarily addresses its military aspects.

All military strategies can in turn be decomposed into four conceptual components,[3] which, though rarely made explicit in practice, are always present. The first is a description of the general political-military objectives that the strategy is intended to serve. Second and closely related is an estimate of the threats that make the use of military means necessary to attain those ends. Third is a statement of the military responses that are deemed most effective or appropriate for addressing the threats that have been identified. It is this third component that has established most explicitly the role of conventional forces in NATO strategy. The range of actions that are actually feasible, however, is determined by the military means available. Appropriately, the final component of military strategy is an indication of the types and numbers of forces that are required to implement the strategy, including conventional forces.

Objectives of Military Strategy. The most basic objective of any military strategy is to ensure the security of a state, or an alliance in this case, in a potentially hostile external environment. This universal goal may be pursued in several ways, however, each of which has very different implications for the other components of military strategy. Some states may seek security through territorial aggrandizement or through control over an expanding share of the world's resources. To achieve such revisionist ends, these states may turn to military strategies that emphasize coercive diplomacy—the threat to use force if certain demands are not met—and even to offensive military operations intended to demonstrate the seriousness of their demands or simply to seize what they desire.

More often than not, states and alliances—including NATO—are content to pursue security by preserving and strengthening the territorial status quo. The military strategies employed by such states are typically intended to serve three closely related functions: to prevent coercion and political intimidation, to deter military aggression, and, if necessary, to defend their populations and territories.[4] These basic objectives have been constant elements of NATO military strategy.

The Military Threat. Until the end of the Cold War in Europe, two fundamental aspects of the military threats addressed by NATO strategy were also constant. First, the danger was viewed as emanating primarily if not exclusively from the Soviet Union and its Warsaw Pact allies. Second, NATO military strategy was principally concerned with the threat to Western Europe, and especially to the countries in the Central Region. The Soviet Union posed a direct military threat to the United States as well during much of the postwar era, but that peril was the exclusive province of U.S. military strategy.

In other respects, however, the threats addressed by NATO strategy were far from static. First, the perceived likelihood of Soviet aggression and thus the imminence of war in the Central Region varied considerably. Second, the precise forms that aggression was expected to take ranged widely in terms of both the scale of the attack and the military means likely to be employed. At times, for example, NATO planners have been primarily concerned with limited attacks, while at others they have been preoccupied by the possibility of general war. Conventional forces and various types of nuclear weapons have also figured differently in alliance threat assessments over the years.

Military Responses. The overall objectives of a military strategy together with the threats it identifies suggest the military responses that may be necessary to prepare. NATO planners have usually been able to choose among a variety of possible responses, which can be grouped into three conceptually distinct categories. The immediate aim of some military responses would be to deny an aggressor any possible material gains from the use of force. Other responses would seek to punish an aggressor by seizing or destroying objects to which the enemy leadership attached great value, thereby imposing losses that would outweigh any possible gains. Yet others would be primarily intended to raise an aggressor's estimate of the probability of escalation to a more destructive level of conflict and thus of the dangers of continuing hostilities.[5]

The role of conventional forces in NATO strategy has been defined largely by the answers that alliance policy makers have given to two questions. The first has concerned the relative emphasis to be placed on each of the three basic types of responses described above. Conventional forces have the greatest potential role in strategies that emphasize denial. Although they could conceivably contribute to strategies stressing punishment by, for example, seizing the adversary's territory in counteroffensive operations,[6] their role in responses intended to inflict damage and, especially, to raise the risk of escalation to all-out war would necessarily be highly circumscribed in comparison with nuclear weapons. The second question has concerned the relative emphasis to be placed on conventional and nuclear forces for implementing the denial element of NATO strategy.

In principle, conventional forces could perform a wide variety of roles in NATO strategy. For example, the alliance could eschew nuclear weapons altogether and rely exclusively on conventional forces to deter and, if necessary, to halt aggression. More plausible would be a strategy of no-first-use of nuclear weapons, whereby NATO would respond with conventional means alone to conventional attacks but would respond to nuclear attacks in kind. At the other end of the spectrum, the alliance could in theory rely entirely on nuclear weapons, assigning no role whatsoever to conventional forces. Or conventional forces might serve as a mere trip-wire intended to trigger nuclear retaliation in response to all but the most minor incidents. Between these extremes lies a range of other potential roles.

Conventional Force Requirements. A variety of forces and weapons may be necessary to implement the responses indicated by a military strategy. These may range from conventional to strategic nuclear forces, and they may include ground, air, and naval elements. Each force component, moreover, may be useful for more than one type of response.

Ideally, any military strategy should provide an estimate of the numbers and types of conventional forces that are required. In addition, it should suggest where those forces are to be deployed. Finally, it may indicate whether a given conventional force component should be available for combat at all times or whether a lesser degree of readiness, such as that befitting mobilizable reserve units, should suffice, depending on the immediacy of the threat.

CONVENTIONAL CAPABILITIES

Whether NATO could have implemented its military strategy in the event of a conflict would have depended in large part on the actual capabilities of the alliance forces stationed in the Central Region. The magnitude of these capabilities would have determined whether a conventional attack could have been defeated without resort to nuclear weapons and, if not, at what point the alliance would have had to cross the nuclear threshold. Fortunately, the adequacy of NATO's conventional forces was never tested in combat during the Cold War. For the purposes of this book, however, this fact poses a problem because conventional capability is otherwise impossible to measure directly. Instead, we must employ surrogate indicators.

The primary and most easily measured determinant of capability is force structure. This book pays particular attention to two aspects of NATO's conventional force structure: the alliance's overall conventional force levels in the Central Region and the size of the conventional contributions of individual countries with forces in the region. Although the relative size of different national contingents does not necessarily have a direct bearing on an alliance's overall conventional capabilities, it has often been the focus of heated debates over burden sharing within NATO.

Force structure can be measured in several ways. The most common measures are numbers of military personnel; combat units of different types, such as divisions, brigades, and squadrons; and major pieces of combat equipment, such as tanks, artillery, and aircraft. Although there is often a good deal of correlation between these three numerical measures because many combat units contain uniform numbers of personnel and equipment, substantial discrepancies may arise. For example, standard combat units sometimes vary in size from one country to another. Or the number of divisions or brigades fielded by a particular nation may fluctuate independently of the number of military personnel simply as a result of reorganizations that produce little or no change in fighting power. Since the contribution of support elements is usually reflected in the number of military personnel but not the number of combat units, the former is perhaps the best single indicator of actual combat capability.

Of course, force structure is not the only determinant of capability. Other important factors include equipment modernization, readiness, sustainability, personnel quality, and, in the case of an alliance, the degree of coordination and interoperability among the forces of different countries. Consequently, capability may increase or decrease even as force structure remains constant. In practice, however, qualitative factors are much more difficult to measure than is force structure, and their contribution to capability is commensurately harder to assess in the absence of actual combat experience. Accordingly, this study pays less attention to them.

THE RELATIONSHIP BETWEEN STRATEGY AND CAPABILITIES

One might expect to find an intimate relationship between NATO military strategy and force structure. On the one hand, strategy, and the indication it provides of force requirements, should have an important influence on the size of national force contributions and thus overall force levels. On the other hand, evidence of the numbers of forces that alliance members have actually provided in the past should suggest the types of military response that are likely to be feasible and thus shape the military strategies that are adopted.

Explaining Alliance Behavior: Three Perspectives

This book examines the evolution of NATO's conventional force posture from three perspectives drawn from the theoretical literature on international relations. While emphasizing a different set of determinants, each promises to be useful for explaining alliance behavior. The balance-of-power perspective concerns how the alliance as a whole responds to changes in its environment. The intra-alliance perspective focuses on differences among alliance members and how they affect individual and collective behavior. The institutional perspective considers the influence of alliance institutions on the actions of member countries.

It may be tempting to regard the three perspectives as offering competing explanations of the phenomena under investigation.[7] One reason is that the balance-of-power and intra-alliance perspectives represent different levels of analysis. At the same time, they both assign considerable weight to variable structural factors, such as the distribution of military power, while the institutional perspective does not. Consequently, the three perspectives might be expected to offer conflicting hypotheses regarding NATO's conventional force posture.

These perspectives are not necessarily mutually exclusive, however. To the contrary, they may be more usefully viewed as potentially complementary. They often focus on different aspects of alliance behavior and thus seek answers to different sets of questions. Indeed, this study will show that all three perspectives must be drawn upon to develop a satisfactory account of the evolution of NATO's conventional force posture, although each makes a distinct contribution.[8] Thus the central analytical challenge is not to determine which perspective by itself offers the most convincing explanation but to identify how all three can be most effectively combined to provide a comprehensive interpretation.[9]

One might well ask at this point why other perspectives, such as those emphasizing domestic structure, bureaucratic politics, organizational behavior, and even cognitive factors, do not receive equal consideration. To be sure, the use of these perspectives would undoubtedly enrich the analysis.[10] This study seeks, however, to develop an approach to explaining alliance behavior that can be fruitfully applied across a wide range of times and places. In contrast to the three perspectives explored here, which are primarily concerned with international determinants, most other perspectives, by focusing exclusively on factors at the domestic and individual levels, would be of more limited applicability.

THE BALANCE-OF-POWER PERSPECTIVE

The balance-of-power perspective views the alliance as a whole as the unit of analysis, treating it as a unitary, rational actor. For the purposes of the present study, this assumption is not unreasonable as

a first approximation of reality, given the high degree of shared interests and common characteristics among the member states of
NATO. This perspective promises to be especially useful for explaining alliance strategy and overall force levels.

According to the balance-of-power perspective, alliance behavior is
strongly determined by the larger international system within which
the alliance is embedded. The international system comprehends
both structural factors, such as the distribution of military power,
and environmental factors, such as the state of military technology.[11]
From this perspective, an alliance's military posture is primarily a
function of four variables: the military power of the adversary, the
intentions of the adversary, the state of military technology, and the
alliance's own overall ability to provide military forces.

The Military Power of the Adversary. At the heart of the balance-
of-power perspective is balance-of-power theory, which is primarily
concerned with the effects of the distribution of power in the international system.[12] In its simplest form, this theory predicts that states
will typically seek to balance the power of potential adversaries. Individual states may make internal efforts to increase their military
power and to use it more effectively, or they may form alliances with
other states. By extension, the members of an alliance will increase
their collective power in response to an increase in the power of the
state or coalition against which the alliance was formed. Although
power may be measured in several ways, most applications of this
approach have emphasized military power.[13]

Overall, balance-of-power theory suggests a close correspondence
between an alliance's conventional force requirements and its actual
capabilities. It predicts that in a non-nuclear world, alliance force
requirements and force levels will vary with the size of the conventional forces arrayed against them. As an adversary's forces increase
or decrease, so will those of the alliance. In the case of NATO's conventional force posture in the Central Region, however, this direct
relationship has been greatly complicated by the existence of nuclear
weapons, which may be substituted for conventional forces to some
extent and which have often been regarded as more cost-effective
than conventional forces.

The Intentions of the Adversary. Recent scholarship has shown
that alliance formation may be determined as much by perceptions of
a potential adversary's intentions as by its military power.[14] Analogously, we may expect such perceptions to play an important role in
shaping an alliance's military posture. The more hostile an adversary
appears, the greater the possibility of aggression will seem, and the

greater the alliance's military preparations are likely to be. Conversely, if the adversary's intentions are viewed as relatively benign, military preparations are likely to slacken because there is no perceived immediate need to balance fully the adversary's military power. Thus a second hypothesis is that alliance conventional force requirements and force levels will vary with perceptions of the adversary's intentions and the risk of war. In the case of NATO, these have largely been influenced by actual or imputed Soviet actions and by the overall state of East-West relations.[15]

The State of Military Technology. Although a concern with the implications of military technology for strategy and force levels is implicit in balance-of-power theory, previous applications of the theory have failed to explore this question in detail. When military technology has received explicit consideration, moreover, its impact has been judged to be modest.[16] Consequently, the analysis has suggested thus far that alliance force requirements and force levels will vary smoothly with the adversary's military capabilities and intentions. It has likewise implied that the types of responses emphasized by alliance strategy will remain constant. In fact, developments in military technology can trigger dramatic and sometimes abrupt shifts in many of the components of military strategy as well as force structure.

Such shifts may come about in two ways. First, one's own technological progress may increase the military means that are available and thus the number of possible responses. In other words, new military technologies may broaden the range of feasible strategic options and generate new force requirements. Second, technological advances by an adversary may result in revised assessments of its military capabilities and, in some cases, even its intentions. The altered view of the threat may in turn raise doubts about the adequacy of one's existing strategy and forces and thus may dictate the need to prepare new military responses.

Of all the advances in military technology that have occurred in the second half of the twentieth century, developments in nuclear technology have had the most profound impact. Because of their tremendous destructive power, the appearance of first atomic and then thermonuclear weapons significantly increased both the types of forces that were available and the range of possible military responses, setting the stage for successive modifications of NATO strategy and force levels.

Accounting for the presence of nuclear weapons thus leads to two further hypotheses. First, one might expect to find that whenever the United States and its allies enjoyed an advantage in nuclear weapons, they would have placed greater reliance on nuclear responses in NATO strategy and would have substituted nuclear for conventional forces. As a result, NATO conventional force levels would have been considerably lower than if the alliance had had to rely on conventional forces alone for deterrence, and they would have been relatively insensitive to increases in the conventional capabilities of the Warsaw Pact.

Conversely, one might also expect to find that whenever the Soviet Union acquired comparable nuclear capabilities and reduced NATO's nuclear advantage, thereby increasing doubts about the credibility of threats to use nuclear weapons in response to conventional aggression, the alliance would have diminished its nuclear reliance and placed greater emphasis on conventional forces. As a result, NATO's conventional force requirements and force levels would have increased. And if the alliance's nuclear advantage were neutralized, its conventional capabilities would have once again become primarily a function of those of the Warsaw Pact.

Alliance Military Potential. A further determinant of an alliance's military posture is its military potential. Because of its focus on the broad features of the international system within which states must operate in search of security, balance-of-power theory has also paid insufficient attention to the overall ability of states to generate and maintain the forces required to implement particular strategies. In principle, a state or an alliance may choose from among a variety of possible military postures. The choice that it eventually makes will depend critically on its military potential.

Military potential is itself a function of several factors.[17] Most easily measured are the material resources at the disposal of a state or alliance, notably its population, natural endowments, infrastructure, and level of economic development. Equally important but less quantifiable is the ability to translate material resources into usable military power. This more intangible resource, which Raymond Aron has termed the capacity for collective action, involves a combination of administrative and organizational skills as well as political will.

An alliance's military potential may shape its military posture in two ways. Clearly, it has a direct bearing on force levels. Some tension always exists between the political and economic opportunity costs of providing all of the forces required to implement a given strategy and the military risks involved in failing to do so. The more that material resources are constrained or the greater the obstacles to translating them into usable military power, the greater the gap between requirements and capabilities is likely to be. In this connection, military potential also influences strategy, albeit less directly. A general sense of an alliance's ability to generate military forces will inevitably color views of the feasibility of different strategies. Once a strategy is adopted, moreover, the subsequent accumulation of evidence indicating that the alliance is unable to provide the necessary forces will produce pressure for a revised strategy that strikes a better balance between costs and risks.[18]

THE INTRA-ALLIANCE PERSPECTIVE

The second perspective shifts our attention from the alliance as a whole to the states that compose it. This perspective focuses on the ways in which alliance members differ from one another and asks how these differences affect individual state as well as overall alliance behavior. Because of variations in their interests, capabilities, and other attributes, alliance members may respond very differently to the same international context.

The impact of intra-alliance differences on alliance military posture depends critically on the nature of the alliance under consideration. In this regard, it is useful to imagine all alliances as lying along a continuum between two ideal types, with purely coercive arrangements at one end of the spectrum and purely consensual ones at the other.[19] In coercive alliances, the dominant member is able to determine alliance military strategy with little regard for the concerns of the others, and it may even be able to compel its partners to contribute specified quantities of military forces.[20] Thus intra-alliance differences other than gross disparities in power are likely to be of little consequence.

NATO, however, has been based on the principle of consensus. In such alliances, all common policies, including military strategy, must be approved by all members, which are then free to disregard any agreements that are reached. Allies may seek to influence each other by extending or withholding benefits, but they are unable to impose their will upon one another. Each country retains the option to exit and may very well do so if the costs of membership exceed the benefits.

In a noncoercive setting, intra-alliance differences can affect alliance military posture in three general ways. First, they may cause individual countries to have contrasting strategic preferences. Alliance members may hold conflicting views about the precise nature of the threat to be addressed, and even when they do agree, they may favor different responses and place varying degrees of emphasis on different types of forces. Even if we assume that each state will prefer the strategy that minimizes its costs and maximizes its benefits, we must make use of the intra-alliance perspective to understand why member countries assess costs and benefits differently.

Intra-alliance differences will also shape the outcome of the bargaining that takes place among alliance members over their respective strategic preferences. Member countries will enjoy different degrees of influence over these outcomes. The states that will carry the greatest weight will be those that can contribute most to their allies' security and those with the greatest interests at stake. Some countries are more dependent on allies than others and thus may feel a greater compulsion to defer to the wishes of their benefactors.[21] At the same time, alliance members do not typically face precisely the same degree of threat, nor do they share the burden of costs and risks equally. The more a state stands to lose from the adoption of a particular strategy, the firmer the stance that it is able to take. Conversely, the more indifferent a state is to the outcome, the more willing it will be to compromise.

Finally, intra-alliance differences will influence the size of the forces actually contributed by each ally and thus overall force levels. In this connection, the intra-alliance perspective is often equated with public goods theory, which has been used to explain discrepancies in national contributions to a public good in terms of differences in the ability and the willingness of individual countries to make such contributions. Public goods theory has been extensively applied to alliances and to NATO in particular, where the public good is regarded as collective security through deterrence.[22]

Traditionally, the contributions of member countries to collective deterrence have been measured in terms of relative defense burden, defined as military expenditures divided by gross domestic product. There is no inherent reason, however, why NATO conventional force levels in the Central Region cannot be treated as a public good. Indeed, the contribution of these forces to collective security through deterrence is more direct than that of military spending.[23] Thus public goods theory, and the intra-alliance perspective more generally, promises to be useful for explaining NATO conventional force levels, both relative and overall.

Differences in Military Power and Potential. The emphasis placed upon military power and potential by the balance-of-power perspective suggests that we begin our analysis of the intra-alliance perspective by considering differences among alliance members in this regard. In fact, variations in the relative ability of countries to contribute forces may affect the alliance's military posture in each of the three ways outlined above. First, such differences will cause countries to prefer different military strategies. Each will promote the strategy that is most consistent with the forces it wishes to provide. This phenomenon is likely to be particularly striking in the case of nuclear capabilities. Because of the prestige and influence as well as the security benefits associated with nuclear weapons, nuclear-capable countries will want to provide them rather than conventional forces whenever possible and will thus seek to place greater emphasis on nuclear responses in alliance strategy.

Differences in military power and potential may also translate into different degrees of influence over alliance policy, including military strategy.[24] In this case, influence flows from a country's ability to grant or to withhold forces that could contribute to its allies' security. As a general rule, the views of the most powerful members will carry the greatest weight. In this regard, the study of NATO's military posture should pay particular attention to the policy preferences of the United States, which has been likely to wield a disproportionate share of influence by virtue of its unmatched military capabilities.

Finally, the distribution of military power and potential among the alliance members will also shape their relative force contributions and overall force levels. As a first approximation, we might expect that states will provide forces in proportion to their relative ability to do so. Larger, more capable countries will contribute proportionately more forces than smaller, less capable ones. To the degree that conventional force contributions in the Central Region satisfy the criteria of nonrivalness and nonexcludability that must be met by public goods, however, we might expect the less intuitive predictions of public goods theory to hold. This approach hypothesizes that the states with the greatest ability to provide conventional forces will contribute disproportionately large shares, relative to their means. Conversely, states with relatively few resources will make disproportionately small contributions, a phenomenon popularly known as "free-riding." The overall provision of conventional forces, moreover, will be suboptimal.[25]

Differences in Geographical Location. Allies may also differ markedly in their location with respect to the principal adversary, which can also affect alliance military posture in each of the ways discussed above. First, geography colors strategic preferences.[26] Differences in location mean different degrees of vulnerability, which in turn shape perceptions about the nature, magnitude, and immediacy of the threat, resulting in varying beliefs about what military preparations should be undertaken. For example, states that border on an adversary should prefer strategies that emphasize the deterrence of all types of conflict because any hostilities are likely to inflict widespread destruction upon them. In contrast, states that are not directly exposed to invasion should prefer strategies that emphasize the avoidance of all-out nuclear war and the need to contain any fighting that does occur.[27]

Geographical differences may also result in different amounts of influence over alliance policy. For example, location may provide certain countries with additional bargaining chips. In some circumstances, the ability to grant or to deny access to one's territory, bases, or airspace may be a powerful source of leverage. As a result, a small ally in a particularly critical location may be able to wield as much influence as a state with much greater military capabilities.[28] At the same time, geography may be a crucial determinant of a state's resolve in the intra-alliance bargaining process. The countries that feel most exposed are likely to care more about the outcome and thus be more willing to accept the risks to alliance cohesion associated with intransigence.

Finally, an ally's location may influence the size of its conventional forces. Public goods theory hypothesizes that a country's contribution is also a function of the benefit it can expect to receive from the availability of the public good. Consequently, we may expect to find that those states that are most at risk will benefit the most from the security provided by the presence of conventional forces and thus will contribute a disproportionately large share.

Further Sources of Intra-Alliance Differences. At least two other factors may differentiate the members of an alliance in ways that cause them to favor contrasting strategies and to make different force contributions. One is the existence of differences in extraregional interests and commitments. Because alliances tend to promote only a fraction of each member's total interests,[29] such differences are not unusual. As a general rule, members with substantial involvements outside the alliance area will be less willing and less able to provide forces for alliance purposes, and they will be more likely to prefer strategies that minimize their alliance obligations.

Variations in national experiences may also have an important impact on alliance force posture. Because of their unique histories, member countries may hold differing views about the nature, magnitude, and even the identity of the threat. They may also attach more or less weight to the prestige and influence that would be conferred upon them by different strategies and different force contributions.

THE INSTITUTIONAL PERSPECTIVE

The third perspective focuses on alliance institutions and asks how they affect the actions of member states and the alliance as a whole. As a general rule, institutions serve to constrain behavior. Departures from institutional norms are either costly or simply impossible. Once established, moreover, institutions are often difficult to alter. Thus in contrast to the first two perspectives, which suggest that alliance military posture will vary in response to changes in the external environment or among the members of an alliance, this perspective emphasizes the constraining effects of enduring institutional factors, which promote behavioral regularity and policy continuity.[30]

International Institutions and International Regimes. Oran Young describes international institutions as social institutions governing the activities of the members of international society, where social institutions are identifiable practices consisting of recognized roles linked by clusters of rules or conventions governing the relations among the occupants of these roles. He distinguishes between two general types of international institutions: international orders and international regimes. "International orders are broad framework arrangements governing the activities of all (or almost all) the members of international society over a wide range of specific issues. . . . International regimes, by contrast, are more specialized arrangements that pertain to well-defined activities, resources, or geographical areas and often involve only some subset of the members of international society."[31] Thus those institutions that have a bearing on alliance military posture may be more properly thought of as international regimes.

At the core of every international regime is a cluster of norms and rules, or "well-defined guides to action or standards setting forth actions that members are expected to perform (or to refrain from performing) under appropriate circumstances."[32] Such injunctions provide benchmarks against which the behavior of states can be assessed. In the context of this book, relevant rules would concern the process through which alliance strategy is formulated and the provision of conventional forces in the Central Region. The overall hypothesis generated by regime theory is that states will generally act in accordance with these behavioral guides, even when other international and domestic factors might tempt them to do otherwise.

How International Regimes Influence State Behavior. Why do participating states comply with a regime's norms and rules in the absence of a central authority to enforce them?[33] Regime theorists have offered two general answers to this question, each based on a different set of assumptions about the nature of state actors. The first approach regards states as rational, self-interested utility maximizers that act in accordance with regime injunctions on the basis of straightforward calculations of a utilitarian nature. The second explanation treats states as habit-driven actors, which, for a variety of internal reasons, are conditioned to comply with regime injunctions even when the benefits do not exceed the costs.[34]

Despite their differing assumptions, these alternative explanations may be viewed as mutually reinforcing and can be combined to provide a more comprehensive theory of how regimes influence state behavior that has both external and internal components. The external component emphasizes how regimes may modify the incentives and constraints that even rational, self-interested states encounter in the international environment. The existence of a regime may alter the costs and benefits to the state associated with different possible courses of action in ways that increase the likelihood of behavior in accordance with regime injunctions. The internal component captures the ways in which the tendency to comply with regime injunctions may become embedded in domestic structures and even in the belief systems of influential individuals. As a result of such internalization, compliance may become increasingly automatic.

External Incentives for Compliance. Rational, egoistic states have a number of incentives to comply with the rules of a regime, even when other factors might suggest that their interests would be better served by acting otherwise. Such behavior benefits participating states by contributing directly to the achievement of the common purpose served by the regime. In addition, by providing evidence of a state's willingness to compromise and to exercise self-restraint, it may reinforce the tendency—or overcome the reluctance—of other states to comply as well, resulting in even greater benefits both immediately and in the future. Finally, compliance can help establish a reputation for trustworthiness, which may increase the willingness of other states to enter into further mutually beneficial arrangements.[35]

Regime rules are invariably accompanied by expectations of future behavior that conforms to them. Such expectations may spring from a variety of sources, such as states' formal commitments to abide by agreed institutional provisions or their observation of certain behavioral norms in practice. Whatever their origins, these expectations endow regimes with much of their influence. Because participating states expect each other to behave in accordance with established norms and rules, they will feel aggrieved and may seek redress whenever another state fails to comply. In this way, the very existence of regime rules, by raising expectations about the proper conduct of states, may result in stronger reactions when discrepant behavior occurs than if the same action had been taken in their absence.

Consequently, noncompliance may redound to a state's disadvantage in several ways. At a minimum, it is likely to damage the state's reputation, causing its partners to be skeptical of its promises and commitments in the future. In addition, a disregard of accepted standards of behavior may prompt other participants to exclude the state from some of the benefits provided by the regime or to take various other forms of punitive action. More profoundly, noncompliance may cause other states to question the merits of their own continued compliance, thereby weakening the regime. And in the worst case, such behavior may culminate in the disintegration of the regime and the consequent loss of all associated benefits. The collapse of one regime, moreover, may inadvertently undermine others involving the same participants, further raising the costs of breaking the rules.[36]

Internal Sources of Compliance. Thus far, the analysis has assumed that states are rational, self-interested, unitary actors and that decisions to comply with regime norms and rules are the result of careful cost-benefit analysis. Some theorists, however, argue that states often comply with established conventions without making elaborate calculations on a case-by-case basis. To the contrary, behavior in accordance with the injunctions of an international regime may become internalized by participating states and take on a life of its own.[37]

Compliance may become institutionalized at the domestic level through the formation and strengthening of substate actors that promote such behavior. For example, government agencies may come to identify their interests with the preservation of an international regime and thus become staunch advocates of compliance.[38] Nongovernmental interest groups as well may benefit from the regime and lobby for adherence to its rules.[39] Negotiated regimes in particular are likely to command broad support both inside and outside the government.[40]

In addition, the process by which national policy is made may become structured in ways that consistently bias decisions in favor of continued compliance.[41] For example, regime rules may become incorporated into the standard operating procedures of the agencies responsible for setting policy. As a result, the information and policy options presented to senior decision makers will tend to reinforce the status quo. At the same time, individuals or agencies with an interest in compliance may acquire disproportionate influence over the decision-making process, while alternative or dissenting viewpoints may become marginalized or excluded from the process altogether.

Finally, and perhaps most profoundly, the habit of compliance may acquire roots in the very beliefs and values of policy makers and the informed public. Individuals may come to hold strong views about the importance of a particular regime, or they may develop firm convictions about the legitimacy of a regime's injunctions and thus a sense of obligation to promote compliance. Once established, such cognitive structures are often resistant to change.[42]

Hypotheses Regarding Regime Influence on Alliance Behavior. Several hypotheses regarding the military posture of alliances can be derived from regime theory. At the most general level, this approach predicts continuity in both alliance strategy and force structure. Because the military contributions of alliance members will tend to accord with established conventions, force levels, both national and overall, should remain constant as long as the relevant norms and rules do not change. This stability should even characterize situations in which shifts in the variables emphasized by the balance-of-power and intra-alliance perspectives would lead one to expect some variation in the alliance's conventional force posture.

Of course, behavior in conformity with regime injunctions by itself may not constitute conclusive proof of regime influence. Under some circumstances, other explanations may be equally plausible. Conversely, examples of discrepant behavior do not necessarily invalidate the claim of regime influence. As with domestic institutions, a certain level of noncompliance is almost always regarded as normal.[43] What matters is whether the existence of a regime increases the likelihood of behavior in conformity with its norms and rules. Thus the most compelling evidence of the influence of regimes will be found in the calculations of decision makers.[44] Consequently, a second general hypothesis, which follows from the external component of the model developed above, is that regime considerations will carry significant weight in the policy-making process. The existence of the regime will alter decision makers' evaluations of the costs and benefits of different courses of action in ways that favor compliance.

Three more hypotheses can be derived from the internal component of the model. First, the decision-making process is likely to be further influenced by the emergence and strengthening of groups with a direct interest in the maintenance of the regime. Second, the process itself may become structured in ways that result in outcomes consistent with compliance. Third, the beliefs and values of decision makers may evolve under the influence of the regime in ways that cause them to favor adherence to established norms and rules.

These additional hypotheses suggest that we should expect to see few proposals for change in alliance strategy or a country's force contribution, even when shifts in the balance of power, perceptions of the threat, or the distribution of capabilities within the alliance suggest the need for adjustment. Any proposals that do emerge, moreover, are likely to encounter fierce resistance, notwithstanding their objective merits, and, as a result of such opposition, few are likely to become national policy. Those proposals that survive will be considerably diluted during their passage through the policy-making process, and even then, plans for force reductions are prone to be moderated in the face of allied objections.

International Regimes and International Organizations. International regimes need not be accompanied by formal organizations, which are material entities possessing physical facilities, personnel, and budgets.[45] Few peacetime military alliances have involved such arrangements. Over the years, however, NATO has acquired an elaborate organizational structure. And where associated organizations do exist, they tend to reinforce the types of state behavior dictated by the regime.

This phenomenon can be attributed to several factors. Organizations and their representatives may play a central role in creating, articulating, and legitimizing regime rules and, subsequently, in promoting compliance and enforcement. Organizational bodies may also be used to facilitate the collection and dissemination of information about the behavior of participating states, thereby contributing to the transparency of the regime. To the extent that they are involved in the development and revision of regime rules, moreover, international organizations may place limits on the types of rules that are possible. As in the domestic setting, they may become dominated by interest groups that benefit from the existing institutional arrangements, while their procedures will tend to become highly bureaucratized. Thus the existence of an associated international organization will make a regime even more resistant to change.

Methodology

The methodology employed here is straightforward. The first step is to reconstruct the history of NATO's conventional force posture. Because of the paucity of well-documented published works on the subject, the following account relies heavily on primary source material.

The second step involves comparing the details of NATO strategy and force structure over the years with the hypotheses about alliance behavior derived above. This step helps to establish the relative usefulness of the three perspectives for explaining the main features of NATO's conventional force posture at different times in the past.[46]

Merely demonstrating the presence or absence of a correspondence between the predictions of the institutional perspective and the alliance's actual behavior, however, is often not sufficient to establish the impact of institutional factors. Thus a third step is to identify the motives of individual decision makers. This requires a careful examination of the processes by which national policies regarding NATO's conventional force posture have been formulated.[47] Where possible, I provide evidence of the influence of institutional considerations in the calculations of decision makers and of the internalization of behavior in accordance with regime injunctions.

CHAPTER 2

Constructing NATO's Conventional Force Posture, 1949–1952

The basic pattern NATO's conventional force structure was to assume throughout the Cold War was established during the first years of the alliance. The NATO countries began to formulate a military strategy and to determine force requirements soon after the North Atlantic Treaty took effect. Through the middle of 1950, however, they did little to strengthen the modest forces then stationed in Western Europe. Not until after the outbreak of the Korean War was a major effort undertaken to create an effective conventional defense of the region. This effort was guided by the alliance's initial strategic concept, which placed almost the entire burden of defending Western Europe in the event of an attack on conventional forces. Although the United States would immediately launch an atomic air offensive against the Soviet Union upon the outbreak of hostilities, strategic bombing was not expected to have a significant impact on the early stages of a ground war.

This first attempt to erect a strong conventional defense in the Central Region proceeded along four lines. Both the United States and Britain deployed additional forces on the continent. At the same time, the other European allies stepped up their rearmament efforts. The NATO countries also agreed to pursue a German military contribution. Finally, the alliance established an integrated military command and planning structure and appointed a supreme commander, to whom national forces would be assigned. The overall objective was to create a force of some 30 ready divisions in the region that would be backed by a mobilizable reserve of similar size.

These actions resulted in a substantial strengthening of NATO's conventional forces in the Central Region. Indeed, the initial force goals were largely met. As 1952 came to a close, however, the alliance's capabilities stood well below the overall force requirements that had been established, and the buildup was rapidly losing momentum. European rearmament efforts had proceeded more slowly than many officials had hoped and were already leveling off, and it was uncertain when Germany would begin to make a military contribution. Thus a wide and seemingly unbridgeable gap appeared to be opening between the heavy reliance placed on conventional forces in the strategic concept and what the alliance's forces would actually have been able to accomplish in the event of war.

NATO's Conventional Force Posture
Before the Korean War

During the fall and winter following the ratification of the North Atlantic Treaty, NATO officials drafted a strategic concept and prepared a Medium Term Defense Plan (MTDP), which included the first estimate of the alliance's conventional force requirements. NATO's actual conventional capabilities at the time, however, amounted to only a fraction of the forces deemed necessary to mount a successful defense of Central Region, and there was little prospect that they would be increased substantially anytime soon.

NATO STRATEGY AND FORCE REQUIREMENTS

NATO's initial military strategy for Western Europe was set forth in two documents. The first, "The Strategic Concept for the Defense of the North Atlantic Area," approved by the defense ministers in December 1949, amounted to little more than a broad statement of defense principles. The general nature of the document's content owed much to the circumstances of its preparation, the most immediate purpose of which was to enable President Harry S. Truman to release the bulk of the U.S. military assistance that had been authorized for the NATO allies early in the fall. Consequently, allied officials sought to complete the strategic concept as quickly as possible. Somewhat more detailed guidance was provided by the Medium Term Defense Plan, which was drawn up by NATO military planners during the first three months of 1950 based on the strategic concept.[1]

Like U.S. war plans at the time, NATO strategy was predicated on the assumption that any war with the Soviet Union would be total and would require the alliance to use every weapon at its disposal, including nuclear weapons.[2] If deterrence failed, the United States would immediately launch a massive air-atomic counteroffensive aimed at destroying the ability and willingness of the Soviet Union to prosecute its war effort. At the same time, NATO conventional forces stationed on the continent would attempt to hold the expected Soviet ground offensive as far forward as possible until the strategic bombing took effect. Specifically, the MTDP called for establishing a defensive perimeter along the north-south line formed by the Rhine and Ijssel rivers and, if possible, along the Kiel Canal.[3] Because the strategic counteroffensive was not expected to influence the land battle for at least 90 days, however, the defense of Western Europe would depend almost entirely on NATO's conventional capabilities.[4]

The MTDP was accompanied by an initial estimate of the forces required to implement this strategy—a total of 90 ready and reserve divisions, 54 of which would be deployed in the Central Region.[5] The MTDP did not attempt to break down the requirements by country. According to the strategic concept, however, the European allies would initially provide "the hard core of ground forces" and "the bulk of the tactical air support and air defense, other nations aiding with the least possible delay in accordance with over-all plans."[6]

NATO'S CONVENTIONAL CAPABILITIES

NATO's actual conventional capabilities at the time fell well short of these requirements. In the Central Region, the alliance disposed of approximately eleven ready and three reserve divisions, which consisted largely of the occupation forces deployed in West Germany— two British, three French, and the equivalent of two American divisions. France also maintained several additional ready divisions on its own territory (see Table 2.1).

These figures paled in comparison with contemporary estimates of the forces the Soviet Union and its satellites could bring to bear against the Central Region. In East Germany alone, the Soviet Union maintained some 22 divisions, backed up by 7 more divisions in Austria, Hungary, and Poland. These forces in Eastern Europe could be reinforced by another 32 divisions from the three westernmost military districts (Baltic, Belorussian, and Carpathian) of the USSR, possibly followed by the 30 divisions in the adjacent Leningrad, Moscow, Kiev, and Odessa military districts. The nearest satellite countries—Czechoslovakia, Poland, and Hungary—deployed another 30 or so divisions of their own.[7]

The alliance's numerical disadvantage was compounded by a variety of qualitative weaknesses. Most of the NATO units were regarded

TABLE 2.1
NATO Central Region Ground Forces,
August 1950

	Total divisions	Ready divisions in Western Europe
United States	9	2
Belgium	1	0
Canada	1 brigade	0
France	9	7
Luxembourg	2 battalions	0
Netherlands	5 battalions	0
United Kingdom	4	2

SOURCE: Appendix to enc. to JIC 530/3, "Most Likely Period for Initiation of Hostilities Between the USSR and the Western Powers," Aug. 22, 1950, RG 218, CCS 092 USSR (3-27-45), sec. 49.

as poorly equipped, poorly trained, and undermanned, the main exception being the U.S. contingent in Germany,[8] and even the American forces were still deployed in a manner more suited for their original occupation duties than for defense. Finally, the allies had taken few steps to facilitate the coordination of the military actions they would have to execute in the event of hostilities.

OBSTACLES TO STRENGTHENING NATO'S
CONVENTIONAL CAPABILITIES

Before the Korean War, allied officials contemplated three main avenues for closing the wide gap between the MTDP requirements and the size of the forces actually available in Western Europe: stationing additional U.S. divisions on the continent, increasing the forces of the European allies, and obtaining a German military contribution. In each case, however, seemingly insurmountable obstacles stood in the way. Consequently, little progress was made toward strengthening NATO's conventional capabilities prior to the outbreak of the war, and the prospects for any significant improvements in the future remained dim.

U.S. Forces in Western Europe. Notwithstanding the overall military potential of the United States, the Truman administration was in no position to station additional U.S. forces in Western Europe. It had been less than a year since administration spokesmen had sought to ease the ratification of the North Atlantic Treaty by assuring skeptical senators that there was no need to deploy large numbers of American troops on the continent and that there were no plans to do so.[9] The mere suggestion that the administration was reconsidering its position on the issue would have risked unleashing a storm of criticism from Capitol Hill.

This political hurdle was reinforced by economic constraints. U.S. defense resources were already stretched to the limit. Believing that large federal budgets were detrimental to the economy, Truman endeavored to hold defense spending to the lowest possible level, which amounted to only $13.5 billion in his proposal for fiscal year (FY) 1950.[10] Within this austere budget, the U.S. Army was able to maintain only ten divisions, virtually all of which were "badly under-strength, ill equipped, and lacking in logistical support." Of this total, moreover, five divisions were already stationed in Europe or the Far East, leaving none to spare in the U.S. strategic reserve.[11]

Even if more forces had been available, the U.S. Joint Chiefs of Staff (JCS) opposed the deployment of additional American units to Europe on strategic grounds. In the absence of strong West European forces, they believed, any conceivable U.S. military presence would have been unable to stage more than a brief holding action. In the event of hostilities, the American forces would quickly face the stark choice of withdrawing completely from the continent or being overrun.[12]

European Rearmament Efforts. Although the strategic concept assigned primary responsibility for providing the required conventional forces to the European allies, these countries made little effort to increase their military capabilities before the Korean War. Perhaps the most important reason for the slow pace of European rearmament was the absence of any significant fear of Soviet attack. Although the Soviet Union was seen as having a commanding advantage in conventional forces, Soviet leaders were regarded as content to pursue their foreign policy objectives through nonmilitary means.[13] Consequently, the overwhelming U.S. advantage in atomic weapons seemed more than adequate as a deterrent.

Of far greater concern in Western Europe was the danger of domestic turmoil and internal subversion. Following World War II, communist parties had been popular throughout much of the region, and many countries were still beset by serious economic difficulties. This threat was better addressed by raising living standards than by augmenting Western Europe's military power. Indeed, excessive defensive preparations could exacerbate the internal situation by hampering economic progress. Consequently, the European allies insisted that economic recovery should have clear priority over rearmament.[14]

Although many U.S. officials would have liked to see the Europeans make a greater effort to increase their military forces, they were initially willing to accept this ordering of priorities. The American attitude was reflected in the modest initial size of the U.S. military assistance program, which provided less than $1 billion to the NATO allies. The purpose of the program was "less to prepare defense measures against potential attack than to contribute, as [the Marshall Plan] had done, to the political and economic stability of Western Europe."[15] Consequently, the JCS, concerned that only militarily irreducible demands be imposed on the economies of the NATO countries, concluded that the original MTDP force requirements were in need of a "radical" downward revision and recommended that NATO planners develop a new plan "based on a realistic estimate of force availabilities."[16]

Finally, many of the resources the European allies might have made available for the defense of the Central Region were diverted to other uses. Britain was attempting to build the atomic bomb, and both Britain and France continued to maintain far-flung commitments and dependencies, which required them to station large numbers of troops overseas. In some cases, such as Indochina, where France was fighting an increasingly costly war, the threats to these interests seemed much more urgent than that of aggression in Europe.

The Question of a German Military Contribution. Perhaps the greatest obstacles stood in the way of a possible German military contribution. Allied military planners had concluded that even if more American and European forces were made available, it would be impossible to implement NATO strategy without the use of German manpower. In late 1949, the U.S. Army began to lobby strongly for the creation of German divisions, and in May 1950, the JCS agreed that "the appropriate and early rearming of Western Germany [was] of fundamental importance to the successful defense of Western Europe against the USSR."[17]

Officials in the State Department, however, were unwilling to propose German rearmament to the allies. The complete demilitarization of Germany had been a cornerstone of U.S. policy since the end of the war and had been reaffirmed as recently as November 1949. Consequently, although American diplomats looked forward to the eventual integration of German forces into a European defense structure, they were convinced that an early attempt to reverse this policy would have negative political repercussions within NATO and West Germany that would almost certainly outweigh any possible military benefits. They also expected that the Soviet Union would react sharply to any move toward recreating a German military. Thus Secretary of State Dean Acheson refused even to discuss the matter at the meeting of the North Atlantic Council (NAC) in May 1950, despite the insistence of the military.[18]

The State Department position reflected the views of the major European allies. France in particular feared a revival of German militarism and thus was staunchly opposed to German rearmament under any circumstances. The British stance was more nuanced. Because the existing NATO members would be unable to raise the forces required to defend Western Europe by themselves, they recognized that a German contribution would eventually be necessary, but they argued that the countries of the region must be strengthened militarily before German rearmament could be contemplated.[19]

PRESSURE FOR STRENGTHENING NATO'S
CONVENTIONAL CAPABILITIES

Meanwhile, pressure was building behind the scenes for a sharp increase in NATO's conventional capabilities. In particular, the discovery in September 1949 that the Soviet Union had tested an atomic bomb laid the groundwork for a major conventional buildup. During the following months, a growing number of U.S. officials concluded that NATO's conventional inferiority could no longer be tolerated. Nevertheless, until the outbreak of the Korean War, few of the obstacles described above had been overcome, and the policies of the NATO allies continued along their previous courses. Only the shock of the North Korean invasion finally prompted the alliance to initiate a serious effort to improve its conventional capabilities.

The Soviet Atomic Bomb and NSC 68. The Soviet achievement of an atomic capability, which came several years earlier than had been expected, weighed heavily on the minds of Western leaders. Not only did it abruptly terminate the American atomic monopoly, but it heralded the beginning of the end of U.S. strategic superiority. American and allied officials would have to reconsider the degree to which they relied—implicitly or explicitly—on the U.S. nuclear advantage to compensate for NATO's inferiority in conventional forces. In combination with the subsequent communist victory in China, this event prompted a general feeling in the West that the communists were on the move throughout the world and that the Soviet Union was becoming increasingly willing to undertake risky adventures that might lead to war.[20]

The most immediate policy decision facing U.S. officials was whether to proceed with the development of the hydrogen bomb, which Truman approved at the end of January 1950. At the same time, however, he directed the State and Defense departments to undertake a joint review of U.S. policy in light of known and anticipated developments in Soviet atomic capabilities.[21] Spearheaded by the State Department's Policy Planning Staff and its new director, Paul Nitze, the study was completed in early April and designated National Security Council document 68 (NSC 68).[22]

NSC 68 both reflected and stimulated a growing consensus within the Truman administration that the West must make a much greater effort to strengthen its military capabilities. Its underlying theme was that the Soviet acquisition of an atomic capability had revolutionized the world situation. Although Soviet leaders were deterred for the moment from attacking the United States or its allies by the U.S. advantage in atomic weapons, by 1954 they might have enough weapons to disarm the United States in a surprise attack.[23] As that date approached, the Soviet Union would be increasingly able to use its superiority in conventional forces to intimidate and coerce Western Europe.

To avert disaster, NSC 68 called for a substantial American military buildup, an acceleration of European rearmament efforts, and as much U.S. military assistance to the allies as possible. Greatly increased conventional forces would be needed to survive a Soviet surprise attack. They were also required to reduce the West's dependence on atomic weapons because the United States might be reluctant to initiate a nuclear exchange under conditions of atomic parity. Above all, a strong U.S. military posture and an increase in the allies' defensive capabilities would build European confidence and strengthen Europe's determination to counter Soviet political and psychological moves.

NSC 68 made a substantial impact within the State Department. Acheson in particular appears to have been strongly influenced by the report, and at the May meeting of the NATO foreign ministers, he made a determined effort to convince the allies to take steps to increase their conventional capabilities. He called on the North Atlantic Council to "initiate action on an urgent basis to build up forces in accordance with the medium-term defense plan" and to place as much emphasis on rearmament as on economic recovery.[24]

The U.S. position, however, encountered stiff resistance from the allies, who believed that economic recovery should continue to have first claim on national resources and thus remained unwilling to increase their military spending. Consequently, Acheson's proposals were either watered down or omitted from the resolutions adopted by the foreign ministers.[25] Meanwhile, NSC 68 encountered rough sledding within the U.S. government. Recognizing that the report's conclusions flew in the face of the economy principles that had guided his administration until then, Truman withheld his approval and instead referred the document to the National Security Council (NSC) staff for further study. In particular, the president sought a clearer indication of the programs envisaged in NSC 68, including estimates of their probable cost.[26]

An ad hoc interdepartmental committee was established to develop a response by August 1. The State Department proposed that military assistance for Western Europe be increased by over 150 percent, for a total of $12.5 billion over the next five years. The JCS developed a plan for a U.S. military buildup peaking in 1954, which was later estimated to cost $26–$27 billion annually.[27] The committee, however, also provided opponents of increased government spending with an opportunity to question NSC 68's assumptions and conclusions, which ensured that progress would be slow.[28] Thus as the summer of 1950 began, the eventual impact of the report on U.S. national security policy remained uncertain, and its significance for NATO's conventional force posture was even less clear. Pending the completion of the additional analysis, the administration prepared budget requests for defense spending and military assistance in FY 1951 that were almost identical to those of the previous year.

The Impact of the Korean War

The outbreak of the Korean War demolished many of the political obstacles that had previously blocked efforts to strengthen NATO's conventional forces. In some respects, this event merely completed the process of altering Western threat perceptions that had begun with the first Soviet atomic explosion and been accelerated by the preparation of NSC 68. The surprise North Korean invasion, however, underscored the altered nature of the situation in a way that left little room for doubt.

The sudden onset of the Korean War affected the views of NATO leaders in several interrelated ways. First, it caused them to reassess their views of the Soviet threat to Western Europe. Until then, it had still been possible to regard Soviet leaders as content to pursue their political objectives through nonmilitary means. The war, however, seemed to confirm the grim interpretation of Soviet intentions promulgated by NSC 68, which was quickly adopted by virtually all top U.S. officials. There was little doubt that the Soviet Union had instigated the attack. Consequently, most allied leaders concluded that the Kremlin would henceforth be more willing to use force to achieve its aims. The danger of international conflict now overshadowed the threat of internal subversion in Europe.[29]

The war also seemed to demonstrate that the degree to which the West could rely on atomic weapons to deter aggression was sharply limited. U.S. nuclear superiority had not prevented the North Koreans from acting, and these weapons would be of declining utility as the Soviet Union's own atomic arsenal expanded and its reluctance to risk general war diminished proportionately. In July, moreover, the JCS concluded that the Soviet Union might achieve effective atomic parity and thus maximize its military strength relative to the West as early as 1952.[30]

Until then, the Soviet Union might be expected to exercise caution. Although they would be certain to probe U.S. firmness at vulnerable points around the world, Soviet leaders would seek to avoid a global conflict until they were sure they could prevail. The war, however, suggested that they would not hesitate in the meantime to use proxy forces to achieve their ends. In this respect, divided Germany bore a disquieting resemblance to Korea. Indeed, the Soviet Union was busily creating an East German police army, which had reached a total of some 70,000 men by the end of August 1950 and which some Western leaders feared might be used in an attempt to seize Berlin or even all of West Germany.[31]

As a result of these considerations, many U.S. and allied officials concluded that Western Europe's conventional defenses must be strengthened on an urgent basis. Even if war in Europe were not imminent, considerable time would be required to accumulate the forces needed to deter and, if necessary, to repel an invasion. A prompt and rapid buildup was needed, moreover, to reverse the negative impact on allied morale of the initial setbacks in Korea, which had led some people to question the value of the North Atlantic Treaty itself.[32]

Designing NATO's Conventional Buildup

During the six months after the outbreak of the Korean War, NATO fashioned a comprehensive program intended to provide the conventional forces required to defend Western Europe. This program was developed in two stages. In the summer of 1950, the United States devised a multifaceted proposal for strengthening the alliance's conventional capabilities, which was presented to the allies at the September meeting of the NATO Council. A series of intense intra-alliance negotiations over the details of the American proposal then ensued. Most of the wrangling concerned the question of a prompt German military contribution, over which the United States and France were initially deeply divided. Through a series of compromises, however, the allies managed to bridge their differences, enabling the NATO foreign ministers to approve in December the final measures needed to set the conventional force buildup in motion.

FORMULATING THE U.S. PACKAGE PROPOSAL

In the weeks following the beginning of the war, several European countries took unilateral steps to strengthen their conventional forces.[33] But the allies looked principally to the United States to lead the way because they believed that the creation of an adequate conventional defense of Western Europe would require a substantial American contribution. Initially, however, the Truman administration merely offered to increase U.S. military assistance in return for stepped-up European rearmament efforts. Yet this proposal generated little allied enthusiasm, and it quickly became clear that such a limited program would never result in sufficient forces. Most U.S. officials concluded that a dramatic departure from previous policy was required, but the State and Defense departments remained deeply divided over the proper approach to take. The government eventually adopted a package proposal that firmly linked the creation of an integrated defense force, the stationing of additional American forces in Europe, and increased U.S. military assistance to an acceleration of European rearmament efforts and a German military contribution.

The Initial U.S. Proposal and Allied Responses. In the days following the North Korean invasion, U.S. officials concluded that the United States should substantially increase the level of military assistance to Europe to enable the allies to make their maximum contributions toward meeting NATO's force requirements.[34] To this end, Pentagon planners quickly set about devising an expanded aid program that would provide the necessary stimulus while remaining acceptable to Congress.[35] The pace of the work was accelerated by the need to have a formal U.S. position in time for the first meeting of the newly formed NATO Council of Deputies, which was scheduled for late July. This goal was achieved on July 21, when Truman approved an increase in the size of the military assistance program of $4 to $6 billion. To avoid giving the Europeans any incentive to relax their efforts, however, the president was careful to condition the offer on allied willingness to develop forces roughly sufficient to implement the MTDP.[36]

The U.S. proposal was promptly relayed to the allies. In return for "the maximum possible addition by each country to its own military budgets," the United States offered to provide a substantial volume of military equipment and possibly funds to stimulate European military production, while Marshall Plan aid might be made available beyond its scheduled termination in 1952. As a first step, the allies were requested to provide the "firmest possible statement" of the increased effort they were prepared to make. European economic recovery would remain an important objective, but rearmament would require economic sacrifices that would inevitably slow the pace of recovery. All countries would have to accept lower levels of consumption and to divert resources from investment in peacetime production to finance the necessary military effort.[37] Although they would not say so publicly, U.S. officials still expected the allies to bear the brunt of the burden of creating an adequate defense of Western Europe.[38]

The major European allies responded promptly to the U.S. offer. Within days, Britain announced its willingness to undertake a three-year, £3,400 million defense program, beginning in 1951, which constituted an increase of 40 percent over existing plans. This, the British felt, was the most they could do without jeopardizing economic recovery. Similarly, France proposed a three-year, 2,000 billion FF ($5.7 billion) rearmament effort intended to add 15 divisions in Europe, which represented an increase of one-third over the previously planned level of spending.[39]

Despite the magnitude of the proposed increases, U.S. officials regarded them as unsatisfactory.[40] What the allies had offered seemed both inadequate for the task at hand and considerably less than they were capable of. First, the proposed programs would result in little immediate growth in the alliance's force structure. In fact, the British contemplated no increase in the size of their forces. Second, the allied initiatives were predicated on the receipt of substantial U.S. military assistance. Consequently, the total increase in defense spending outlined by the Europeans amounted to only $3.5 billion over three years, far short of the $10–$12 billion in additional effort that U.S. officials calculated as being within their means.[41] The administration feared that Congress would not approve the proposed military assistance legislation unless the allies demonstrated their willingness to undertake far greater efforts.[42] Thus throughout the rest of summer, the United States regularly pressed the Europeans to adopt concrete measures to augment their forces and military production substantially in the near future.[43]

Shifting Views Within the United States. At the same time, however, U.S. officials gradually came to the conclusion that European rearmament efforts alone, even if backed by substantial military assistance, could not provide the forces needed to defend Western Europe. Consequently, they began to give serious consideration to several additional measures for strengthening the region's defenses—notably U.S. participation in a NATO military planning and command structure, the stationing of additional U.S. forces on the continent, and a German military contribution—each of which had been regarded as politically infeasible before the Korean War.

U.S. thinking evolved most rapidly in the State Department, where officials were exposed to a steady stream of cables on the views of the allies. These reports suggested that the commitment of a greater number of U.S. troops and vigorous U.S. leadership through participation in a command structure headed by an American would be essential if the Europeans were to be persuaded to move forward with their own rearmament efforts. France, in particular, was viewed as unwilling to build up its own military strength unless substantial allied forces were stationed on the continent.[44] Consequently, Acheson and Secretary of Defense Louis Johnson agreed in early August that the United States should accept the responsibility of a unified command and should place additional combat units in Europe.[45]

By that time, Acheson was also considering the possibility of creating a European or North Atlantic army as a means of harnessing Germany's military potential without compromising other U.S. objectives in the region, an idea Truman also endorsed. The French had already indicated their willingness to consider a German military contribution as long as there was no risk of a resurgence of German militarism. As the limited nature of the rearmament efforts the allies proposed became evident, moreover, State Department officials began to regard German manpower and resources as essential for an effective West European defense. The successful establishment of a European army, however, was also seen as depending on the appointment of an American commander and the stationing of more U.S. and British troops on the continent.[46]

In mid-August, these ideas were assembled into a comprehensive State Department position paper titled "Establishment of a European Defense Force." The paper warned of a dangerous trend toward fear and resignation in Western Europe occasioned by the weakness of the region's defenses and the inability of the United States promptly to turn back the threat to South Korea. To boost the morale of the allies and to ensure that they made their maximum military contributions, the United States would have to participate fully in the European defense effort. Because the material resources of the European countries would still be insufficient to erect an effective defense, moreover, German participation would be essential. To this end, the paper called for the formation of a European defense force composed of increased U.S., British, and continental forces and operating under an American "Supreme Commander," who would be served by an international military staff. Such a force would make possible a German military contribution and eventual German membership in NATO while avoiding the creation of German national forces by integrating German ground units of division strength into larger non-German formations. The paper was silent, however, on the delicate issues of the timing and scope of German rearmament.[47]

Development of the U.S. Package Plan. State Department officials hoped to reach a quick agreement with the Defense Department on the creation of a European defense force so that Truman could approve a U.S. position well in advance of the North Atlantic Council meeting that was scheduled for mid-September in New York, and preferably before the NATO deputies resumed their deliberations in late August. Because of "the difficulty of the problems and the far-reaching importance of the answers," however, Johnson replied that the Pentagon could not complete its own studies and furnish comments on the proposal on such short notice.[48]

The two departments remained deeply divided on several critical issues. A report prepared for the JCS flatly rejected the State Department concept and proposed instead a "NATO defense force" composed of national contingents operating under their own national commanders. It also recommended the immediate incorporation of German ground units into this entity on a national basis, arguing that the creation of adequate German forces should not be delayed for any reason. The JCS contemplated an eventual German contribution of 10 to 15 divisions. Finally, while agreeing that there should eventually be an American supreme commander, the report concluded that he should not be appointed until sufficient forces existed to enable him to carry out his responsibilities.[49]

With time growing short, Truman sought to bring the matter to a head. On August 26, he asked Acheson and Johnson to develop by the beginning of September joint recommendations on a number of issues regarding the related problems of rapidly strengthening Western Europe's defenses and determining the most appropriate German contribution. The key questions concerned the commitment of additional U.S. forces to the defense of Europe; support for the concept of a European defense force, including German participation on a non-national basis; the immediate creation of a combined staff and the eventual appointment of a supreme commander for the European defense force; and full U.S. participation in European defense organs, including an American supreme commander.[50]

Initially, this added pressure did little to narrow the gap between the two departments. Indeed, the Chiefs stiffened their position, insisting that the deployment of further U.S. forces be linked to adequate steps by the European allies to increase their own forces, and they continued to emphasize the immediate initiation of German rearmament.[51] Consequently, an extra week of intense negotiations was needed to reach a common State-Defense position, which was finally submitted to the president on September 8. The departments agreed that additional U.S. forces should be committed to the defense of Europe at the earliest feasible date and that the overall size of the U.S. contribution should be about five and one-half divisions and eight wings of aircraft. They also agreed to the early creation of an integrated European defense force with an international planning staff and the appointment of a supreme commander at the soonest suitable date.[52]

The views of the JCS prevailed on virtually all of the issues that had divided the two sides.[53] The commitment of additional U.S. forces was made contingent on firm allied programs for the development of the balance of the required forces. The European defense force would be composed of national contingents operating within overall NATO control but under immediate national commanders. An American would be appointed supreme commander only upon the assurance of sufficient forces "to constitute a command reasonably capable of fulfilling its responsibilities." The European defense force would include nationally generated German units of up to division strength, and even larger formations might eventually be authorized. Finally, the establishment of the German units should proceed without delay once the planning staff had been formed.

THE QUESTION OF A GERMAN MILITARY CONTRIBUTION

Despite misgivings about the rigidity of the package, Acheson formally presented the administration's new position at the September meeting of the NATO foreign ministers in New York. Not surprisingly, the French delegation opposed the U.S. proposal for a German military contribution, and given American insistence that the package of recommendations be adopted as a whole, the ministers were unable to reach an agreement. Consequently, further discussion was postponed until late October, by which time it was hoped that France would moderate its stance or offer a viable alternative. The resulting French counterproposal for the creation of a supranational European army that would include a German component, however, was not acceptable to the Americans. Once more, NATO leaders found themselves deadlocked, and the question of a German contribution was referred to subordinate bodies for further study. It was not until late November that altered circumstances narrowed the differences between the two sides to the point that a compromise solution could be worked out.

Allied Responses to the U.S. Proposal. Although the United States had attempted to provide Britain and France with some indication of the emerging American position in advance of the New York meetings,[54] U.S. officials were unable to furnish all the details until the foreign ministers were actually assembled. In repeated presentations to the allies, Acheson emphasized that the United States was now prepared to join them as a full partner in the defense of Western Europe. Specifically, the United States would send additional forces at the earliest possible date and was willing to participate in an integrated force headed by a supreme commander. Even with much greater U.S. involvement, however, this force would be inadequate unless it were augmented by German resources. To be successful, moreover, the defense of Western Europe would have to be staged on German territory. For these reasons, the full support and participation of the Germans would be essential. Acheson assured his colleagues that the incorporation of German military units into the integrated force would ensure against any revival of German militarism.[55] He repeatedly stressed, however, that U.S. willingness to proceed with this far-reaching program would depend on allied agreement to a substantial German role.[56]

In general, the allies reacted favorably to the U.S. proposals, especially the offer to station additional forces in Europe. The majority favored accepting the American plan in its entirety. The British were particularly eager to expedite the appointment of a supreme commander as a stimulus to European action.[57]

The British representatives nevertheless questioned the necessity of immediately including German units in the integrated force. They feared that German rearmament might provoke a serious reaction by the Soviets and their allies. Instead, priority should be given to strengthening the forces of the other West European countries. They felt, moreover, that asking the Germans for a military contribution would put them in too strong a bargaining position, and they doubted that the Germans would agree in any case. The British preferred to proceed cautiously, beginning with the establishment of an armed German police force of 100,000, as Chancellor Konrad Adenauer had recently proposed.[58] As it became clear that the U.S. position would have to be accepted as a whole, however, they decided to accept the principle of German participation, although they insisted that the integrated force be established before German units could be created.[59]

The French representatives were strongly opposed to the American proposal for a German contribution and refused to agree even in principle. Like the British, they felt that it would be necessary to create a European army and to achieve a minimum level of forces in Western Europe before any German units were formed. They warned, moreover, that French public opinion would not tolerate German rearmament until France itself was fully rearmed. The immediate need was not to obtain a German contribution but to equip the West European allies, which had untapped reserves of manpower. Consequently, the French felt that it was premature even to discuss the inclusion of German units in the integrated force, much less to make a decision.[60]

Although the New York meetings continued intermittently for nearly two weeks, the United States and France were unable to overcome their differences, and the U.S. delegation withheld approval of the other elements of the package proposal. To avoid the appearance of disarray, the NAC announced that it had approved the concept of an integrated force under centralized command and control. Finalization of the arrangement, however, would have to await recommendations by the committee of NATO defense ministers as to the steps necessary to establish such a force and how Germany could make its most useful contribution. The scheduled meeting of the Defense Committee was postponed from mid- to late October to give the French time to reconsider their position.[61]

The Pleven Plan. The second attempt to forge a NATO agreement on the question of German participation was no more successful than the first. During the weeks following the NATO meeting, the French government hastily formulated an alternative to the American proposal, which was publicly unveiled on October 24 by Prime Minister René Pleven. The so-called Pleven Plan would enable Germany to make a military contribution through the creation of a European army within which German units of the smallest possible size could be merged into larger integrated units. This army would eventually be incorporated into the NATO integrated force with the other national forces of NATO. Like the coal and steel community envisaged by the French-sponsored Schuman Plan, the European army would form part of a larger European institutional framework, in this case consisting of a European minister of defense, a Council of Ministers, a European parliamentary assembly, and a common European defense budget.[62]

The Pleven Plan was intended to serve several interrelated purposes. First, it would postpone actual German rearmament for as long as possible. More important, by obviating the need for a national German army with a General Staff and supported by a Ministry of Defense, it would prevent a revival of German militarism and German domination of Western Europe. In fact, France would enjoy considerable influence over the size and use of the resulting German forces. At the same time, however, the proposal reflected a willingness on France's part to sacrifice a portion of its sovereignty in return for closer association with the other European countries, which would ultimately benefit French security.[63]

Allied reaction to the Pleven Plan was almost universally negative. Both the United States and Britain regarded the plan as completely unworkable. First, it would postpone the final resolution of the question of German participation for many months, if not years. Formal negotiations to establish the European army would not begin until the treaty establishing the coal and steel community was signed, and the actual formation of German units would have to await the completion of the series of complicated political steps leading to the appointment of a European defense minister. As a result, the development of an adequate European defense would be seriously delayed, and Congress might balk at appropriating the necessary military assistance. Second, the plan would be rejected by the West Germans. German units would be limited to company or battalion size, relegating them to a permanent second-class status within the European army. Yet it had already become clear that West Germany would be willing to rearm only on the basis of full equality.[64]

For the second time, agreement on the terms of a German contribution proved impossible, and the late October Defense Committee meeting ended in an impasse. The French refused to discuss German participation in the defense of Western Europe except in the context of the Pleven Plan, which the majority of the allies found unacceptable. Because the United States continued to insist that the alliance agree at least in principle to German rearmament before moving ahead with the other elements of the original package proposal, the new American secretary of defense, George Marshall, requested that a decision on the establishment of an integrated force and the appointment of a supreme commander be deferred once again and proposed that the entire matter be referred to subordinate bodies for further consideration.[65]

The Spofford Compromise. Despite their seemingly irreconcilable differences, the United States and France managed to work out a compromise in late November and early December. Several developments forced the two sides to moderate their positions. In the case of the United States, these changing circumstances loosened the grip of the JCS over U.S. policy, allowing control to pass back into the hands of the State Department, which took a more flexible stance on the issues that divided the allies.

One factor was the presence in the Pentagon of Marshall, with whom Acheson enjoyed an extremely good working relationship. The war hero and former secretary of state quickly became convinced that rigid adherence to the original package proposal was counterproductive and worked to soften the U.S. position.[66] Another consideration was simply the need to reach an agreement so as to maintain alliance cohesion. In mid-October, Acheson noted that a moment could arrive when further delay in creation of the integrated force would not be in the best interest of the United States. That point may have been reached in late November, when the Chinese intervention in Korea signaled a further sharp deterioration in the world situation and a heightened danger of war elsewhere. Consequently, U.S. officials concluded that the construction of a strong defense in Western Europe could not tolerate further delay.[67]

In the meantime, developments in West Germany demonstrated that the original U.S. position on German rearmament was unrealistic. Initially, American diplomats had believed that German support for rearmament could be gained in only two to four months, allowing the first units to be ready by the end of 1951.[68] Following the New York meetings, however, it had become increasingly clear that German agreement would not be automatic. Many Germans were concerned about a possible revival of militarism, which would threaten the development of democracy in their country. They also feared that rearmament would foreclose the possibility of reunification, which remained an overriding long-term objective, and that it might even provoke a Soviet attack. Thus they preferred to wait at least until substantial NATO forces were in place on the continent.[69]

In late October, U.S. officials reluctantly concluded that it would take a year to prepare Germany politically and psychologically for rearmament and thus it made no sense to press for an immediate decision.[70] During the following month, moreover, opposition in the Federal Republic continued to mount so that by December, it appeared that obtaining willing German participation would be a major problem.[71] As a result, the Truman administration decided to adopt a much more restrained approach in dealing with the Germans.

The compromise was embodied in a plan devised by the U.S. representative to the Council of Deputies, Charles Spofford, which sought to separate the political and military aspects of the problem of German participation. According to this plan, the alliance would proceed immediately with the recruitment and training of German soldiers, subject to strong controls and German approval. These rearmament measures would be termed provisional, however, and the relaxation of the controls would be linked to progress toward the development of permanent political and military arrangements. Thus the Spofford compromise set the stage for two distinct sets of negotiations: one between the three occupying powers and the Federal Republic on the immediate remilitarization of West Germany, and another to consider the French proposal for the establishment of a European army and its attendant political institutions. The final element of the compromise was an agreement to limit German formations to regimental combat teams, which would be smaller than the divisions the JCS had wanted but larger than the battalions France had originally proposed.[72]

EUROPEAN REARMAMENT EFFORTS AND U.S.
MILITARY ASSISTANCE

Progress toward the resolution of U.S. and French differences over the question of German participation was paralleled by increases in the scope of the rearmament efforts that the European allies proposed to undertake. In early August, the Council of Deputies had recommended that the member governments take immediate steps to augment the combat forces to be available by mid-1951. In reply, the European countries proposed to expand their forces from 22 to some 43 divisions, of which at least 23 would be stationed in the Central Region.[73]

At its October meeting, the Defense Committee adopted a revised set of requirements for the MTDP, which was labeled DC 28. The new requirements amounted to 49 divisions on D-Day, 79 on D+30, and 95 on D+90 by mid-1954. For the Central Region alone, the respective figures were 32, 54, and 58 divisions. At the same time, the NATO countries indicated that they planned to contribute no less than 46 ready divisions and 83 divisions by D+90 (see Table 2.2). Thus it appeared that the alliance might actually be able to achieve its ambitious rearmament objectives.

TABLE 2.2
*Medium Term Defense Plan Requirements
for 1954 (DC 28)*
(Divisions)

	Central Region	Total
D-Day		
Requirements	32	49⅓
Planned contributions	30⅔	46⅓
Gap	1⅓	4[a]
D+30		
Requirements	54	79⅓
Planned contributions	44⅓	68⅓
Gap	9⅔	11
D+90		
Requirements	58	95⅓
Planned contributions	55⅓	83⅓
Gap	2⅔	12

SOURCES: JCS 2073/157, May 22, 1951, RG 218, CCS 092
Western Europe (3-12-48), sec. 81; and D. Condit, *History of the
OSD*, 370.
[a]Planned contributions exceeded requirements by one division in the southern region.

The anticipated increase in forces in the Central Region was the result of enhanced efforts by several countries, although those of Britain and France were most important. Britain's original proposal of a three-year, £3,400 million defense program (subsequently raised to £3,600 to finance a pay increase) fell well short of the estimated £6,000 million necessary to meet all the requirements of the MTDP and was expected to provide for only 75 percent of the equipment needed by the British army.[74] The British government nevertheless extended the period of conscription from eighteen months to two years and, acknowledging that the buildup of allied forces on the continent should receive top priority, agreed to station two more divisions in Germany by late 1951, for a total of four. British leaders were eager to match the additional efforts proposed by other countries so they could maintain a leadership role in NATO.[75]

When Prime Minister Clement Attlee visited Washington in early December, U.S. officials pressed him to expand the size of the British military effort substantially. Expecting that the United States would ask its European allies to double their pre–Korean War rates of expenditure at the forthcoming Brussels meeting of the NATO Council, the British government decided to announce a dramatic increase in its defense budget. The new figure, which was eventually set at

£4,700 million over three years, amounted to nearly twice what had been planned just six months before. Although the program would entail considerable sacrifice, British leaders felt it was necessary to demonstrate that Britain was wholeheartedly fulfilling its alliance commitments and thereby to preserve the special Anglo-American partnership.[76]

The proposed French rearmament program grew apace. In October, the French government presented plans to raise a force of 20 divisions over three years based on an annual military budget of $1.74 billion. It set an interim target of 10 full-strength divisions by the end of 1951. At the same time, France would spend $686 million in Indochina and other overseas territories. Thus the French government now proposed to increase its overall military expenditures from $1.4 to $2.4 billion, or nearly 70 percent.[77]

France nevertheless expected the United States to equip most of the new divisions through the military assistance program. Of the total budget, moreover, $770 million would have to be provided by other countries. The Truman administration, however, was able to commit no more than $200 million for production and military construction for the first half of 1951 in addition to transfers of military equipment (end items), although it hoped to be able to offer a similar amount during the second half of the year. The French government replied that such a small contribution would force it to scale back its rearmament program and would make it impossible to raise the number of divisions planned. Subsequently, the French National Assembly approved only $2.1 billion in military spending for 1951. This figure still represented an increase of nearly 50 percent, however, and in January, the assembly passed additional legislation providing for the creation of 20 divisions for European defense in 1953 and extending compulsory military service from twelve to eighteen months.[78]

THE COMPLETED PROGRAM: THE DECEMBER MEETING
OF THE NATO COUNCIL

The final critical elements of the NATO program for strengthening Western Europe's defenses fell into place at the December meeting of the North Atlantic Council in Brussels.[79] The foreign ministers approved the creation of an integrated force for the defense of Western Europe and requested that the United States designate an American officer as Supreme Allied Commander Europe (SACEUR). The NAC recommended that Truman appoint Dwight Eisenhower to the position, which the president promptly did. Acheson informed the allies that the United States would place all of its forces in Europe under the supreme commander, and Truman reiterated the U.S. intention to station additional combat units on the continent.[80] In response, France declared that it would immediately put three divisions at Eisenhower's disposal, shortly to be followed by another two divisions, and Britain announced its plans to increase further its production of military equipment.

Finally, the council approved a Defense Committee report, based on the Spofford compromise, regarding the political and military aspects of a German military contribution.[81] The report, which concluded that the formation of small German units should begin in the immediate future, also proposed safeguards to ensure that Germany could not "again plunge Europe into war." German ground force units would initially be limited to regimental combat teams or brigade groups, and their number could not exceed one-fifth the total number of allied units assigned to the supreme commander. In approving the report, the foreign ministers invited the occupying powers to discuss with the German government the question of participation in the defense of Western Europe along these lines, and they acknowledged France's intention to call a conference of the countries, including the Federal Republic, that might wish to participate in a European army.

Notwithstanding the high spirits that marked the Brussels meeting, much ground would still have to be covered before NATO's ambitious force goals would be achieved. There were good reasons to question whether a substantial increment in the size of the U.S. contribution would be soon forthcoming. The Truman administration had not yet decided how many divisions it would send to Europe— or when they would be dispatched—and the prospects of a prompt and substantial deployment were growing dimmer. The few available combat-ready units had been rushed off to Korea, leaving only one division in reserve in the United States at the end of 1950, and the Chinese intervention dashed any hopes that the war would be brought to a quick conclusion. Given the resulting demands for additional manpower, the JCS even opposed providing the allies with a specific figure for the number of reinforcements they could expect in the near future.[82]

The ability of the Europeans to carry out the ambitious rearmament efforts they had proposed also left considerable room for doubt. But perhaps the greatest uncertainty surrounded the issue of German rearmament. Rather than taking any concrete steps toward making German participation in the defense of Western Europe possible, the Brussels meeting had merely set the stage for another round of discussions and negotiations. By then it had become clear, moreover, that future progress would be complicated by the need to gain German consent. In mid-December, Adenauer insisted that any German units should enjoy complete equality, rejecting the restrictions embodied in the Spofford compromise.[83] To discourage German intransigence, the three occupying powers would be forced to adopt a circumspect approach in their talks with the German government. Seeking to play down the importance of German rearmament, they would have to refrain from making any concrete proposals, waiting instead for the Germans to take the lead.[84] This approach, although necessary under the circumstances, was sure to delay the realization of a German military contribution.

Implementing the Buildup, 1951–1952

During the next two years, the NATO countries attempted to implement the program to strengthen the alliance's conventional capabilities that had been so arduously developed in the fall of 1950. The results of their efforts were mixed. The integrated defense force was quickly established under Eisenhower's command. Despite strong initial congressional opposition, the United States deployed four additional divisions to Germany by the end of 1951. The implementation of the other elements of the program, however, left much to be desired. European rearmament efforts proceeded much more slowly than U.S. officials had hoped. Although in May 1952 six continental countries signed a treaty establishing a European Defense Community (EDC) that would enable West Germany to make a military contribution, moreover, the document remained far from ratification by the end of the year. Thus substantial progress was made, but as the Truman administration prepared to leave office at the beginning of 1953, it was far from clear whether NATO's conventional capabilities would ever match the alliance's force requirements.

THE ESTABLISHMENT OF THE INTEGRATED
MILITARY STRUCTURE

The rapid creation of an integrated NATO force under Eisenhower was perhaps the brightest feature of the alliance's effort to erect a strong conventional defense in Western Europe. Previously, there had been no adequate arrangements for the coordination of military planning in peacetime or for the command of combat operations in wartime. The United States, in particular, stood outside the limited existing arrangements, which had been largely adopted from the Brussels Treaty organization.

TABLE 2.3
Revised MTDP Requirements: SHAPE and MC 26/1
(Divisions)

	Central Region		Total	
	SHAPE	MC 26/1	SHAPE	MC 26/1
D-Day	31	31	46	46
D+30	65	65	97	98
Aircraft			8,500	9,212

SOURCES: JCS 2073/201, Sept. 7, 1951, RG 218, CCS 092 Western Europe (3-12-48), sec. 93; Poole, *History of the JCS*, 276; and D. Condit, *History of the OSD*, 373. Other sources give the overall MC 26/1 figures as 46⅓ and 98⅓. See Poole, *History of the JCS*, 305; Watson, *History of the JCS*, 282; and MC 33, Nov. 10, 1951, PRO, PREM 11/369.

With his appointment as SACEUR, Eisenhower was given a broad mandate to establish a Supreme Headquarters, including an integrated planning staff, and the necessary subordinate operational commands. By mid-January, an advance planning team was already busy in Paris devising the appropriate organizational structures, and by late March it had worked out most of the details. Supreme Headquarters Allied Powers Europe (SHAPE) was activated on April 2, at which time Eisenhower formally assumed command of the national forces that had been placed at his disposal and issued the first General Orders.[85]

One of Eisenhower's first actions was to review NATO's conventional force requirements. During the spring and summer, SHAPE planners prepared a revised estimate, which was designated SG 20/32. The SHAPE requirements were based on a more forward strategy than that embodied in the MTDP, whereby NATO would attempt to establish a strong defensive zone east of the Rhine. In addition, they assumed that the critical initial phase of hostilities would occur during the first 30 days. Thus the alliance's reserve forces would have to be made ready much sooner than under DC 28, which assumed a 90-day period. Consequently, SG 20/32 called for 31 ready and 65 D+30 divisions in the Central Region alone and 46 D-Day and 97 D+30 divisions overall. The SHAPE requirements were considered by the NATO military and defense committees at their fall meeting in Rome and approved with only slight modifications as MC 26/1[86] (see Table 2.3).

THE DEPLOYMENT OF ADDITIONAL U.S. FORCES
TO EUROPE

Like the establishment of SHAPE, the augmentation of U.S. forces in Western Europe was carried out with considerable dispatch. During the course of 1951, the United States deployed an additional four divisions along with six wings of tactical aircraft in West Germany, raising its contribution to European defense to a total of six division-equivalents and seven wings. Although many of these units had to be called to active duty, the main obstacle was strong opposition in the Senate, which precipitated what became known as the "Great Debate." The deployment proposed by the administration was ultimately approved in a nonbinding resolution, but the Senate action served notice that the president would not be free to increase the size of the U.S. military presence in Europe at his own discretion in the future.[87]

Public opposition to the administration's deployment decision surfaced immediately after the close of the Brussels meeting. On December 20, former president Herbert Hoover, in a radio broadcast, insisted that the United States should not send "another man or another dollar" to Europe until the allies had demonstrated the ability to defend themselves.[88] Most of the Great Debate, however, took place during the following three months on the floor and in the hearing rooms of the Senate, where a number of critics of the administration's policies were to be found.

The opponents of the administration condemned the decision to dispatch additional troops to Europe on three grounds. First, they criticized the wisdom of the action. Some questioned the importance of Western Europe to American security and the need to station forces on the continent, given the natural advantages afforded by the United States's geographical position. Some also doubted the feasibility of defending Western Europe, even with substantial American help, in view of the size of the Soviet army. Because any U.S. forces in Europe might be quickly overrun, they maintained the United States should limit its contribution to the defense of the region to strategic bombing. The process of rapidly building up NATO's conventional forces, moreover, could provoke the very attack it was intended to deter. And once the United States had dispatched additional troops to Europe, it might never be able to extricate them. In fact, the allies would only press for even more U.S. help.[89]

This last point was closely related to the second set of criticisms, which concerned the appropriate size of the U.S. deployment. The administration's opponents were concerned lest the allies relax their rearmament efforts. In their view, the responsibility for the defense of Europe should rest primarily with the Europeans themselves; American forces should only supplement, not substitute for, indigenous forces. Although few senators went as far as Hoover in insisting that the commitment of additional troops should only follow the allies' development of an adequate defense, they argued that the size of the U.S. contribution should be strictly conditioned on European efforts.[90]

Finally, some senators questioned the authority of the president to station troops in Europe without congressional approval. The Senate had ratified the North Atlantic Treaty on the understanding that it did not involve the deployment of any additional U.S. forces. Because the administration was now proposing a fundamental change in the nature of the U.S. commitment, the Congress should be involved in the decision-making process. This line of criticism was embodied in a resolution introduced by Senator Kenneth Wherry of Nebraska stating that "no ground forces of the United States should be assigned to duty in the European area for the purposes of the North Atlantic Treaty pending the formulation of a policy with respect thereto by the Congress."[91]

In response, Truman steadfastly maintained that his constitutional role as commander in chief of U.S. armed forces gave him the authority to send troops anywhere in the world without congressional permission,[92] and a parade of high-level administration witnesses spelled out the rationale for deploying additional troops to Europe. To begin with, they contended, the preservation of an independent Western Europe was of unique importance to U.S. security, in view of its tremendous resources. Consequently, the United States had a vital interest in deterring Soviet aggression in the region.[93] Successful deterrence, however, required the presence of adequate "forces in being" in Western Europe in addition to strategic air power and the West's substantial war-making potential. Strong conventional forces were needed to defend the region in the event of hostilities until strategic bombing took effect. In fact, by protecting forward air bases from which many of the bombers used in a nuclear attack would have to be launched, given that only a small number of aircraft with intercontinental range existed at the time, these forces would play an indispensable role in any large-scale air campaign and thus bolstered the threat of strategic retaliation. Furthermore, only conventional forces in place in Western Europe could deter and defend against lesser forms of aggression, such as attacks by satellite countries perpetrated at Soviet urging. Finally, conventional forces would assume increasing importance as a deterrent as the U.S. lead in atomic weapons diminished in the coming years. Thus it was essential that NATO use its present advantage to build up sufficient ground forces for the future.[94]

Nor did the administration doubt that Western Europe could be de-

fended. Fresh from a whirlwind tour of the region in January, Eisenhower assured Senate skeptics that NATO could protect "rather significant portions" of Western Europe with far fewer forces than the
Soviet Union possessed. In his judgment, forty divisions—far less
than the existing MTDP requirement—could offer effective resistance. In the short run, however, the countries of the region would be
unable to generate the required forces by themselves. A significant
U.S. contribution would be necessary for the time being.[95]

In fact, by boosting morale in Europe, the deployment of additional U.S. troops would stimulate—rather than dampen—European
rearmament efforts and inspire the allies to provide the remaining
required forces. The bulk of the ground forces would eventually be
generated by the Europeans themselves. Once the allies had built up
sufficient forces, the United States might be able to withdraw its
troops. The Europeans would not act, however, unless the United
States took the lead.[96]

It was not until mid-February that the administration specified the
number of units that it intended to deploy in Europe. This delay
reflected the continuing uncertainty over just what forces would be
available, given the still fluid situation in Korea. As a result of this
silence, however, unofficial estimates proliferated, ranging as high
as 15 additional divisions. To allay congressional fears that the administration was contemplating the dispatch of a significantly larger
contingent, Marshall revealed on February 15 that only four more
divisions would be sent.[97]

The Great Debate came to an end in early April, with the passage
(69–21) of Senate Resolution 99. The resolution represented a compromise between the administration's supporters and its critics. On
the one hand, it approved the assignment of the four additional divisions to Europe, the appointment of Eisenhower as SACEUR, and the
placement of all U.S. forces in Europe under his command. On the
other hand, it clearly stipulated that the allies should shoulder a fair
share of the burden of their defense by making "the major contribution to the ground forces" under Eisenhower's command, and it
expressed the "sense of the Senate" that no additional U.S. troops
should be sent to Europe without further congressional approval.
Thus SR 99 reasserted Congress's right to play a role in the decision-
making process and placed a further constraint on future U.S. policy.[98]

Following the Great Debate, U.S. deployments to Europe proceeded rapidly. The first reinforcements set foot in Germany in late May;
the last combat units arrived in December. Thus by the end of the

year, the newly formed U.S. Seventh Army was at full strength, although the troops would need some training before they would be able to fight as a cohesive whole.[99] Meanwhile, however, SR 99 had already begun to inhibit the administration from further increasing the size of the U.S. military presence in Europe short of a major crisis. A May 1951 U.S. Army proposal to dispatch an additional division in early 1952 was deferred until the following February and then indefinitely.[100] Any future growth in NATO's conventional capabilities would first have to come from the European members or through German rearmament.

EUROPEAN REARMAMENT

Although considerable progress was made, European rearmament efforts proceeded much more slowly than U.S. officials had hoped. During the first half of 1951, it became clear that the allies would be able to provide far fewer forces than they had promised the previous fall, a proposed contribution that already fell somewhat short of the alliance's force requirements. In an attempt to narrow these differences, the North Atlantic Council established a high-level committee to review both NATO requirements and the ability of the member countries to generate the necessary forces. The resulting report, which was adopted at the Lisbon meeting of the council in February 1952, took an optimistic view of the situation and largely reaffirmed the program to strengthen NATO's conventional forces that had been developed in late 1950. Almost at once, however, Britain and France scaled back their rearmament efforts, and by the end of the year the gulf between force requirements and likely future capabilities seemed unbridgeable.

The Emerging Gap. Although the force contributions pledged by NATO countries substantially met the requirements established for 1954,[101] at least two developments threatened to result in a significant widening of the remaining gap. First, the United States wanted to advance the date by which the MTDP would be completed from mid-1954 by as much as two years.[102] Second, additional forces would be required if NATO plans were amended to include the defense of West Germany.[103] The anticipated gap also widened somewhat as a result of the analysis conducted at SHAPE that summer. NATO planners now estimated an overall shortfall of 15 ready and 23 D+30 divisions against the new requirements of 46 and 97 divisions, respectively. Most of the deficit—10 ready and 17 D+30 divisions—was expected in the Central Region.[104]

Of equal if not greater concern to U.S. officials was the growing gap

between the estimated cost of implementing the MTDP and anticipated levels of European defense spending. At the beginning of the
year, they estimated the four-year cost of meeting the MTDP requirements for Western Europe at $57.4 billion. The countries of the re-
gion were expected to contribute approximately $32 billion of the
total over the same period, with the United States making up the
difference.[105] To achieve this goal, the European allies would have
to double their combined pre–Korean War defense budgets from $5
billion to $10 billion, 30 percent of which would be needed to meet
their major equipment requirements. As of that time, however, annual European defense spending had risen to only $6.6 billion, while
spending on equipment had reached only $1.5 billion per year.[106]

Such concerns only intensified as the anticipated cost of implementing the MTDP skyrocketed during the following months. By
mid-1951, the estimate for the European share had ballooned to approximately $72 billion, but the countries of the region were ex-
pected to be able to provide no more than $40 billion. Even if the
United States continued to provide assistance at the rate of about $5
billion per year, there would be a financial shortfall of some $12.5
billion by mid-1954.[107]

Several reasons existed, moreover, for doubting whether even the
force contributions promised by the European allies would be forthcoming. One was the intrinsic political and economic weakness of
many of the countries in the region, which, after all, had been the reason for subordinating rearmament to recovery before the Korean War.
The standard of living was still very low in Europe, where per capita
income averaged less than a third of that in the United States.[108] Some
European governments were unstable, moreover, and hence reluctant to make unpopular decisions.[109] In light of these problems, the
rearmament programs undertaken by most of the allies could be regarded as extremely ambitious.[110]

These intrinsic weaknesses were quickly compounded by the severe economic strains produced by the military buildup, which effectively resulted in a state of partial mobilization.[111] Increased outlays
for defense reduced the amount of money available for civilian consumption, further depressing living standards, and in some cases,
governments imposed rationing.[112] At the same time, imports of raw
materials jumped while exports declined as productive capacity was
redirected to the manufacture of military equipment, worsening Europe's already chronic balance-of-payment deficits. The heightened
worldwide demand for raw materials, moreover, caused their prices
to rise much faster than those of manufactured exports, adversely

shifting the terms of trade and putting additional pressure on trade balances.[113] Rising import costs also stimulated inflation. In the first 12 months after the Korean War began, the cost of living increased 10 percent in most countries, and it climbed 20 percent in France, prompting the restoration of some wage and price controls.[114]

At the same time that the sacrifices entailed by rearmament were making it increasingly unpopular with domestic audiences, the very argument used to justify the effort in the first place was losing its force. The sense of urgency that had animated the allies following the outbreak of hostilities in Korea diminished quickly on the continent. Although tensions rose briefly following the Chinese intervention in November, the ensuing stalemate soon resulted in a palpable decline in the sense of threat. European fears of a Soviet offensive were largely dispelled as early as the summer of 1951, when armistice talks began. Thus the allies became "more impressed by the burden of rearmament than by the danger of aggression."[115]

The allied buildup was also slowed by competing security concerns. The British were reluctant to undertake additional commitments on the continent, which they regarded as incompatible with the maintenance of the Commonwealth and their other international involvements. What commitments they did make were largely intended to cement Britain's special relationship with the United States and to prevent the Americans from once again withdrawing into isolationism.[116] The expansion of French forces in Europe was hindered by France's deepening involvement in Indochina, which diverted scarce military personnel and resources. Although conscripts were not used in the conflict, it absorbed a high percentage of France's officers and noncommissioned officers and impeded the training of draftees in Europe. The dispatch of twelve battalions in early 1951, moreover, meant that France would be unable to raise the ten divisions planned for Europe by the end of the year. Meanwhile, the financial cost of the war effort mounted to more than $1 billion per year, although a growing portion was assumed by the United States.[117]

Finally, European defense efforts were hampered by the slow arrival of U.S. military assistance, which the allies depended on to achieve their rearmament objectives, especially since many governments were reluctant to call up men for military service without assurance that adequate equipment would be available. U.S. policy was to provide only those items of equipment that each European country could not produce for itself or procure elsewhere within the limits of its financial resources. Initially, however, the amount of end-item assistance was further restricted because the United States assigned higher priority to prosecuting the wars in Korea and Indochina and to meeting the requirements of its own military buildup. Thus most early equipment transfers were limited to what was available in U.S. reserve stocks located in Europe. Efforts were made to circumvent this bottleneck by appropriating increasing amounts of assistance in the form of "counterpart funds," which could be used by recipients to offset the costs of their own military production, and "offshore procurement," which involved placing orders for equipment in European rather than American factories. Notwithstanding such innovations, at the end of 1951 only $1.5 billion in equipment had been delivered while billions of additional dollars in appropriated military assistance funds remained unexpended.[118]

The Temporary Council Committee. Through the summer of 1951, U.S. officials continued to believe that NATO could achieve the ambitious MTDP objectives. They viewed all the European countries, with the possible exception of France, as capable of steadily increasing their efforts in 1952 and the following years. The allies, however, felt that their economies were already stretched to the limit and that the gap could be filled only with additional U.S. assistance.[119]

Intra-alliance differences over the size of the gap and what to do about it reached a head at the September meeting of the North Atlantic Council in Ottawa. To avoid a divisive debate, the NAC deferred discussion of the subject and instead established a high-level Temporary Council Committee (TCC) to analyze the issues involved in reconciling the requirements of fulfilling a militarily acceptable plan for the defense of Western Europe with the realistic political-economic capabilities of the member countries. For the first time NATO would review jointly the alliance's military requirements, economic and financial resources, and political constraints before recommending a concrete course of action.[120]

The TCC was composed of officials, often ministers, appointed by each of the member governments. Most of the analysis, however, was conducted by a three-man Executive Bureau, which was chaired by the U.S. delegate, Averell Harriman, and included the British and French representatives, and a Screening and Costing Staff (SCS), which was also headed by an American. These two bodies worked in two directions. The members of the Executive Bureau, also known as "the Wise Men," focused their attention on identifying the political and economic obstacles that most hampered each country's rearmament effort and determining how to reduce those obsta-

TABLE 2.4
TCC Force Goal Recommendations

	Forces-in-being			
	1951	1952	1953	1954
Total divisions				
M-Day	19	25	36⅔	41⅓
M+3	20⅔	31⅓	43⅔	53⅔
M+30	34	54⅔	69⅓	86⅔
Aircraft	2,907	4,230	7,005	9,965

SOURCES: Poole, *History of the JCS*, 276; and D. Condit, *History of the OSD*, 375.

cles so as to maximize the resources that could be devoted to defense. The SCS screened the alliance's military requirements and examined each country's actual capabilities to devise an effective but economical program for increasing NATO's military strength.[121]

The TCC issued its findings in mid-December. The committee's report largely confirmed the necessity and feasibility of the existing rearmament efforts. It proposed a detailed program for "the most rapid practical buildup of balanced effective combat forces" over the next three years that would result in the fulfillment of roughly 90 percent of the MC 26/1 force requirements that had been adopted in Rome the previous month. Specifically, the TCC recommended an increase by the end of 1952 in the number of effective divisions available by M+30 from 34 to 54⅔ followed by a further buildup to a total of 86⅔ ready and mobilizable divisions during the next two years (see Table 2.4).[122]

These goals represented a reduction in the demands that had been previously placed on the Europeans in several other respects as well. In contrast with the MTDP, the target date for the completion of the buildup had been postponed by at least six months. More important, the figures for 1954 included a German contribution of eight ready and twelve M+30 divisions that the other European countries would no longer have to provide. In addition, the screening of NATO's force requirements had reduced the most recent estimate for the cost of the European buildup from $92.7 billion to $73.9 billion overall and to $66.5 billion for the next three years.[123]

To pay for even this less demanding program, however, the TCC recommended that the European allies (including Germany) raise their planned defense expenditures by $2.8 billion over the 1952–54 period, for a total of $41.8 billion, and it spelled out specific spending increases for each country ranging from zero to as high as 40 percent in the case of Belgium. Notwithstanding these additional financial contributions, the allied effort was expected to fall short of the target by $6.1 billion, even when $18.6 billion in U.S. end items was taken into account. Because of the continuing low level of European production, moreover, the TCC predicted an equipment gap of some $10 billion, which could be filled only by increased U.S. military assistance.[124]

Further Setbacks. The force goals established by the TCC were adopted by the North Atlantic Council at its late February meeting in Lisbon with only slight modifications (see Table 2.5).[125] By that time, however, the detailed rearmament program that Harriman and his associates had so painstakingly devised had already begun to unravel. The TCC's estimates of the likely size of the financial and equipment gaps had been based on several favorable assumptions. In particular, they assumed that the shortages of manpower and raw materials that had hitherto hamstrung the European military buildup would be overcome, allowing an overall increase in the region's GNP of 14 percent, and that inflation would be halted.[126]

That these assumptions were overly optimistic became clear when it came time for individual countries to increase their efforts along

TABLE 2.5
Lisbon Force Goals
(Divisions)

	1952 (firm commitments)		1953 (provisional)		1954 (planning goals)	
	M-Day	M+30	M-Day	M+30	M-Day	M+30
United States	5⅔	7⅔	5⅔	7⅔	6⅔	9⅔
France	5⅓	12⅓	5⅓	17⅓	7⅓	22⅓
Germany	—	—	6	6	8	12
Italy	6	11⅔	9	15⅓	9	16⅓
United Kingdom	4⅔	6⅔	4⅔	6⅔	4⅔	7⅔
Subtotal	21⅔	38⅓	30⅔	53	35⅔	68
Other	3⅓	15⅓	6	19⅓	6	21⅔
Total	25	53⅔	36⅔	72⅓	41⅔	89⅔
Central Region					31	63⅓
Aircraft	4067		7005		9965	

SOURCES: Poole, *History of the JCS*, 293; Watson, *History of the JCS*, 282; Combs, "From MC-26 to the New Look"; and author's estimates. Other sources provide lower figures for the 1952 M+30 goal. Poole, *History of the JCS*, 305, and D. Condit, *History of the OSD*, 376, place it at 51⅔ divisions. *FRUS, 1952–54*, 5: 147, gives 50⅔ divisions.
NOTES: Other includes Belgium, Canada, Denmark, Luxembourg, Netherlands, Norway, and Portugal and excludes Greece and Turkey.

the lines the TCC had recommended. Although the size of the additional expenditures requested of the European allies was slightly reduced, to $2.4 billion, as of February 1952 they had agreed to spend no more than a further $1.2 billion on defense. In addition, the allies failed to make firm commitments to increase their production of defense items after 1952, as the TCC had hoped, asking instead that the United States place more production orders in their countries.[127] Thus a supplementary report prepared by the TCC noted that the readiness of as many as 25 percent of the divisions programmed for 1952 was uncertain, depending on the resolution of economic and financial problems, and that the prospective shortage of major items of military equipment for the subsequent stages of the buildup was likely to be greater than anticipated.[128]

Part of the gap was caused by slippage in the British defense effort. In October 1951, the ruling Labour party was defeated in a general election, largely on the issue of its handling of the economy, and replaced by the Conservatives, setting the stage for a revision of British policy. The new government, headed by Winston Churchill, held more conservative fiscal attitudes than those of its predecessor and felt less bound to persevere with the rearmament program that Labour had set in motion.

Churchill and his cabinet inherited a difficult economic situation. Britain, which had enjoyed a substantial positive balance of payments as recently as 1950, was now running a deficit at the rate of approximately $2 billion per year, and its dollar and gold reserves were falling rapidly.[129] Some British officials feared that confidence in sterling as an international reserve currency would be shaken, necessitating a further devaluation of the pound, and that Britain might be unable to purchase necessary imports.[130]

To eliminate the deficit, the government sought to reduce overseas expenditures by $1 billion and to increase exports. Because the metal-using industries were responsible for a high percentage of Britain's exports, however, the latter measure required that British military production be scaled back.[131] The Conservatives also decided to reduce the defense budget for the 1952–53 fiscal year from £1,667 million to £1,470 million. As a result, Britain would be unable to complete its rearmament program in three years, as previously planned, and it would not be able to place two additional divisions on the continent by M+30, as called for in the TCC force goals for 1952 and 1953.[132] Despite these efforts, however, Britain's economic problems persisted. To stem the continuing loss of reserves, government officials pressed for a further increase in exports, requiring additional restrictions on defense output, and in the fall of 1952, the cabinet once again agreed to a substantial reduction in the defense budget for the following year, this time from £1,759 million to £1,610 million, and to limit sharply the level of military production by the metal-using industries.[133]

The French buildup suffered from an even greater disparity between promise and performance during 1952. The TCC had recommended that France budget $3.4 billion for defense, of which the United States would contribute $257 million in the form of direct financial assistance, to raise fourteen divisions in Europe by the end of the year.[134] At U.S. urging, the recently installed government of Edgar Faure promised to spend a total of $4.0 billion, which represented a further $357 million increase in French outlays. In return, the United States agreed to provide $500 million in budgetary aid.[135]

U.S. officials estimated that the expanded budget would ensure that France could meet its TCC force goal for 1952, although some of the divisions would be at a reduced state of readiness.[136] The French government, however, argued that because of its effort in Indochina, even a twelve-division force was out of the question and that ten would be a more reasonable target. The additional expenditures would be needed simply to cover inflation and the rising cost of the war.[137] Eventually, the two sides agreed that France should plan to field twelve divisions that year, even if two were not complete, and the force goals adopted for France at Lisbon were reduced accordingly.[138]

The achievement of even this more modest rearmament objective was immediately called into question. To pay for the revised defense budget, Faure would have to increase taxes by approximately $900 million, or 15 percent. But France's dire economic situation made this next to impossible. Thus only days after the Lisbon conference, Faure's request was rejected by the National Assembly, and the government resigned.[139]

Having inherited a severe economic crisis and not wanting to repeat the mistakes of its predecessor, the new government of Antoine Pinay slashed the defense budget by nearly 10 percent and conditioned the achievement of even this reduced target on the receipt of adequate U.S. aid. As a result, the remainder of the year was marked by constant squabbling between France and the United States over the amount of military assistance France could expect to receive and a steady decline in the size of the forces it promised to provide.[140] In May, France asked for $616 million in U.S. production orders over the next three years, and in June it requested $150 million in additional aid for Indochina.[141] In response, the Truman administration offered only $185 million in offshore procurement during the next fiscal year, which France claimed would result in insufficient military production, although the United States would increase its contribution to the war effort by the amount requested.[142] In the meantime, the French government had developed its 1953 defense budget on the assumption that the United States would provide a total of $650 in financial aid—the $500 million promised at Lisbon plus the additional $150 for Indochina—and in October it announced plans to spend a total of $4.18 billion the following year.[143] Because Congress had cut the military assistance proposed for Europe by more than 20 percent, however, the United States could offer only $525 million, and it was also forced to reduce the amount of military hardware that it could provide from $690 million to $400 million.[144] France reduced its proposed budget accordingly. In November, the French government informed NATO that it would not be able to increase its military contribution to fifteen divisions the following year, and by the end of 1952 only six French divisions had achieved a reasonable degree of readiness, rather than the ten that had been promised.[145]

PROGRESS TOWARD A GERMAN MILITARY CONTRIBUTION

Although the efforts of the European allies fell short of their original rearmament goals, the greatest setback to the accomplishment of the conventional force buildup set in motion in 1950 was the failure to achieve a German contribution.[146] The Brussels meeting of the NAC had set the stage for two independent sets of negotiations, yet the direct talks between the three occupying powers and West Germany concluded after less than six months without success. Consequently, most hopes were pinned on the parallel French-sponsored European army conference in Paris, which the United States belatedly supported in mid-1951. The participants eventually reached agreement on the details of a treaty to establish a European Defense Community, which was signed in May 1952. During the following months, however, little progress was made toward ratification. Thus by the end of the year, the prospects for raising German military units in the near future seemed little brighter than they had two years before.

The Petersberg Talks and the European Army Conference. The negotiations between the United States, Britain, France, and West Germany on the question of a German contribution to Western Europe's defense formally began in early January at the Petersberg near Bonn. The three occupying powers opened the meetings by authorizing the German government to prepare plans for raising 100,000 ground troops in 1951.[147] Fearing that the French proposal for a European army would result in lengthy delays, U.S. officials hoped that the talks would quickly result in an agreement allowing the first German units to be created and assigned to NATO while permanent institutional arrangements were devised.

These hopes were soon dashed. From the outset of the negotiations, the German government took a firm stand, setting forth several political conditions that would have to be met before it would agree to make a military contribution. These conditions included the stationing of adequate allied forces in Germany, the replacement of the imposed Occupation Statute with a contractual relationship, the treatment of German units on a basis of full equality, and financial assistance.[148] In response to these demands, the allies revised the Occupation Statute to enable West Germany to conduct diplomatic relations and to place additional curbs on the high commissioners' control over domestic legislation and trade.[149] Britain and France, which remained unenthusiastic about promptly rearming Germany, however, were unwilling to make substantial concessions on other issues, notably the size of the largest allowable German ground force unit and many of the additional discriminatory safeguards that had been devised the previous fall. As a result, the talks adjourned in early June without agreement.[150]

Following this setback, U.S. officials reluctantly pinned their hopes on the European army conference in Paris, which France had convened in mid-February. The Truman administration's initial impulse was to continue to seek some way of forming German military units without delay, even if this meant pressing France to accept the German position on the issues left unresolved by the Petersberg talks.[151] By this time, however, the U.S. ambassadors in Europe as well as Eisenhower were coming to the conclusion that, given the views of the allies, a German military contribution could be obtained at least as rapidly through the mechanism of a European army as by any other means and thus began to urge Washington to throw its support behind the Paris conference.[152] Consequently, the administration decided to press for the rapid creation of what it termed a "European Defense Force," while still insisting on the development of a specific plan for raising German contingents at the earliest possible moment and a political arrangement restoring substantial German sovereignty, which was considered necessary to obtain German agreement.[153]

Even this approach faced obstacles. In its first six months, the European army conference had made only modest progress. Only five countries—France, West Germany, Italy, Belgium, and Luxembourg—had agreed to participate, and because national elections were scheduled in France for June, the French government had sought to postpone discussion of the more controversial issues, especially the size of the basic military units and the level of integration.[154] Nevertheless, the participants issued an interim report in late July, which proposed the creation of a European Defense Community by 1953. The EDC would consist of a force of 20 divisions and some 600,000 to 700,000 men and would be governed, like the coal and steel community, by an Assembly, a Court of Justice, a Council of Ministers, and a European Authority endowed with supranational powers.[155]

In Pursuit of a Treaty. U.S. endorsement of the concept of a European army gave a strong boost to the conference, which the Truman administration hoped would be completed by November. Much work remained to be done before a treaty would be ready, however, and unforeseen hurdles soon arose. In the effort to establish a new contractual relationship between the occupying powers and Germany, for example, France insisted that the Occupation Statute should not be replaced until the EDC had been ratified. At the same time, West Germany requested that the allies extend formal security guarantees and commit themselves to maintain their forces in the country. This request raised the question of German membership in NATO, which France staunchly opposed.[156]

As a result of such conflicts, the deadline set by the United States passed without an agreement, even though France had made considerable concessions on the size of the units that would be allowed. In November, the EDC conference approved a plan for a force equivalent to 43 divisions—14 French, 12 German, 12 Italian, and 5 Benelux[157]—which Eisenhower subsequently pronounced as adequate. Instead of being limited to regimental combat teams, each country, including Germany, would contribute division-size units, or "Groupements," of 13,000 men, although support functions would be concentrated at the corps level under multinational control.[158]

At this point, the prospects for ratification of any treaty that might emerge from the conference seemed highly doubtful.[159] Considerable opposition to the project existed in the French parliament, and the Benelux countries held strong reservations about the proposed supranational arrangements, which they feared would force them to surrender too much control over their armed forces, budgets, and economies.[160] Thus a meeting of the foreign, defense, and finance ministers of the six participants at the end of December left many issues unresolved.[161]

Although these differences were narrowed considerably at a subsequent meeting in late January 1952,[162] yet another conflict erupted between France and Germany, which threatened to derail the whole enterprise. In early February, the German Bundestag formally approved the government's handling of the negotiations but conditioned its assent on the elimination of "every vestige of legal, economic, and financial discrimination." One week later, the French National Assembly endorsed the EDC while recommending the imposition of certain restrictions and limitations on West Germany.[163] The positions of the two bodies were clearly in conflict, and a NATO endorsement of the EDC at the upcoming Lisbon meeting, which was sought by U.S. officials, was consequently jeopardized. In last-minute negotiations in London, however, Adenauer and French Foreign Minister Robert Schuman, with the help of Acheson and British Foreign Secretary Anthony Eden, managed to overcome nearly all of their differences. As a result, the NAC was able to adopt a resolution approving the principles underlying the EDC, urging the early signing of the treaty, and affirming the necessity of creating a European defense force as soon as possible.[164]

Despite this apparent progress, the signing of the treaty was delayed for several more months by demands for additional assurances. France was concerned about Germany's possible secession from the EDC once its forces had been built up and thus sought guarantees by the United States and Britain that they would not stand idly by should the organization's integrity be threatened. Germany wanted an automatic commitment on the part of EDC members to provide military assistance in the event of an attack. And the Netherlands pressed for a guarantee of automatic British involvement.[165] To remove these last hurdles, Britain agreed to enter into a treaty with the EDC pledging mutual assistance and, jointly with the United States, to issue a declaration stating that the two governments would regard any threat to the integrity of the EDC as a threat to their own security and would act in accordance with the consultative provisions of the North Atlantic Treaty.[166] With the way finally cleared, the EDC treaty and the new contractual agreements with Germany were signed in late May 1952.

In Pursuit of Ratification. The signing of the EDC treaty marked the high point of the Truman administration's efforts to obtain a German military contribution to the defense of Western Europe. Although the U.S. Senate quickly approved the contractual agreements and a protocol to the North Atlantic Treaty extending its guarantee to the EDC countries, new obstacles promptly arose that would prevent ratification of the EDC in 1952. Paris was increasingly concerned that because of the continuing diversion of French forces and resources to Indochina, Germany would eventually dominate the EDC. Alternatively, if France sought to maintain parity with Germany in the EDC while continuing the war effort, its resources would be stretched to the breaking point. In addition, many French leaders feared that France's involvement in a supranational organization would threaten the country's relatively privileged position in NATO, placing it at a decided disadvantage vis-à-vis the United States and Britain. Any remaining hopes for early French ratification were dashed in October, when the president of the National Assembly, Édouard Herriot, attacked the treaty, triggering the French equivalent of the Great Debate that had raged the previous year in the United States.[167]

In Germany, the opposition Social Democratic party immediately attacked the EDC as unnecessarily perpetuating the division of the country. In addition, there was a danger that the treaty and the contractual agreements would conflict with the German Basic Law. In that case, a two-thirds majority in the Bundestag would be needed to amend the constitution. Uncertain of even a simple majority, Adenauer postponed a final vote on the treaty until the question could be clarified.[168]

NATO'S CONVENTIONAL FORCES AT THE END OF 1952

By the end of 1952, NATO's conventional capabilities were much greater than they had been before the outbreak of the Korean War. In fact, the Lisbon force goals for that year had been substantially achieved. The alliance now fielded some 25 ready divisions, including approximately 18 in the Central Region, and could mobilize a total of about 45 divisions by M+30, leaving a shortfall of only 8 mobilizable divisions. In contrast, NATO had been able to muster only 21 reasonably effective divisions, of which no more than 14 could be rated as ready, just two years before. Thus in the judgment of the Supreme Allied Commander, both the quantity and the quality of the alliance's conventional forces had doubled.[169]

This increase in conventional strength meant that the Red Army could no longer count on being able to overrun Western Europe with the 22 divisions in East Germany. NATO's forces in the Central Region were no longer significantly weaker. To assure victory, the Soviet Union would henceforth have to reinforce these units with the 40 additional divisions stationed elsewhere in Eastern Europe and in the western military districts of the USSR. This effort, however, would provide the West with considerable advance warning. Of course, what NATO could do in the event of a large-scale Soviet mobilization was unclear because the number of reserve divisions was small compared with the forces the Soviet Union and its allies could potentially bring to bear. Nevertheless, it was no mean accomplishment to have eliminated the possibility of a successful Soviet surprise attack.[170]

Despite this achievement, however, the prospects for reaching the Lisbon force goals for 1953 and 1954 seemed increasingly dim.[171] European rearmament efforts were clearly faltering, especially in France, which had been expected to make by far the largest contribution to NATO's ground forces in the Central Region. Despite its earlier commitment to field as many as twenty divisions by the end of the following year, France disposed of at most ten effective divisions, and its plans for increasing the total in 1953 had been all but scrapped. Progress toward obtaining a German military contribution had been even more disappointing. U.S. officials had originally hoped that the first German units would be organized in 1951. Although the EDC treaty had been signed, substantial political obstacles remained to be overcome in both France and West Germany before it could become operative.

As a result of these setbacks, NATO planners began to prepare for a downward revision of the alliance's force goals during the first Annual Review of requirements and capabilities, which had been authorized at Lisbon and began the following summer. In early December, U.S. officials agreed to support a 10 percent reduction in the Lisbon goals for 1953 and 1954, which the JCS stated was the largest that could be allowed without "seriously jeopardizing" the defense of Western Europe. And at their mid-December meeting in Paris, the NATO foreign ministers concluded that the pace of the buildup would have to be slowed. Emphasis would henceforth be placed on seeking qualitative improvements, such as providing necessary support elements, rather than creating additional combat forces. Because of the outgoing Truman administration's reluctance to make any important commitments, however, the ministers deferred final consideration of the revised force goals until the following spring, when the new Eisenhower administration would be in office.[172]

Analysis

Why did NATO's initial strategy place so much emphasis on conventional forces at a time when the United States enjoyed decisive nuclear superiority over the Soviet Union? Why did the allies nevertheless begin to make a concerted effort to strengthen their conventional forces in the Central Region only after the outbreak of hostilities in Korea? What influences shaped the form and the objectives of the resulting buildup, and why did the allies' efforts begin to flag well before these objectives were met?

The evolution of NATO's conventional force posture during this period can be explained largely in terms of the determinants emphasized by the balance-of-power perspective. The central role of conventional forces in NATO strategy and the alliance's conventional force requirements were dictated primarily by estimates of the size of the Soviet forces facing Western Europe and the absence of alternative military means for resisting aggression. Although U.S. strategic bombing would ultimately be the decisive determinant in any conflict with the Soviet Union, its impact would be felt too late to obviate the need for substantial conventional forces in the Central Region if Western Europe were to be successfully defended.

Similarly, the effort to provide conventional forces was greatly influenced by balance-of-power factors, especially Western perceptions of Soviet intentions. Initially, most NATO leaders believed that an attack was extremely unlikely, notwithstanding the seemingly serious military imbalance on the continent. Therefore, it made little sense to jeopardize political and economic recovery in Western Europe by attempting to raise the forces called for by NATO strategy or by reversing the previous policy of demilitarizing Germany.

The development by the Soviet Union of an atomic capability, the outbreak of hostilities in Korea, and China's subsequent intervention there caused Western leaders to revise their assessments of the likelihood of war in Europe. As long as there was a distinct possibility that the Soviet Union might resort to force, rearmament could not be subordinated to other objectives. Similarly, the arguments against pursuing a German military contribution lost much of their force in the face of the new imperative to erect an adequate defense. The intensive effort to strengthen NATO's conventional forces that resulted, however, could be sustained only as long as a sense of urgency prevailed, and when the perception of threat receded, it was only logical that the buildup would falter.

Variations in the overall level of effort were amplified by the political and economic conditions prevailing within the alliance at the time, which sharply curbed its military potential. The ability of the NATO countries to generate the forces asked of them was highly constrained during this period, coming so soon as it did after World War II. As a result, the trade-offs between meeting the external and internal requirements of security was especially acute. If West European governments had been more stable, economic recovery more advanced, and Germany's integration into the West further along, meeting NATO's conventional force requirements would have been much less problematic.

Nevertheless, there were important differences in the responses of the NATO countries that the balance-of-power perspective alone cannot explain. The efforts to strengthen the alliance's conventional forces undertaken by the various members were not all equally ambitious or equally successful. And although the allies eventually agreed to seek a German military contribution, there was no consensus on how and when that contribution should be obtained. To account for these differences, we must consider the determinants emphasized by the intra-alliance perspective.

Perhaps most fundamentally, the allies differed in their perceptions of the threat to Western Europe. Two sets of differences were particularly important. First, the United States and its European allies did not always see eye to eye on the risk of war in Europe and thus the urgency of fulfilling the alliance's force requirements, especially after the Korean War stalemated in 1951. U.S. officials continued to be anxious to complete the NATO buildup by mid-1954, if not earlier. The West Europeans, however, resumed their previous reluctance to subordinate economic recovery to rearmament.

Because of their differing geographical locations and historical experiences, the allies also held conflicting views about the likely sources of trouble within Europe over the long run. While the United States tended to see the Soviet Union as the primary problem, Germany's immediate neighbors were more inclined to regard that country as a source of considerable concern. That they did not view Germany as an immediate threat at the time owed more to the fact that the former foe was disarmed than to any fundamental change of heart. Thus the United States and its European allies approached the German question from different perspectives. For the latter, the timing, manner, and circumstances of German rearmament were at least as important as whether the erstwhile enemy ever made a military contribution to the defense of Western Europe. Although France was most responsible for delaying the formation of German units through its insistence on erecting strong barriers against a revival of German militarism and continental hegemony, Britain also advocated moving more slowly than the United States desired.

In addition to their differences over the immediacy and even the identity of the threat to Western Europe, the allies did not always assign the same priority to the task of erecting an effective defense of the region because of their competing interests and commitments elsewhere in the world. Notwithstanding the profound concern of the United States about the security of its European allies, the war effort in Korea received first call on U.S. military resources, resulting in delays in the dispatch of equipment and American forces to the continent. Similarly, because of its global interests and its commitment to develop a nuclear capability, Britain limited its NATO contribution to the minimum it regarded necessary to ensure continuing U.S. engagement in the defense of Europe, and it refused to participate in the EDC, thereby diminishing that project's chances of success. Of greatest consequence for NATO's conventional force posture, however, was France's deepening involvement in Indochina. Largely because of the war effort's growing demand on French military resources, France was unable to provide the forces it had promised in the European theater. By demonstrating the need for France to retain the flexibility provided by national control of its forces and by raising the specter of German domination of the EDC, moreover, the conflict sapped French enthusiasm toward the project, which France itself had proposed for obtaining a German military contribution, further slowing progress toward that goal.

Finally, variations in the allies' responses must also be understood in terms of differences in their intrinsic abilities to generate military forces at the time. Ironically, the countries that might naturally have made the largest contributions in view of their greater proximity to the Soviet threat were often the least able to do so because of political instability and economic debility. France, in particular, proved unable to make the sacrifices required by rearmament because of the poor economic conditions and the political and social divisions prevailing in the country. Thus French officials were forced to assure the public that the increased defense effort would not be allowed to lower living standards,[173] and they insisted that France would not be able to complete the ambitious rearmament program that it had undertaken without substantial U.S. assistance. More generally, it was only natural that once the fear of imminent hostilities had subsided, the European countries, which labored under greater domestic constraints than did the United States, would turn their attention back to potential internal threats.

Interestingly, the much greater military potential of the United States translated into only modest influence over intra-alliance outcomes. The promise of additional troops, military assistance, and U.S. participation in an integrated force was instrumental in securing allied promises to increase their conventional forces and agreement, at least in principle, to a German contribution. Yet once these commitments had been made, the United States could do relatively little to maintain the pace of the European buildup or to hasten German rearmament.[174] In fact, because their endorsement of any mechanism for raising German forces would be essential, France and West Germany enjoyed the greatest influence on the issue.

In contrast, the causal factors emphasized by the institutional perspective played only a minor role in shaping NATO's conventional force posture during this period. It could be argued that the realization of a German military contribution was greatly impeded by the institutionalization of German disarmament both among the allies and within Germany itself, which helps to explain the tremendous resistance engendered by the U.S. proposal. As recently as late 1949, the occupying powers had reaffirmed the prevailing regime of demilitarization. This regime, moreover, had become deeply rooted in West Germany, where there was no constituency favoring rearmament. The interest groups that would traditionally have supported such a course of action had been delegitimized if not dissolved altogether. Cognitive factors were also at work. Many Germans were convinced that rearmament so soon after the war would undermine their country's fledgling democratic institutions and revive German militarism.

For the most part, however, few institutions were in place to shape the evolution of NATO's conventional force posture during this early period. More important from this perspective are the new institutions that were created and that would be capable of exerting considerable influence in the future. In this regard, the establishment of SHAPE and the position of Supreme Allied Commander were significant. Because of his considerable independence and authority, SACEUR would embody and give voice to the values of the alliance as a whole and thereby reinforce existing norms of behavior. At the same time, SHAPE could serve as a presumably impartial source of new norms, regarding conventional force levels, for example, that would in turn generate powerful expectations of compliance.

A second potentially important institutional development was the passage of SR 99. This measure established an effective upper limit on the size of the U.S. military presence in Europe that could not easily be exceeded without congressional approval. As a result, future administrations would be reluctant to dispatch additional forces for fear of reopening the contentious issue of the president's authority to deploy troops abroad and perhaps even undermining the consensus in support of the existing U.S. contribution to NATO. At the same time, the deployment of six U.S. division-equivalents in Germany set a precedent that would not be easily undone. This substantial military presence became the most visible symbol of the U.S. commitment to European security, and over time, the allies came to regard it as a prerequisite for their own collaboration. Such expectations would make it difficult for the United States to reduce its forces on the continent without triggering allied remonstrances.

Finally, this period saw the development of a ceiling on the size of any future German contribution, which was set forth in the EDC treaty and its annexes. Although the treaty had not been ratified as 1952 came to an end, the EDC figure of twelve divisions represented a consensus among many of the NATO countries and West Germany that would not be easily renegotiated.

CHAPTER 3

The New Look and NATO's Conventional Force Posture, 1953–1955

NATO's conventional force posture underwent significant change during the first years of the Eisenhower administration. In December 1954, the alliance's military strategy for Western Europe was revised when the North Atlantic Council approved Military Committee document 48 (MC 48), "The Most Effective Pattern of Military Strength for the Next Few Years." Reflecting the growing availability of tactical nuclear weapons as well as the increased reliance on them embodied in the administration's "New Look," MC 48 authorized the NATO military authorities to base their plans and preparations on the assumption that large numbers of these new weapons would probably be used from the outset of any future conflict in Europe. Because the tremendous destructive power of tactical nuclear weapons would compensate for the alliance's presumed inferiority in conventional forces, MC 48 offered the hope that a truly forward defense of the region might finally be realizable.

As tactical nuclear weapons grew in importance, the role of conventional forces in NATO strategy declined commensurately. No longer would the alliance have to rely exclusively on conventional forces to hold Western Europe until the U.S. strategic counteroffensive took effect. Nevertheless, conventional forces would continue to serve a vital auxiliary function: they would oblige an attacking force to concentrate sufficiently so as to present lucrative targets for tactical nuclear weapons. Consequently, although MC 48 allowed for some reduction in NATO's overall conventional force requirements, particularly in mobilizable reserve divisions, the success of the new strategy hinged on the availability of substantially the same number of ready forces as had been deemed necessary under the old strategic concept, including a full West German contribution.

NATO's conventional capabilities, however, continued to fall well short of these military requirements. The buildup initiated after the outbreak of the Korean War slowed even further and eventually leveled off, as was reflected in sharp reductions in the alliance's force goals. Indeed, France's effective contribution in the Central Region declined as increasing numbers of units were dispatched to North Africa beginning in the second half of 1954. In the prevailing atmosphere of retrenchment, it was perhaps no small achievement that American force levels on the continent remained constant, notwithstanding reductions in the defense budget and in the overall force structure of the United States resulting from the implementation of the New Look.

This period also witnessed the belated achievement of an agreement enabling West German rearmament to proceed. The drawn-out attempt to secure ratification of the EDC treaty eventually ended in failure when France rejected the project in August 1954. But thanks to strong British leadership, an alternative arrangement—involving German membership in both NATO and an expanded Brussels Treaty organization—for harnessing West Germany's military potential on terms acceptable to its neighbors was quickly devised and went into effect the following year. Despite this progress, however, it would be some time before the first German units could be formed and add their weight to Western Europe's defenses.

Changes in NATO Strategy and Force Requirements: MC 48

MC 48, which completed the passage of NATO strategy into the nuclear age, was adopted only after more than two years of heated discussion within the alliance. The inherent difficulty of achieving the objectives set at Lisbon and a declining sense of threat had stimulated considerable interest in finding an economically less burdensome alternative strategic concept, and hopes that the introduction of tactical nuclear weapons might reduce the alliance's need for conventional forces emerged as early as 1952, especially in Britain. Although the new Eisenhower administration promptly spearheaded a significant reduction in NATO's force goals for the next two years, however, it was initially unwilling to countenance a fundamental revision of NATO's planning practices, which derived the alliance's military requirements from Soviet capabilities. The first systematic

NATO analysis of the potential impact of tactical nuclear weapons, which was completed in mid-1953, moreover, suggested that they would not allow as great a relaxation of effort as some had hoped. Only late in the year did the Eisenhower administration decide to seek greater NATO reliance on nuclear weapons in line with the New Look it had recently adopted as a means of reducing U.S. military expenditures over the long term.

The principal source of MC 48 was an alternative study, conceived and conducted at SHAPE, of how NATO could most effectively use the forces likely to be available several years hence. The "New Approach" study concluded that a future war in Europe would invariably be nuclear. NATO forces would need both to withstand a Soviet nuclear surprise attack and to launch a devastating counterattack immediately upon the outbreak of hostilities. Yet given sufficient nuclear and conventional capabilities, the alliance would be able to mount a truly forward defense. This effort to put NATO strategy on a fully atomic footing left several critical issues unresolved, however, and questions immediately arose about the feasibility of its implementation.

PRESSURES FOR CHANGE IN NATO STRATEGY

By early 1953, two factors in particular argued for a revision of NATO's strategy and conventional force requirements in Western Europe. One was the need to close the stubbornly persistent gap between the existing requirements and the forces that the allies were actually willing and able to provide. As discussed in Chapter 2, the carefully elaborated Lisbon program quickly ran aground. Indeed, the gap threatened to widen when the familiar economic and financial constraints were aggravated by weakening public support for high levels of defense spending, the continuing diversion of British and French resources to alternative commitments and programs, and delays in the achievement of a German military contribution.[1] The second factor was the development of tactical nuclear weapons. The growing prospect of substituting these new weapons for conventional forces raised hopes that NATO's seemingly unattainable requirements could finally be reduced.

The British government was the first to recognize the potential implications of the multiplication and diversification of nuclear weapons for NATO strategy. During the spring of 1952, the British Chiefs of Staff conducted a comprehensive review of British global commitments and strategy, taking into account the rapid development of atomic weapons and of U.S. strategic air power as well as the tremendous financial burden imposed by rearmament. They concluded that the forces NATO could afford to acquire by 1954—which amounted to only a fraction of those required under the existing strategic concept—would, if backed by the use of tactical nuclear weapons, be sufficient to resist a Soviet invasion of Western Europe until the impact of the strategic counteroffensive was felt.[2]

In mid-1952, the British presented these findings to the U.S. Joint Chiefs of Staff, who strongly disagreed with the assertion that NATO's force requirements could be reduced in the near future. The Americans pointed out that tactical nuclear weapons would not be available in quantity for several years and argued that the British had exaggerated the effect of strategic bombing on a Soviet offensive. They feared, moreover, that European rearmament efforts would dip sharply if any excuse were given to relax.[3] The JCS nevertheless agreed that the presence of sufficient numbers of tactical nuclear weapons might allow a reduction in NATO's conventional force needs as early as 1956, and at British request, they authorized General Matthew B. Ridgway, Eisenhower's successor as Supreme Allied Commander Europe, to estimate the force requirements for that year assuming the availability of a certain number of the new weapons.[4]

REDUCTION OF NATO FORCE GOALS

The British government and subsequently the other allies hoped that Ridgway's study would result in considerably lower conventional force requirements, allowing the burden of rearmament to be reduced. In part because of such hopes, the NATO Council postponed completion of the 1952 Annual Review, originally scheduled for December 1952, until the following spring.[5] It soon became evident, however, that Ridgway would not be able to provide even preliminary results until the summer of 1953. Consequently, the alliance would have to update its force requirements and goals based on the same assumptions it had relied on since 1950.[6]

Despite the cool reception their ideas had received in the United States the previous summer, British leaders had continued to entertain the notion of proposing a new approach to NATO military planning that would take into account possible changes in Soviet intentions, Western economic difficulties, and technological developments.[7] In early 1953, when word leaked out that Ridgway's study might not result in any narrowing of the gap, they gave serious consideration to abandoning the prevailing method of determining force requirements, which looked primarily at the adversary's capabilities and resulted in consistently high estimates, in favor of a political approach that was more sensitive to the economic situation and the actual risk of war. Rather than focusing narrowly on what was required to defeat an attack, the latter method would emphasize the forces needed to constitute an effective deterrent.[8]

By this time, the change of government in Washington had made U.S. officials more receptive to the British approach. Eisenhower believed that the Truman administration had unduly compromised America's economic strength for short-term military expediency.[9] Consequently, he sought to strike a better balance between maintaining adequate military forces and preserving a sound economy, both of which he deemed essential to long-term security. The new president felt particularly strongly about the need to hold government spending to a level that could be maintained over a period of years, and during the campaign he promised to eliminate the federal budget deficit, largely by cutting national security programs. The necessary savings could be attained, however, without sacrificing U.S. military strength, Eisenhower reasoned. Rather, the United States could achieve "both security and solvency" simultaneously.[10]

The new administration wasted little time in seeking to put these ideas into practice. In early March, the National Security Council agreed to explore the possibility of approaching a balanced budget in fiscal year 1954, which began in mid-1953, and then achieving one in FY 1955. Because the realization of this goal would require finding substantial savings in the budget proposed by Truman shortly before he left office, the Department of Defense and the Mutual Security Agency (MSA), which was responsible for the military assistance program, were asked to examine the effects on national security programs of cuts totaling $6.8 billion and $14.0 billion in FY 1954 and FY 1955, respectively.[11] These spending limitations were strongly opposed by the U.S. military, however, and were subsequently dropped.[12]

Despite this initial setback, Secretary of Defense Charles E. Wilson quickly devised an alternative approach for reducing the anticipated deficit. The military programs inherited from the Truman administration, he noted, were aimed at achieving certain force and readiness goals by 1956, requiring $45 billion in each of the next three years and $40 billion thereafter. By working on the assumption that these programs could be related to a "Floating D-Day," however, he estimated that defense expenditures could be cut to $43 billion in FY 1954 and reduced progressively thereafter without affecting U.S.

TABLE 3.1
NATO Force Goals, April 1953
(Divisions)

	End 1953 (firm commitment)		End 1954 (provisional)	
	M-Day	M+30	M-Day	M+30
Western Europe				
Belgium/ Luxembourg	3	5	3	5⅔
Denmark	⅔	4	⅔	4
France	5⅓	14⅓	5⅓	14⅓
Italy	9	14⅓	9	15⅓
Netherlands	1	3	1	3
Norway	⅔	3	⅔	3⅔
Portugal	—	1	—	1
United Kingdom	4⅔	4⅔	4⅔	6⅔
Subtotal	24⅓	49⅓	24⅓	53⅔
United States	5⅔	6⅔	6⅔	8⅔
Canada	⅓	⅓	⅓	⅓
Total	30⅓	56⅓	31⅔	63
Central Region	20	34	21⅓	39
Greece and Turkey	23⅓	41	23⅓	39⅓

SOURCES: Memorandum for the Joint Chiefs of Staff, Apr. 16, 1953, and Memorandum, Mautz to Vass, June 1, 1953, both in RG 59, 740.5; Watson, *History of the JCS*, 287; and author's estimates.
NOTE: Excludes Germany.

combat strength, at least for the first two years.[13] In other words, U.S. programs would no longer be geared toward preparing for war at a specific date.

Wilson's ideas were quickly worked into a statement of proposed national security policies and programs, which was discussed by the National Security Council several times during April and, only slightly modified, was adopted as NSC 149/2 at the end of the month.[14] Shortly thereafter, Eisenhower publicly described this new philosophy of defense planning, which became known as the "long haul" concept.[15]

The administration followed a similar approach in preparing the U.S. position for the April 1953 NATO Council meeting. To be sure, it believed that the European allies continued to face a serious military threat from the Soviet Union.[16] But their economies, still recovering from World War II, had been put under considerable strain by the hurried effort to rearm following the Korean War, and as a result, many European countries seemed seriously overextended and in grave danger of collapse.[17] Thus by the end of March, U.S. officials had concluded that the NATO buildup would have to be stretched out to avoid exceeding the economic capacities of the allies. Instead, alliance resources should be increasingly devoted to making the available forces fully combat ready. This shift of emphasis, they believed, would result in the most effective deterrent to aggression.[18] Although the Americans had gone a long way toward embracing the view held by the British, however, they remained unwilling to repudiate the use of force requirements as the ultimate basis of NATO planning.[19]

The U.S. delegates vigorously promoted the new American position at the NATO meeting. A general war with the Soviet Union was unlikely in the near term, but international tension would continue for the foreseeable future. Consequently, they argued, national security expenditures should be spread out over a longer period to make them bearable by the Western economies on a sustained basis. Otherwise, the alliance might undermine the ultimate source of its strength.[20]

Not surprisingly, the U.S. delegation had little difficulty in securing the allies' approval of the "long haul" concept.[21] Consistent with this shift of emphasis, the NATO ministers simultaneously agreed to reductions in the force goals for Western Europe established at Lisbon (see Tables 3.1 and 3.2). The largest cuts were made in the goals for the number of divisions to be mobilized within 30 days, which were reduced by 10 for 1953 and 14⅔ for 1954. In contrast, the goals for the number of ready divisions were reduced only slightly, by ⅓ and 2 divisions, respectively, reflecting the fact that the majority of these forces were already provided for in national defense plans. Largely at U.S. urging, the NATO Council assigned more importance to the

TABLE 3.2
*Comparison Between Lisbon and April 1953 Force Goals
(excluding Germany, Greece, and Turkey)*
(Divisions)

	End 1953		End 1954	
	M-Day	M+30	M-Day	M+30
Lisbon	30⅔	66⅓	33⅓	77⅔
1953	30⅓	56⅓	31⅓	63
Reduction	⅓	10	2	14⅔

SOURCES: Tables 2.4 and 3.1.

objective of raising the combat effectiveness of existing units.[22] As a result, American officials confidently predicted a steady increase in NATO's actual military strength in 1953 and 1954, lower force goals notwithstanding.[23]

ORIGINS OF THE NEW APPROACH

Although the revised force goals offered some economic and financial relief to the Europeans, they still represented a challenge that most of the allies would not be able to meet.[24] Nor did the new targets resolve the problem of the gap between force requirements and prospective NATO conventional capabilities, which promptly reemerged as a central alliance issue.[25] The British feared that the persistence of the gap, together with the planned cutbacks in U.S. defense spending and assistance to the allies, might induce a sharp decline in the defense efforts of some NATO members, a view that came to be shared in the United States.[26]

At this point, the one remaining hope for narrowing the gap lay in Ridgway's study of the effect of new weapons on NATO force requirements. As expected, however, the preliminary results, which were submitted in July 1953, confirmed that these requirements were not likely to decline. Indeed, because of the heavy losses expected to result from initial Soviet atomic attacks, NATO would need a substantially larger number of aircraft and ten more combat-ready divisions—seven in the Central Region—than were already called for, although the required number of mobilizable reserve divisions could be substantially reduced.[27]

These initial findings were nevertheless extremely tentative, being based on the questionable assumptions that the alliance would have as many as seven days to mobilize and to deploy its forces before hostilities began, that tactical nuclear weapons would be available in adequate numbers and would have the anticipated effects, and that SACEUR would indeed have the authority to retaliate when attacked and to use nuclear weapons on NATO territory if necessary. In addition, the report failed to consider the effects of the U.S. strategic air offensive or how tactical and organizational changes might reduce losses suffered by allied ground forces. To develop firm force requirements, Ridgway's successor, General Alfred Gruenther, requested further guidance.[28] Instead, however, the Standing Group of the NATO Military Committee instructed him to discontinue work on the project for the time being and to prepare an estimate based solely on conventional weapons.[29]

The Ridgway study's failure to project a lower estimate of NATO force requirements prompted the British government to make a further attempt to put NATO planning on a different footing. In most respects, the proposal it developed was fully consistent with the approach that had been adopted by the Eisenhower administration. The British felt strongly, however, that there was no chance of ever attaining the existing force requirements. Given that most NATO countries were already doing all they could short of undermining their economies and that the level of U.S. assistance was declining, the gap, if anything, was likely to widen. The resulting sense of frustration coupled with the widespread belief that the danger of attack had decreased, they feared, might lead some countries to curtail sharply their military efforts.[30]

The British therefore concluded that planning on the basis of requirements must be pushed decidedly into the background. Instead, NATO's primary objective should henceforth be to provide a force that would constitute an effective deterrent to Soviet attack and yet that countries could afford to maintain and to equip with modern weapons over a period of years. Arguing that the NATO buildup had already resulted in a force that met these alternative criteria, they proposed that alliance planning should proceed on the assumption that roughly the existing allied contributions plus the expected German contribution—but nothing more—would be available over the next few years.[31]

Gruenther had decided independently to try a similar approach. As a substitute for the suspended requirements study, he proposed to develop a plan for the defense of Western Europe based on the forces, both conventional and nuclear, that were likely to be available several years hence. Because it would focus on actual capabilities rather than requirements, this so-called Supplementary Planning Project would not require strategic guidance from the Standing Group and was soon given informal approval.[32]

To pave the way for NATO agreement to its "new approach" at the next ministerial meeting, the British government felt that it first had to gain U.S. backing.[33] U.S. officials were generally sympathetic toward the British proposal. They agreed that any substantial further buildup was unlikely and that emphasis should be placed on preserving the gains that had been achieved to date and securing such improvements as were economically feasible rather than attaining the existing requirements.[34] But they disputed the idea that force requirements could be essentially disregarded and that military planning could be based on the notion of an effective deterrent—or that such a concept could even be defined. They feared, moreover, that merely admitting the inability to narrow the gap would foster complacency. Consequently, the Eisenhower administration continued to believe that the gap would eventually be closed by reducing—not abolishing—the force requirements, chiefly by accounting for the effects of tactical nuclear weapons, and they agreed that NATO should once again undertake a study of the question the following year.[35]

The U.S. response to the British proposal was strongly colored by the conclusions of the sweeping review of basic national security policy conducted by the Eisenhower administration, which was completed in October 1953. Many administration officials had long hoped that further substantial reductions in U.S. defense spending could be effected by substituting atomic weapons for conventional forces. Through September, however, formal U.S. policy had not explicitly acknowledged this possibility.[36] This omission was corrected in the new policy, NSC 162/2, which stated: "In the event of hostilities, the United States will consider nuclear weapons to be as available for use as other munitions."[37]

To reap the full benefits of greater reliance on nuclear weapons, especially in the European context, however, the United States would first have to gain the approval of its allies for this policy and, where necessary, their consent to the use of nuclear weapons from their territory.[38] Consequently, U.S. officials raised these issues repeatedly at the tripartite Bermuda conference with Britain and France in early December and the subsequent meeting of the NATO Council.[39] The allies agreed that NATO strategy would have to take account of such new weapons and showed particular interest in the possibility of reducing conventional forces. But they resisted the idea that nuclear weapons would be used automatically in the event of hostilities. As a result, the U.S. attempt to achieve a consensus on the need to authorize tactical nuclear operations in advance was stymied.

In addition, the resolution adopted by the NATO Council to guide the 1954 Annual Review represented a compromise between the different force planning approaches preferred by the United States and Britain.[40] Formally endorsing the principle of the long haul and reflecting the British emphasis on deterrence, the resolution agreed that the NATO countries would have to support over a long period of time forces that would "be a major factor in deterring aggression" and acknowledged that no increases in defense spending could be expected in the next several years. More significantly, it instructed the military authorities to review "the size and nature of the forces required to defend the NATO area, taking account of developments in military technology, Soviet capabilities, and the overall strategic sit-

TABLE 3.3
NATO Force Goals, December 1953
(Divisions)

	End 1954 (firm commitments)		End 1955 (provisional)		End 1956 (planning goal)	
	M-Day	M+30	M-Day	M+30	M-Day	M+30
United States	5⅔	8⅔	5⅔	8⅔	5⅔	8⅔
Other	41⅓	93⅓	42	95	41	95⅓
Total	47⅔	102⅓	48	104	47	104⅓
Germany					12	12
Central Region					31⅔	53⅔

SOURCES: Watson, *History of the JCS*, 299–300; and Combs, "From MC-26 to the New Look."
NOTE: Excludes Germany. In most cases, further breakdowns are unavailable. The 1954 M-Day total includes 18 Greek and Turkish divisions. Thus it represents a reduction of 5⅓ Greek and Turkish divisions and approximately one other West European division from the April 1953 force goals.

uation," and "to press on with their reassessment of the most effective pattern of military strength for the next few years within the resources which it is anticipated may be made available." Thus the resolution endorsed both the further requirements study desired by the Americans and the capabilities study proposed by Gruenther and backed by the British.[41]

THE DEVELOPMENT AND APPROVAL OF MC 48

Believing that allied thinking about nuclear weapons lagged far behind that of the United States, the Eisenhower administration refrained from directly confronting the sensitive issue of prior authorization in the December meetings.[42] During the following months, however, U.S. officials became increasingly concerned that certain allies continued to draw a sharp distinction between conventional and atomic weapons. In their view, it no longer made sense to refuse to prepare to use the new weapons, given recent Soviet advances in the field, or "to maintain two separate military establishments," which resulted in unnecessarily inflated defense expenditures.[43] Consequently, the administration decided to press its views at the next meeting of the NATO Council in late April 1954.[44]

The agenda of the meeting was dominated by the forthcoming Geneva conference on Indochina and the reply that the three occupying powers were preparing in response to the latest Soviet demarche regarding Germany. Secretary of State John Foster Dulles nevertheless took advantage of the occasion to offer his colleagues an unprecedentedly detailed description of new "U.S. official thinking regarding nuclear weapons," in which he stressed the important role that nuclear-armed forces based in Europe could play in defending the NATO area. "It should be our agreed policy," Dulles argued, "to use atomic weapons as conventional weapons . . . whenever and wherever it would be of advantage to do so."[45] Yet he made no concrete proposals requiring prompt consideration by the alliance. Thus the evolution of NATO strategy that occurred during the following months was driven primarily by the analysis being conducted at SHAPE rather than by U.S. policy.

In July, Gruenther submitted the results of the Supplementary Planning Project—now called the Capabilities or "New Approach" Study—that he had begun the previous fall. The SHAPE paper was combined with complementary studies by the other NATO commanders into a single draft Standing Group report, which was approved largely unmodified by the NATO Council in December as MC 48. The only significant alteration made during the intervening review process concerned the still sensitive issue of the freedom of NATO military commanders to use nuclear weapons.

The Capabilities Study concluded that SACEUR must have sufficient authority to ensure there would be no delay in responding with nuclear weapons to an attack. To this end, the original Standing Group paper stated that the very commitment by NATO countries of their forces to action would constitute authorization of the use of nuclear weapons in their defense.[46] Both U.S. and British officials feared, however, that the inclusion of such language would seriously jeopardize the document's chances of approval by the alliance. Although the Joint Chiefs of Staff continued to hold a view similar to that of the NATO military authorities, by mid-1954 State Department officials had begun to back away from the original U.S. goal of gaining prior allied agreement to the immediate use of nuclear weapons. Instead, they hoped to create a de facto situation in which there could be only one decision in the event of war.[47]

By October 1954, even Gruenther had decided not to push for express agreement on the right to use nuclear weapons.[48] As a result, the controversial section of the Standing Group paper was modified simply to state the basic military requirement that NATO forces should be able to use nuclear weapons in their defense from the very outset in an attempt to make clear that the NATO ministers were not thereby granting allied military commanders the authority to do so.[49] And in preparing the U.S. position for the December meeting of the NATO Council, the JCS finally acquiesced in the State Department's decision not to press the allies at that time to commit themselves to the use of nuclear weapons in war or to grant operating rights for nuclear-armed U.S. forces. Instead, the United States would be content to gain NATO acceptance of the new strategic concept embodied in the report.[50]

Nevertheless, as the NATO meeting approached, the British began to fear that even this considerably weakened language would imply that under certain circumstances SACEUR could use nuclear weapons without high-level political approval. They hoped to develop a formula that would provide NATO commanders with the authority to respond automatically only when attacked with nuclear weapons.[51] To avert an awkward impasse at the meeting, Dulles and his British and Canadian colleagues devised a last-minute accompanying resolution clearly stating that the council's approval of MC 48 as a basis for planning and preparations did "not involve the delegation of responsibilities of governments for putting plans into action in the event of hostilities."[52] As a result of this behind-the-scenes effort, the council approved MC 48 unanimously after little discussion.[53]

THE CONTENT OF MC 48

MC 48 described the nature of a future war in Europe in terms strikingly different from those of previous NATO strategy documents. Superiority in atomic weapons and delivery capability would be the most important factor in determining the outcome. Both sides would possess a large number of nuclear weapons, which would inevitably be used. Indeed, the war was likely to open with an intensive atomic exchange of short duration, which was almost certain to be decisive, although a second, less intense phase might follow. Because of the importance of getting in the first blow, the Soviet Union would seek to achieve surprise, and NATO would receive little or no warning of an attack.[54]

The radically different nature of atomic warfare had profound implications for NATO military preparations and planning. To deter and, if necessary, defeat aggression, NATO forces required the capability both to withstand a Soviet nuclear attack and to deliver an immediate and effective counterblow. Even in the event of a full-scale non-nuclear attack, the alliance would be unable to prevent the rapid overrunning of Europe without immediately using both strategic and tactical nuclear weapons. Consequently, it was necessary to provide NATO forces with a fully effective early warning system, passive defense measures, and an integrated atomic capability.[55]

Because any war was likely to be extremely brief with little opportunity to deploy reinforcements, priority would have to be given to providing forces-in-being that could contribute to the initial phase instead of planning to mobilize maximum strength at some time after D-Day. These standing forces, moreover, would have to be reorganized and retrained for atomic war, and because there might be no time to make emergency preparations before an attack, they would have to be maintained at a high state of readiness. Finally, it was essential that NATO forces be able to use nuclear weapons from the outset and that the NATO military authorities be authorized to plan on this assumption.[56]

This new emphasis on atomic preparations, however, did not mean that substantial conventional capabilities would be superfluous. To the contrary, U.S. and SHAPE studies indicated that to realize the potential benefits from the large-scale tactical use of nuclear weapons, combat-ready ground forces would have to be deployed in sufficient strength to force an enemy to concentrate so as to present lucrative atomic targets and to be available to exploit the effects of the atomic counterattack before the enemy could recover.[57] Thus NATO military authorities repeatedly argued that any changes in alliance defense programs that might result from the new studies would be evolutionary in nature, not revolutionary. Rather than replace conventional forces, the new weapons would merely be integrated with and supplement the old ones. There was no basis, they maintained, for delaying the implementation of the force goals adopted in 1953.[58]

Similarly, the assumption of a fully effective German military contribution was an integral part of the SHAPE study. Indeed, NATO military leaders regarded the presence of the long-awaited German divisions as no less important to the success of the new strategy than the ability to use nuclear weapons. Consequently, the Joint Chiefs of Staff refused to comment formally on the Standing Group's draft report until the prospects for German rearmament had been clarified following the French rejection of the EDC.[59]

Despite these qualifications, MC 48 had at least one clear virtue. In contrast to the previous NATO strategy, the full implementation of which had long been doubtful, it held the promise of a truly forward defense that would protect the territory of all of Western Europe in the event of war. Earlier NATO war plans had envisaged a retreat to the Rhine, which would serve as the main line of defense, for lack of sufficient conventional forces. Provided that the measures needed to adopt NATO forces for nuclear warfare were implemented and that a full German contribution was forthcoming, SACEUR would be able to stop a full-scale Soviet attack much farther to the east.[60]

Nevertheless, MC 48 left three critical issues unresolved. The primary purpose of the document was to permit the NATO military authorities to plan on the assumption that they would be able to use atomic weapons.[61] The resolution that accompanied its adoption, however, highlighted the tension involved in trying to put NATO strategy on a nuclear footing. As a contemporary analysis pointed out, there was no question about the importance of launching an instant atomic counteroffensive in response to any major attack, either nuclear or conventional, especially once SACEUR's forces had been redesigned for the atomic strategy. In the case of a surprise attack, the outcome of the war might depend on immediate atomic retaliation.[62] Yet the NATO Council's action left open the issue of what would actually happen if SACEUR was unable or had no time to obtain authority to use nuclear weapons or if the necessary approval was slow in coming, raising the awkward possibility that the NATO commanders would not be able to implement the plans they had devised under the new strategic concept.[63]

In addition, MC 48 left unresolved the issue of the role of the European allies in the new strategy. It said nothing about equipping allied forces with nuclear weapons, and the United States had no intention of doing so at the time.[64] Thus in practice, the nuclearization of NATO would be limited to the American—and at a somewhat later date, the British—forces stationed on the continent. Such an arrangement, however, could only aggravate the distinction between the nuclear "haves" and "have-nots" within the alliance, and it could only cause the smaller countries to question the value of maintaining forces that had an increasingly marginal role in NATO strategy. Finally, this limitation might leave NATO vulnerable to a Soviet breakthrough because nuclear-armed American ground forces would not be evenly distributed along the length of the central front.

A third major question raised by MC 48 was whether it could actually be implemented with the resources that were likely to be available. NATO conventional force requirements continued to be based on the size of the forces likely to be arrayed against Western Europe, which had not diminished.[65] Thus while reducing the need for mobilizable reserve divisions, which had accounted for by far the largest share of the gap, the new strategy allowed for almost no reduction in ready forces, most of which had already been created and which were causing the greatest strain on defense budgets. In the Central Region, NATO ground force requirements were set at approximately 30 ready divisions under MC 48, not significantly lower than the MC 26/1 requirement of 31 divisions that had been established in 1951.[66]

In addition to sustaining already burdensome force levels, the NATO countries would have to pay for the many "minimum" measures intended to adapt their forces for the conditions of nuclear warfare. These included reorganizing and retraining units, raising manning levels and maintaining forces in a state of constant readiness, building additional facilities such as airfields so as to disperse aircraft and reduce their vulnerability to attack, improving NATO's early warning system, and increasing the level of military stocks on the continent. To carry out the forward strategy, moreover, pipelines, storage sites, headquarters, and other infrastructure would have to be constructed east of the Rhine.[67]

Thus even before MC 48 was approved, it had become clear that the "new pattern of military strength" the document envisioned would not produce any significant economies. To the contrary, the implementation of all of the measures it recommended would require greater expenditures that might well exceed the available resources. To avoid an overall increase in defense costs, it would be necessary to reduce reserve forces sharply and even to make cuts in the existing active forces, although such actions would undermine the new strategy.[68]

NATO Conventional Force Levels and Capabilities

In 1953 and 1954, before the adoption of MC 48, NATO's conventional force levels continued to rise, although not nearly as rapidly as they had during the previous two years, and then leveled off at roughly two-thirds of the force requirements stipulated under the original strategic concept (see Table 3.4). Thus a wide gap remained between NATO's actual capabilities and those deemed necessary to implement the alliance's strategy. The Central Region benefited from only part of this growth, moreover, since much of it occurred in the forces of the countries on the northern and southern flanks.

To make matters worse, the alliance's conventional capabilities in the Central Region then declined toward the end of this period because an increasingly large share of France's NATO contribution was withdrawn for duty in North Africa. And although a mechanism permitting German rearmament to proceed was finally agreed on, it would still be some time before German divisions would be able to take up their positions in the NATO line. The principal good news was that the American and British conventional force contributions remained stable. The number of U.S. combat units in Europe stayed roughly constant during this period, notwithstanding the pressure for

TABLE 3.4
Actual Forces Available to NATO, 1952–1954
(Divisions)

	End 1952		End 1953		End 1954	
	M-Day	M+30	M-Day	M+30	M-Day	M+30
United States	5⅔	6⅔	5⅔	6⅔	5⅔	8⅔
Western Europe and Canada	19⅔	38⅓	23⅔	48⅔	23⅓	51⅔
Subtotal	25⅓	45	29⅔	56	29	60⅔
Greece and Turkey	21	30	17⅓	41⅓	17⅓	38⅔
Total	46⅓	75	47	97⅓	46⅓	99⅓

SOURCE: Watson, *History of the JCS*, 319.

reductions inherent in the Eisenhower administration's New Look. Similarly, there was no change in British force levels on the continent, despite equally strong desires on the part of British leaders to reduce their defense burden.

THE NEW LOOK AND U.S. FORCES IN EUROPE

According to the Lisbon force goals and those adopted in April 1953, the United States was scheduled to station an additional armored division in Europe in 1954. By mid-1953, however, Eisenhower and his advisers had concluded that fiscal and political considerations ruled out any increase in the size of the U.S. contingent on the continent. Thus the defense budget submitted by the administration to Congress for FY 1954 provided no funds for the planned deployment, and later in the year the United States informed its allies that the additional division would not be forthcoming.[69]

Nevertheless, the real issue during this period was whether the United States would reduce its contribution to NATO. The Eisenhower administration was eager to lower federal outlays and to eliminate the government's budget deficit, in large part by cutting spending on defense. The New Look adopted by the administration in the fall of 1953, moreover, provided a strategic rationale for withdrawing forces from overseas. Despite strong pressure for reductions, however, the size of the U.S. deployment in Europe remained unchanged.

Eisenhower had always regarded the presence of U.S. forces on the continent as only a temporary expedient. When they assumed office, however, both he and Dulles believed that the time was not yet ripe to pull even a single U.S. division out of Europe. Such a move, they feared, might have a disastrous effect on European morale and thus the overall strength of the West.[70] Nevertheless, there were strong financial reasons for reducing the size of the U.S. military contingent in Europe, notwithstanding the political and psychological arguments against doing so. This tension was manifested most sharply in a study conducted during the summer of 1953 by the recently appointed members of the Joint Chiefs of Staff. In July, Eisenhower asked the new Chiefs, who would not formally assume their posts until August, to use the interim period to conduct "an intensive, full-time analysis of U.S. military problems and strategy." Although the president did not fix any arbitrary budgetary or personnel limitations, he urged them to consider the importance of cutting the federal budget in the hope that they might devise a strategy that would allow a reduction in forces and military expenditures.[71]

The new Chiefs submitted their report in early August. After reviewing the entire range of American commitments, they concluded that the United States was militarily overextended. As a result, its freedom of action was seriously limited, and its state of military readiness was deteriorating. The only way substantial savings in defense expenditures could be achieved, they suggested, would be to give less priority to overseas commitments. U.S. forces stationed abroad could then be redeployed to the continental United States, where they would be less expensive to maintain, and some of the withdrawn units might even be disbanded. They noted, however, that this course of action could have a dangerous effect on public and official opinion in allied countries.[72]

The JCS report was discussed at the end of the month by the National Security Council, which became deeply divided over the issue of withdrawing U.S. forces from Europe.[73] The advocates of redeploying those troops argued that U.S. national security priorities had changed dramatically since the American military contribution in Europe had been established. In particular, the growth in Soviet nuclear capabilities had made the defense of the continental United States much more important than it had been several years before. They noted, moreover, that the continued presence of U.S. troops overseas posed political problems and that there was considerable congressional sentiment in favor of troop withdrawals. The opponents of redeployment countered that simply discussing the issue might prompt allied countries to slacken their defense efforts and that insofar as it could be construed as U.S. abandonment of Europe, such talk would threaten the alliance's cohesion.

The issue was largely resolved in early October, when the council met to consider a draft of a new basic national security policy document, NSC 162. On the controversial question of redeployment, the draft offered two alternatives. The first position stated that although a partial redeployment of U.S. forces from Europe might be desirable, any major withdrawal "would seriously undermine the strength and cohesion of the free world" and urged continued study of the matter. The second position, which was based on the JCS report, noted the dangers posed by the overextended U.S. military posture and called for a determination of whether the United States should withdraw the bulk of its land forces stationed overseas during the next few years.[74]

After further debate, the council agreed on a compromise that combined sections from each of the two original formulations. The resulting statement emphasized the present overextension of U.S. forces while affirming the dangers of any major withdrawal. In a final paragraph, it suggested using diplomatic means to obtain allied consent for redeployments at some indefinite date in the future. The potentially most controversial section, which called for an early decision, was omitted. Eisenhower himself concluded that it was vital that there be no hint of a U.S. redeployment until the allies had come to realize that such a move would make military sense. Although the stationing of so many troops in Europe had not been intended to last indefinitely, it was a small price to pay, given the importance of the region to U.S. security and the presumed likelihood that any abrupt change in U.S. policy would destroy European morale.[75]

Nevertheless, the decision to defer U.S. troop withdrawals for the time being seemed at variance with the general thrust of the new basic national security policy, which placed greater reliance on nuclear weapons so that U.S. force levels and military expenditures could be reduced. This policy was given its sharpest expression in early 1954, when Dulles articulated what became known as the doctrine of massive retaliation.[76] The secretary of state quickly clarified, however, that the United States would not rely primarily on the threat of strategic nuclear reprisal to deter Soviet attacks against Western Europe. Because of the region's unique importance to the security of the United States and the temptation to aggression it seemed to offer, it was the one area in the world that required an adequate defense-in-being.[77]

The adoption of NSC 162/2 did not end the debate over the redeployment issue within the administration, however. Indeed, just days before the formal approval of the new basic national security policy, Wilson had "set off a flurry of anxious queries from Allied capitals" when he remarked publicly that "new" weapons might reduce the need for U.S. forces overseas. Both the president and Dulles had found it necessary to announce that the United States had no plans to withdraw troops from Western Europe.[78] And in early December, as the United States prepared its position for the NATO meeting, advocates of redeployment raised the issue again, only to be muzzled once more by the president.[79]

Nevertheless, the contradictions contained in the New Look could not be avoided. In mid-October, Wilson instructed the JCS to prepare an estimate of U.S. force requirements assuming that nuclear weapons would be used "whenever it is of military advantage to do so." In response, the Chiefs agreed that the size of the U.S. Army could be reduced from 1.5 million men and twenty divisions to an even 1 million men and fourteen divisions by mid-1957. Such a sharp reduction, however, would require a global "regroupment" of U.S. forces. As a result, a maximum of six army divisions would be available for deployment abroad. How this limit would be reconciled with the administration's decision to maintain the American presence in Europe at its existing level as well as other U.S. commitments was left unresolved.[80] Thus the issue continued to smolder, reemerging repeatedly in high-level discussions within the administration during the following years, until it once again burst out in public in mid-1956.[81]

EUROPEAN REARMAMENT AND U.S. ASSISTANCE

There was virtually no chance that the European allies would significantly increase their conventional force contributions to NATO during this period. To the contrary, the real question was whether they would be able to preserve the gains they had made since rearmament had begun in earnest several years before. The allies found themselves increasingly squeezed by two opposing tendencies. On the one hand, the considerable political and economic pressures they faced to reduce their military efforts had intensified. On the other hand, the prospect of diminishing levels of U.S. assistance suggested that even the maintenance of defense spending at existing levels would not prevent force reductions.

The sense of threat that had triggered and sustained the original NATO buildup had diminished palpably by 1953. In most corners of Western Europe, the possibility of a Soviet attack now seemed remote. Much more likely was a prolonged period of Cold War, and even that level of tension was not universally regarded as inevitable. The British in particular felt that Soviet intentions had moderated and that NATO defense planning should reflect this fact. In the prevailing atmosphere, European governments faced increasing difficulty in maintaining public support for rearmament.[82]

The diminishing sense of threat was paralleled by a growing consensus that the rearmament effort had reached its economic limits and must level off, if not decline. European defense spending had more than doubled since before the Korean War, from $5.3 billion to approximately $12 billion per annum. In contrast, the combined GNP of the allies had grown by less than 10 percent. Per capita GNP in Europe, moreover, remained less than a third of that in the United States. Thus although the United States was dedicating a somewhat higher percentage of its national income to the military, each additional dollar the allies devoted to defense entailed a much greater sacrifice. Consequently, by mid-1953, U.S. officials joined their British counterparts in concluding that European defense expenditures had reached the upper limits imposed by total budgetary resources.[83]

To make matters worse, the level of U.S. military assistance to Europe had peaked and was starting to decline. In addition to making available substantial amounts of military equipment, the United States had contributed directly to European defense budgets and had provided indirect financial support through the mechanism of offshore procurement (OSP). The latter program was especially useful because it helped to strengthen and expand the still recovering European defense industries while accelerating the delivery of essential items of equipment at the same time. Beginning with the FY 1954 budget, however, all of these programs were sharply curtailed.

Eisenhower initially sought a maximum of $5.8 billion in new budgetary authority for the FY 1954 Mutual Security Program, well below the $7.4 billion Truman had proposed.[84] The new administration subsequently trimmed its request to $5.45 billion, of which $3.37 billion was intended for the NATO allies. Congress was in even more of a budget-cutting mood, however, and appropriated only $4.5 billion in new funds for FY 1954, of which just $2.7 billion was destined for Europe.[85] Of this reduced amount, $1.5 billion was programmed for equipment and material, slightly more than half of which was earmarked exclusively for the EDC countries. An important amendment to the appropriations bill, moreover, stipulated that no more than 50 percent of the EDC subtotal could be delivered until the treaty was ratified.[86]

This downward trend continued the following year. For FY 1955, the administration requested only $3.5 billion in new military assistance. Of this total, less than $900 million was intended for Europe, mostly in the form of finished military equipment. In contrast, some $1.33 billion was earmarked for the conflict in Indochina. Congress, moreover, set the final amount at only $2.8 billion. Thus in two years, Mutual Security Program appropriations had been reduced by more than half.[87]

Because the impact of many of these cuts would not be felt until sometime in the future, a more immediate cause of concern in Europe continued to be the slow pace with which the bulk of the U.S. assistance—in the form of military equipment—arrived. By mid-1953, only $4.5 billion of the $14 billion that Congress had appropriated for end items for the NATO buildup had been delivered, and a year later, roughly $7.5 billion in military assistance remained unspent.[88] The cumulative effect of these delays, in combination with the sharp decline in new appropriations, could not help but undermine the resolve of the allies to persevere with their rearmament efforts, even if the level of actual U.S. spending on military assistance would remain roughly constant through mid-1955.

The danger of allied force reductions became apparent in 1953, when Belgium announced that it planned to phase back one of its three ready divisions to reserve status in 1956. The following year, Belgium began to consider shortening the period of military service, which U.S. officials feared would lead to similar cuts in the Netherlands and possibly elsewhere.[89]

BRITISH FORCE LEVELS ON THE CONTINENT

Nor could the possibility of reductions be excluded even in the case of Britain, which had been the United States's staunchest ally. By early 1953, any further increase in Britain's contribution to NATO seemed out of the question. Achievement of the British force goals for the end of that year depended on the receipt of some $400 million in U.S. economic assistance and of full compensation for the expenses incurred in stationing forces on the continent. British officials were particularly concerned about the heavy burden that the failure to secure adequate support costs (then estimated at £80–90 million per year) from Germany once the EDC came into existence would impose on the balance of payments. If it appeared that these costs would not be met in 1954, they would have to reconsider the British defense program without delay. Indeed, British military planners began to consider the withdrawal of an armored division from Germany.[90]

Although the prospects for early ratification of the EDC remained dim, British plans were upset by the Eisenhower administration's decision to terminate general economic aid, which was revealed at the April 1953 NATO Council meeting, because the British rearmament effort was much more dependent on financial assistance than on the provision of U.S. military equipment.[91] Thus in the middle of the year, the government asked the Chiefs of Staff to estimate the adjustments that would be needed to hold military spending at the existing level for the next two years, forgoing a planned increase of more than £300 million, even if this required making cuts in Britain's NATO force contribution.[92] In proposing the new approach to NATO planning later that summer, moreover, British officials were careful to avoid implying any commitment not to reduce the defense budget.[93]

Although the British government continued to condition its acceptance of the NATO force goals,[94] intra-alliance considerations dampened whatever hopes it may have entertained for actually reducing the size of Britain's contribution to the alliance. A constant concern was that any British cutbacks would encourage other countries to follow suit.[95] In the fall of 1953, this concern was considerably heightened by Wilson's statement that the introduction of new weapons might lead to a reduction in U.S. manpower requirements. Subsequently, British leaders actively sought to dampen any suspicion that their proposed new approach was intended to allow any relaxation of effort and, especially, to avoid providing the United States with any excuse for withdrawing forces from Europe.[96]

Most important, the hope for reductions was incompatible with the goal of securing early approval of the EDC. British leaders realized that a major obstacle to ratification was the French fear of being dominated by the Germans on the continent. Consequently, they sought to reassure France that Britain would maintain its forces at the same level over the next few years, although this commitment was not intended to preclude later cuts.[97] The British search for a sufficient guarantee, however, eventually led them to pledge to maintain the equivalent of four divisions and a tactical air force on the continent and not to withdraw those forces without the approval of a majority of the Brussels Treaty powers.

FRANCE'S NATO CONTRIBUTION

The greatest change took place in France's conventional force contribution to NATO. During the first part of this period, the number of French forces in the Central Region stayed fairly constant. By early 1953, the size of the French army in Europe had reached approximately five ready and seven reserve divisions, where it remained for the next year and a half.[98] Still, this number fell well short of the goal of 22 divisions by the end of 1954 that had been established at Lisbon.

The prospects for an increase in the French contribution brightened considerably in 1954, when France withdrew from Indochina. Hopes that the substantial military resources thus liberated would be devoted to European defense were short-lived, however. Beginning in September of that year, the capability of the French forces in Europe was sharply reduced as increasing numbers of units were redeployed for combat duty in North Africa.

Several factors contributed to the premature termination of the French military buildup in 1953. First, political instability made it impossible to develop the durable consensus that was necessary to sustain high levels of defense spending. One government after another was brought down, paralyzing French policy making. The Pinay government, which had taken power in March 1952, resigned at the end of the year. It was replaced by a government headed by René Mayer, which lost a vote of confidence in May 1953. Mayer's fall precipitated the longest ministerial crisis since the war, which did not end until Joseph Laniel was invested as prime minister in late June. Although Laniel remained in office for nearly a year, he was unable to govern decisively because of deep divisions within his cabinet.[99]

The French effort was also constrained by severe economic and fiscal problems. In 1953, the government was running a budget deficit of 20 percent—which was expected to reach 25 percent in 1954— yet the French economy was stagnating.[100] A major cause of these problems was the high level of defense spending, which was consuming approximately 10.5 percent of France's GNP.[101] A large proportion of the total, however, was going to finance the military effort in Indochina, which was estimated to cost more than $1 billion per year in 1953.[102]

The magnitude of the war effort and its rearmament objectives in Europe as well as the severity of its political and economic constraints made France highly dependent on U.S. assistance. The United States was supplying some 50 percent of the equipment needed to meet France's military requirements in Europe, and it was providing more than $500 million in direct financial aid per year in addition to end items and OSP. By 1953, moreover, the Americans were picking up approximately half of the cost of the Indochina war.[103]

Yet the French government considered this assistance inadequate and insisted that the United States would have to provide $650 million in financial aid in 1953 if France were to meet its force goals for that year. The Truman administration had allocated only 80 percent of this amount, however, and the incoming Eisenhower administration decided not to make up the difference.[104] Subsequently, the French government indicated that it would require an even higher level of U.S. assistance in 1954. But as of late April 1953, the United States was planning to offer only a further $525 in financial aid in FY 1954.[105]

For all these reasons, French defense spending declined steadily during the following months. Shortly after taking office, Laniel decided to reduce military outlays by some $286 million in Europe and $143 million in Indochina, even though these cuts would inevitably reduce France's contribution to NATO.[106] And at the end of the year, his government announced plans to spend only $3.17 billion in 1954, including the approximately $525 million in U.S. assistance, well down from the 1953 total of $3.55 billion.[107]

The resolution of the Indochina war in mid-1954 should have enabled France to devote greater military resources to its NATO commitments. The growing insurgency in French North Africa, however, soon caused France to weaken its forces in Europe even more than had the previous conflict. By September of that year, the French government had become deeply concerned about the situation in Algeria. At that time, it created two new divisions out of battalions taken from the forces on the continent, which were promptly dispatched to North Africa. As a result of this transfer, six NATO-committed divisions—including two of the five ready divisions in Germany—were suddenly reduced in strength by a third. Additional battalions quickly followed, and the next year, this initial trickle became a flood as regiments and even entire divisions were withdrawn from Europe.[108]

By late 1955, the equivalent of only about three French ready divisions remained on the continent, and only three of the original seven French reserve divisions could be made ready by M+30. The combat effectiveness of the remaining units was extremely low. U.S. officials estimated that France's capacity to discharge its NATO missions had been reduced by at least 50 percent, and Gruenther commented that the piecemeal disintegration of the French forces in Europe had made an orderly mobilization of the French army in an emergency almost impossible.[109]

The hollowing out of the French contribution to NATO neutralized most of the gains in conventional capability that had been made in the preceding two years. In early 1954, Gruenther had been confident that the forces then or soon to be available to him would be adequate to handle an unmobilized attack by the Soviet forces stationed in Eastern Europe. If the Soviet Union were to have any chance of overrunning Western Europe, it would have had to bring up additional divisions from its own territory, which would have provided NATO with time to increase its own forces. The higher concentration of Soviet forces would also have made them more vulnerable to atomic attack.[110]

At that time, however, NATO's conventional forces were not yet strong enough to resist an all-out attack. An early 1955 SHAPE analysis, moreover, indicated that "the combat effectiveness of those forces allocated or earmarked for SACEUR's command is considerably less than it was a year ago," and a subsequent study noted many instances of declining NATO combat capability, particularly in the ground forces.[111]

Clearing the Way for a German Military Contribution

The principal cause of the persistent gap between NATO's conventional force requirements and its actual capabilities was the absence of a German military contribution. NATO planners continued to regard German rearmament as an essential element of the alliance's defense plans, notwithstanding their growing nuclear emphasis. The assumption of a German contribution of the size contemplated under the EDC treaty was an integral part of the SHAPE Capabilities Study and of MC 48. Indeed, Gruenther repeatedly described the presence of the long-awaited German divisions as no less important than the ability to use nuclear weapons if NATO were to be able to defend Western Europe against all-out aggression. The failure of the other Europeans to provide the forces expected of them only gave greater urgency to the achievement of a German military contribution.[112]

The way to German rearmament was finally cleared during this period, albeit in a different manner than anyone had expected at the outset. After taking office, the Eisenhower administration reaffirmed U.S. support for the EDC and, along with Britain, made a series of efforts to bring the project to fruition. Although progress continued to be slow, by the spring of 1954, four of the six EDC signatories had ratified the treaty. Nevertheless, a string of adverse internal and external political developments prevented successive French governments from acting on the matter until after the resolution of the Indochina conflict, and when the National Assembly finally did vote on the EDC in August 1954, it effectively scuttled the project. During the following weeks, however, British leaders devised an alternative formula for securing a German military contribution that was quickly accepted by all the allies. As a result, the Federal Republic was admitted to both NATO and an expanded Brussels Treaty organization in May 1955, although it would still be some time before the first German units were ready to take the field.

DEVELOPMENTS THROUGH EARLY 1953

A striking feature of this period was the continuity in U.S. policy toward the EDC. Following the 1952 election, Eisenhower and Dulles had been largely silent on the subject. Both were strong supporters of the EDC, however, and once installed in office, they immediately sought to secure its ratification. The president, of course, had advocated the creation of a European army as the most promising way to achieve a German military contribution since mid-1951. Dulles favored the project as much for its promise to bring about European unity and Franco-German reconciliation as for its potential for increasing Western Europe's military strength.[113] Thus notwithstanding the change of administration, the United States continued to back the EDC, which was widely viewed as the only available instrument for achieving the entire complex of U.S. policy objectives in Europe.[114]

The new administration wasted little time in setting out in pursuit of this objective. Within days after assuming his post, Dulles publicly warned that the United States would have to rethink its policy toward Western Europe if effective unity was not forthcoming.[115] And at the end of January, he and the new director of the Mutual Security Agency, Harold Stassen, left on a whirlwind ten-day visit to seven West European capitals that was largely intended to make clear the U.S. desire for prompt ratification of the EDC treaty.[116] Although U.S. officials considered France to be the main stumbling block, they also sought to persuade the other five signatories to ratify so as to put as much pressure as possible on the French to follow suit.[117]

Initially, the policy of the Eisenhower administration was frustrated because the French seemed no more favorably disposed toward the treaty than they had been the year before.[118] Indeed, the obstacles to ratification in France had grown noticeably since late 1952. Parliamentary elections in December had resulted in the new Mayer government, which depended for a majority on the conservative Gaullists, who opposed the EDC. Thus although Mayer supported the treaty, he felt compelled to impose four new conditions that would have to be met by the allies if the treaty were to be approved by the National Assembly. These were acceptance of protocols to the treaty intended to guarantee greater French freedom of action and to maintain the integrity of the French armed forces; closer association between Britain and the EDC; a satisfactory resolution of the status of the Saar region, which France had sought to separate from Germany; and increased U.S. economic assistance to Indochina so as to allow France to make a military contribution in Europe that would at least be equal to that of Germany.[119]

The first of these conditions was satisfied fairly easily. Although the other signatories had initially feared that the protocols would slow down the process of ratification and might even require renegotiation of the treaty, the EDC countries quickly agreed to regard the protocols as "interpretive texts" that would not affect the "letter and intent" of the treaty. This compromise resolved the matter by late March.[120]

To fulfill the second condition, the French government proposed in mid-February that Britain agree to maintain its armed forces in Europe at their present levels in return for the right to participate in the EDC institutions. The British government replied that it could not make such a commitment, though it was prepared to declare Britain's willingness to consult with the EDC on the level of British forces on the continent as well as to extend the North Atlantic Treaty for 30 years to make it coterminous with the EDC.[121] In May, France accepted the substance of the British counterproposal, although a final agreement on the terms of British political and military association would not be ready until the following April.[122]

In contrast, no progress was made toward the resolution of the Saar issue. Although German Chancellor Konrad Adenauer initially seemed disposed to begin negotiations with the French, he eventually decided against considering the matter seriously until after the German general elections scheduled for September. Thus it was clear that French ratification of the EDC would not be possible until late in the year even before the Mayer government fell in May, leaving a political vacuum that was not filled for over a month.[123]

FURTHER SETBACKS AND THE AGONIZING REAPPRAISAL

By this time, the prospects for ratification had become tightly bound up with the state of East-West relations as a result of Stalin's death in early March 1953. Although the Eisenhower administration took the position that the fundamental objectives of the Soviet Union had not changed, notwithstanding conciliatory gestures from the Kremlin, other Western governments felt that the new Soviet leadership should be provided an opportunity to demonstrate its peaceful intentions. On May 11, Churchill called for a summit conference with the Soviet Union to discuss the outstanding issues of the Cold War, including the division of Germany. Whether or not it was reasonable to hope that such a meeting might result in an overall postwar settlement, the immediate effect of Churchill's appeal was to deepen the policy-making paralysis in France.[124]

One of Mayer's last acts as prime minister was to request an early meeting of the Big Three heads of government to concert Western policy toward the Soviet Union.[125] Because of the ensuing governmental crisis in France, however, the allies did not convene until mid-July, and then only at the foreign minister level. The principal issue on the agenda at the Washington meeting was the timing of a four-power conference on the German question. The British position was that a summit should occur only after ratification of the EDC. The French argued, however, that it would be impossible to gain parliamentary approval until the infeasibility of a four-power solution had been convincingly demonstrated. Because Adenauer believed that his electoral prospects would be enhanced by an announcement that such a conference would be held sometime after the German national elections scheduled for the end of the summer, the United States agreed to back the French approach, and the three governments duly invited the Soviet Union to meet with them in late September to discuss the organization of free all-German elections and the establishment of an all-German government.[126]

The prospects for ratification of the EDC seemed to brighten when Adenauer gained a strong mandate at the polls, which freed him to resume negotiations on the Saar, and the new French prime minister, Laniel, expressed his determination to secure his country's approval of the treaty by the end of the year.[127] The process of putting the EDC into operation soon ran into further snags, however. The escalation of the French war effort in Indochina in September drew yet more troops from Western Europe, thereby sharpening concerns about France's ability to maintain military equality with Germany on the continent.[128] The French government also continued to link EDC ratification to the outcome of the four-power talks on Germany, to which the Soviets had agreed only in late November and which would not begin until early the following year in Berlin.[129] Backing away from his previous pledge, Laniel indicated that he could not bring the EDC to a vote until after the French presidential election in December.[130]

These reversals strained U.S. patience to the breaking point. The

EDC was a cornerstone of American policy toward Europe and was strongly supported by virtually every responsible high official in the Eisenhower administration. In mid-August, the president had approved NSC 160/1, which concluded that the United States should pursue the ratification of the treaty "with all available means." If ratification did not seem imminent "within a reasonable period," the government would review other possible courses of action. But no deadline was specified, and there seemed to be no satisfactory alternative.[131]

As the year came to an end with the prospect of an agreement nowhere in sight, the resulting frustration prompted the administration to take unprecedentedly strong action. At the tripartite Bermuda conference in early December, the members of both the U.S. and U.K. delegations pressed their French counterparts at length on the question of ratification.[132] Much more dramatic was Dulles's intervention at the subsequent meeting of the NATO Council, in which he warned that the United States would be compelled to undertake an "agonizing reappraisal" of its policy toward Europe should the EDC not become effective.[133]

THE REJECTION OF THE EDC

This stepped-up U.S. pressure had little effect on France. In any case, further progress toward French ratification had to await the conclusion of the four-power conference in Berlin, which began in late January. Because of the great differences between the two sides over how to bring about German reunification, the meeting ended without an agreement on the issue. The only concrete achievement was a decision to hold another conference in Geneva beginning in late April to discuss problems in the Far East.[134]

Recognizing that the prospect of the Geneva conference could provide the French with an excuse for yet further delays, Dulles urged that they bring the EDC treaty to a vote before the Easter holidays in mid-April. The achievement of this goal required, in turn, assuring the French government that its three outstanding conditions for ratification—closer British association with the EDC, American assurances regarding the U.S. commitment to Western Europe, and settlement of the Saar issue—would be met by the end of March.[135]

The first two conditions were promptly fulfilled.[136] In March, the British cabinet agreed to issue a declaration that Britain had no intention of withdrawing its troops from the continent as long as a threat to the security of Western Europe existed, and it approved the integration of British army units and EDC contingents where desirable and practicable if requested by SACEUR. In mid-April, Britain signed a formal treaty of association with the EDC.[137] Several days later, Eisenhower offered similar assurances, declaring that the United States would consult with the EDC countries on the level of U.S. forces in Europe and that it would regard the North Atlantic Treaty as being of indefinite duration.[138]

In the meantime, the EDC treaty had been ratified in the Netherlands (January 20), Belgium (March 11), Germany (March 29), and Luxembourg (April 7), and the Italian parliament had nearly completed its procedural work. Thus further progress depended almost entirely on France. The U.S. and U.K. assurances, however, had little impact on French attitudes.[139] The Saar issue remained unresolved, and new divisions within the French government dealt a further blow to hopes for early ratification.[140] As Dulles had feared, moreover, the beginning of the Geneva conference absorbed the attention of the French government, ensuring that the treaty would not receive parliamentary consideration in the near future.

The sharpest setback, however, was caused by the French surrender at Dien Bien Phu in early May, which changed the entire character of the EDC debate. Most immediately, it made a prompt settlement of the Indochina conflict imperative. When this was not forthcoming, Laniel fell and was replaced by Pierre Mendès-France, who assigned top priority to ending the war but had shown little interest in the EDC. As a national humiliation, moreover, Dien Bien Phu made France much more sensitive to its status as a great power. Henceforth, the central issue surrounding the EDC would not be whether the treaty would grant Germany too much influence but whether it would deprive France of its special position within NATO and of an independent national army.[141]

Mendès-France turned his attention to the EDC in late July, following the Geneva conference. In response to pressure from the allies, he promised to bring the treaty to a vote before the National Assembly adjourned in September. To secure a parliamentary majority in support of the EDC, however, Mendès-France felt he had to gain acceptance of several additional protocols to the treaty, which he discussed with the heads of the other five EDC countries in late August in Brussels. An important purpose of the protocols was to limit the restrictions that would be imposed on French sovereignty while maintaining the controls over German rearmament. The overall effect, however, was to emasculate the supranational aspects of the treaty, which the other countries found unacceptable. The conference ended with the parties unable to reach even a compromise agreement. Consequently, Mendès-France submitted the unmodified treaty to the French parliament, which effectively defeated it by a substantial majority on August 30 on a procedural motion.[142]

THE SEARCH FOR AN ALTERNATIVE

The demise of the EDC caught the allies almost totally unprepared. In early July, U.S. and British experts had met in Washington to study what course of action their two countries might take should France fail to ratify the treaty. These talks focused on the possibility of separating the contractual agreements from the EDC treaty to restore German sovereignty at once while deferring the question of German rearmament. When the United States and Britain approached Adenauer with their proposal following the French action, however, the chancellor insisted that solutions to both problems would have to be found simultaneously, and the allies were left with no agreed approach.[143]

The United States had been particularly insistent in its refusal to consider alternatives to the EDC. In part, this position reflected tactical considerations, especially the need to keep pressure on France to ratify. U.S. officials feared that any public discussion of alternatives would have suggested a lack of firm support for the EDC and resulted in further delays. More fundamentally, however, they steadfastly viewed the EDC as the only possible solution.[144]

The defeat of the EDC presented the administration with the difficult choice it had often threatened to make but had long sought to avoid by nursing the project along even when its prospects seemed dim. Rather than revising U.S. policy, as he had previously threatened, however, Dulles immediately reaffirmed the U.S. commitment to Western Europe. As a first step toward devising an alternative framework for restoring German sovereignty and enabling the Germans to make a military contribution, he called for a special meeting of the NATO foreign ministers. The secretary of state and many other U.S. officials nevertheless continued to believe that a viable solution would have to include a measure of supranationality.[145]

The British government chose a different approach. In addition to admitting West Germany into NATO, it would include the Federal Republic and Italy in an expanded Brussels Treaty organization, which would establish and oversee safeguards on German rearmament. Rather than a full meeting of the NATO Council, Britain proposed to begin with a conference in London of the nine countries most directly concerned—the six EDC signatories, the United States, Britain, and Canada.[146]

Initially, the United States reacted negatively to the British formula, which lacked any explicit provision for furthering European integration.[147] The British government, however, was firmly opposed to infusing the Brussels Treaty organization with any supranational authority, having forsaken the opportunity to join the EDC largely because it contained such features. During a hurried tour of the continent in mid-September, moreover, Eden had obtained varying degrees of support for the British approach from all the EDC countries.[148] Thus Dulles eventually conceded that the British plan "seemed the best that could be devised to meet the situation which confronts us."[149]

The London meeting, which began in late September, lasted nearly a week and consisted of fourteen plenary sessions as well as five special meetings of the three occupying powers and the Federal Republic. The high point occurred on the second day, when Eden declared his country's intention to maintain four divisions and a tactical air force—or the equivalent fighting capacity thereof—on the continent and not to withdraw those forces against the wishes of a majority of the Brussels Treaty powers.[150] This announcement represented an important shift in British policy, which until then had sought to avoid any firm commitments, and thus went far toward addressing the arguments that had been used by French opponents of the EDC.[151]

French concerns were further mollified when Adenauer voluntarily renounced the manufacture of atomic, biological, and chemical weapons as well as the production of certain conventional weapons by Germany without allied approval. Following these unilateral gestures, the remaining elements of a comprehensive agreement quickly fell into place. The revived Brussels Treaty organization, to be termed the Western European Union, would regulate the level of each country's forces so that they would not exceed the maxima that had been established within the EDC framework. The conferees created a special agency to control armaments production on the continent, and they granted SACEUR greater authority over the integration and deployment of NATO forces. Finally, full German sovereignty would be restored except with regard to Berlin, reunification, and the achievement of a final peace settlement.[152]

The United States and Britain spent the remainder of the year ensuring that France would ratify the understandings reached at the London meeting. The principal remaining obstacle was the French government's insistence on a settlement of the Saar issue as a precondition for introducing the agreements into the National Assembly.[153] This long-standing problem was finally resolved in late October, when France and Germany agreed to a popular referendum in the disputed territory in 1955, allowing the final texts to be signed at a special meeting of the NATO Council in Paris.[154] And at the end of December, following constant U.S. and British prodding, a majority of the French deputies gave grudging approval.[155]

The Paris Agreements went into effect the following May, finally opening the door to German rearmament. Thus the stage was set for a major improvement in NATO's conventional capabilities in Western Europe. Even the limits on Germany of twelve divisions and 500,000 men would allow an increase of more than 50 percent in the number of allied units and personnel then in the Central Region. Nevertheless, NATO officials estimated that it would be at least three to four years before the full German military contribution would be available.[156]

Analysis

Why did NATO's military strategy undergo such a fundamental change during this period, and why did the new strategy take the form it did? Why did the alliance's conventional capabilities continue to fall well short of its military requirement? And why were the allies finally able to agree on a set of arrangements that would enable West Germany to make a military contribution?

As was true for the previous period, the evolution of NATO's conventional force posture from 1953 to 1955 can be explained largely by the variables emphasized by the balance-of-power perspective. First, internal difficulties continued to limit the alliance's military potential and thus made it impossible for many countries to meet the force goals that had been established at Lisbon. During the first half of 1953, even U.S. officials, who had been the most forceful advocates of rearmament, realized that the limits of the Europeans' political and economic capabilities had been reached, even though the alliance still fell well short of meeting the established force requirements.[157]

These constraints were reinforced by a further decline in Western fears about Soviet intentions. By early 1953, most NATO governments assumed that the Soviet Union was unlikely to embark upon premeditated aggression in Europe in the next few years, notwithstanding the continuing imbalance of conventional capabilities on the continent.[158] And as long as hostilities did not seem imminent, the achievement of NATO's force goals assumed less urgency. These views became more pronounced following Stalin's death in March of that year as a result of the subsequent Soviet "peace offensive" and the Korean armistice.

The persistent inability to close the gap between NATO's conventional force requirements and its actual capabilities bred frustration and served as a constant source of intra-alliance acrimony. Consequently, many officials began to search for a new approach to alliance military planning that not only would be more amenable to full implementation in the near future but would call for a level of military preparation that could be sustained for a prolonged period. It was primarily the increasing availability of tactical nuclear weapons, however, that made the adoption of a new strategy feasible rather than merely desirable. This more than any other development enabled the alliance to reduce its previously high degree of reliance on conventional forces, even if, as soon became clear, it allowed for less of a reduction in conventional requirements than many had hoped.

Although the intra-alliance perspective is less helpful for explaining the evolution of NATO strategy during this period, differences in allied strategic preferences did shape MC 48 in one important respect. The NATO countries by and large welcomed the introduction of tactical nuclear weapons into alliance strategy, but they disagreed on the restrictions that should apply to their use. The U.S. position was that NATO commanders should be given the authority to retaliate with nuclear weapons at their discretion. Many European leaders, however, wanted to reserve to political authorities the decision to initiate tactical nuclear warfare, given the potential consequences for their countries, a view that was embodied in the important accompanying resolution to MC 48.

The intra-alliance perspective is more useful for explaining the continuing marked discrepancies in the magnitude and success of national efforts to provide conventional forces. As a general rule, the European allies, but especially France, were both less willing and less able to achieve their force goals than was the United States. Some of them still suffered from internal political problems, which limited the demands governments could impose on unreceptive societies. In other cases, economic weakness was the principal source of constraint. In addition, the Europeans tended to be more sanguine about the Soviet threat and thus felt less of a need to achieve NATO's force requirements than did the United States.[159]

At the same time, both Britain and France continued to devote substantial resources to non-NATO interests and commitments. In 1953 and 1954, France remained deeply mired in Indochina, which received first call on the French defense budget and officer corps. France had hardly disengaged from that conflict, moreover, when it stepped up its military activity in North Africa, resulting in the diversion of NATO-committed units. Britain continued to maintain substantial forces overseas and to devote a considerable share of its military budget to its growing nuclear program during this period.

The intra-alliance perspective also accounts for the further delay in achieving an agreement that would allow German rearmament to proceed. Through most of 1954, allied views concerning the timing of a German contribution and the way it should be realized continued to be marked by profound disagreement, which reflected underlying differences in their geographical positions, relative power, out-of-area interests, and historical experiences. And as before, the U.S. preponderance in military power and potential translated into minimal influence over the outcome of the issue. As long as the Eisenhower administration was unwilling to withdraw American forces from Europe or even to terminate military assistance, it could do little to compel ratification. Because of the magnitude of the existing U.S. contribution to NATO, moreover, it was in no position to offer substantial additional enticements—as the Truman administration had been able to do—to gain French acceptance of the EDC.

France and Germany continued to possess the greatest leverage over the issue simply because they were prospective members of the EDC and thus had the greatest interests at stake. As long as their ratification of the treaty was necessary, moreover, they would exercise veto power. Of the NATO countries outside the EDC, only Britain was able to shape the outcome substantially. Because of its geographical proximity to and its historically closer relationship with the continent as well as its relatively considerable military resources, Britain alone was in a position to offer inducements sufficient to overcome France's opposition. And although the British government did not choose to exercise this potential leverage to save the EDC, it did use its influence to ensure the success of the alternative formula that was devised for facilitating German rearmament.

The intra-alliance perspective also captures the narrowing of allied differences that eventually allowed a compromise to be struck. Before an agreement could be reached, each of the principal countries involved had to modify its position significantly. Most important, France became much less concerned about a revival of German military power per se and thus more willing to accede to German rearmament, as long as the institutional mechanism chosen for this task included certain safeguards and did not infringe unduly on French sovereignty. At the same time, the United States abandoned its hopes for substantial progress toward European political unity, Germany accepted severe restrictions on military production, notwithstanding its previous insistence on full equality, and Britain, departing from tradition, committed itself to maintain a fixed level of forces on the continent. Each of these important shifts is difficult to comprehend solely on the basis of the factors stressed by the intra-alliance perspective, however. In this instance, a fully satisfactory account may require a more detailed consideration of national decision making.

During this period, institutional factors played the smallest role in shaping NATO's conventional force posture. In this regard, perhaps the most important development was the emergence of SACEUR as an independent actor with a potentially significant influence over events. Gruenther and his staff at SHAPE were instrumental in conceiving and conducting the studies that served as the basis for MC 48. Without that essential groundwork, the adoption of a new strategy would have taken much longer.

In addition, several institutions that could have an important bearing on future alliance behavior were created or strengthened. Perhaps most significant were the constraints that Britain accepted on its ability to reduce the forces it maintained on the continent, as indicated in the agreements reached in London and signed in Paris. Because of this commitment, Britain's freedom to reduce the size of its NATO contribution would henceforth be much more circumscribed. In addition, an upper limit of twelve divisions for the future size of the German military contribution was formally established as the plans devised for the EDC were incorporated into a protocol to the Brussels Treaty. Even the size of the U.S. military presence in Europe showed signs of further institutionalization, as manifested in European expectations that the United States would not reduce its contribution to NATO and in the beliefs of numerous high-level officials in the Eisenhower administration, including Dulles and the president himself, who consistently argued that any reductions would have a disastrous effect on the morale and thus the defense efforts of the allies.[160]

CHAPTER 4

NATO Retreats from Massive Retaliation, 1956–1960

NATO's conventional force posture underwent a further significant shift less than three years after the approval of MC 48. In May 1957, the North Atlantic Council adopted a new strategic concept, MC 14/2, which formally revised the alliance's military strategy and the role of conventional forces in it. In contrast to MC 48, the new strategic concept and the December 1956 Political Directive from which it was derived offered a more complex view of the threat and of how NATO should respond. While still largely addressed to the problem of deterring a general war with the Soviet Union and emphasizing the role of nuclear weapons in preventing and, should deterrence fail, determining the outcome of such a conflict, these documents formally recognized the possibility of limited forms of aggression, against which limited and even non-nuclear responses would be most appropriate. Thus the development of the Political Directive and MC 14/2 represented an important step away from the exclusive concern with all-out aggression and all-out responses that had previously characterized NATO planning and U.S. declaratory policy. Well before the adoption of flexible response in 1967, conventional forces were assigned a potentially autonomous, albeit still limited, role within alliance strategy.

Like flexible response, moreover, MC 14/2 was ambiguous, amenable to a range of conflicting interpretations. Consequently, the debate over strategy within the alliance continued for another year while the NATO military authorities sought to determine the alliance's force requirements under the new strategic concept. As time passed, the view that NATO forces in Western Europe should be capable of responding flexibly and at an appropriate level to any military incident short of an all-out attack became increasingly dominant. As a result, reductions in the alliance's conventional force requirements were precluded.

The departure from massive retaliation and the consequent greater emphasis on conventional forces in NATO strategy had little impact on the alliance's overall non-nuclear capabilities, however. These forces failed to increase commensurately with their enhanced importance, showing instead little or no net growth. Although the first of the long-awaited German contingents were finally formed, the German buildup was considerably scaled back and slowed down. The bulk of the French army remained tied down in Algeria, moreover, and the British and American forces on the continent were actually reduced. Finally, the belated nuclearization of allied forces that began during this period further limited their conventional strength.

Pressures for Change in NATO's Conventional Force Posture

Pressures for modifying NATO's military posture yet again emerged even before the ink was dry on MC 48. On the one hand, strategic developments bred a growing concern in both the United States and Europe about the possible dangers inherent in the West's deepening dependence on nuclear weapons. On the other hand, political and economic considerations dictated taking NATO's version of the New Look to its logical conclusion by placing even greater emphasis on nuclear weapons, both tactical and strategic.

CONCERNS ABOUT NUCLEAR OVERRELIANCE

Doubts about the wisdom of the high degree of nuclear reliance in U.S. and NATO strategy were gaining strength in the United States as early as 1954. It was becoming clear to many American observers that the days of meaningful U.S. superiority in nuclear weapons and delivery vehicles were numbered. As strategic parity approached, the Soviet Union would increasingly pursue its objectives by threatening to use means short of general war. If hostilities resulted, the United States might be faced with the difficult choice of doing nothing or responding in a way that would invite Soviet nuclear retaliation. In addition, more specific concerns emerged about the usefulness of tactical nuclear weapons in combat. In particular, there was growing evidence that the use of these weapons would not confer any decisive military advantages on NATO once the Soviet Union had acquired a tactical nuclear arsenal of its own.[1]

Such views were initially voiced outside the government, largely in response to the Eisenhower administration's declaratory policy of massive retaliation.[2] They nevertheless quickly found a receptive audience within the administration as well. Not surprisingly, these concerns were most strongly held within the army, whose organizational interests were deeply threatened by an exclusive focus on nuclear weapons.[3] As early as November 1954, however, Secretary of State John Foster Dulles wrote, "The U.S. and NATO should explore urgently the possibility of maintaining sufficient flexibility in NATO forces to avoid exclusive dependence on atomic weapons."[4]

These growing doubts within the U.S. government were reflected in the basic national security policy documents for 1955 and 1956, NSC 5501 and NSC 5602/1, respectively.[5] Adopted only a year after Dulles had first articulated the doctrine of massive retaliation, NSC 5501 explicitly recognized the possible development of a condition of "mutual deterrence," which would strongly inhibit the United States from initiating general war and even from taking actions that it regarded as increasing the risk of such a conflict. In particular, the threat to use nuclear weapons in response to lesser provocations would lose its credibility. As a result, the Soviet Union might be increasingly emboldened to pursue its objectives through limited forms of aggression.

NSC 5602/1 carried this logic one step further. Noting that as nuclear parity approached, "the ability to apply force selectively and flexibly will become increasingly important," it concluded that U.S. forces "must not become so dependent on tactical nuclear capabilities that any decision to intervene against local aggression would probably be tantamount to a decision to use nuclear weapons." In a further departure from previous U.S. policy, NSC 5602/1 recognized that a general war could be the unintended result of a series of escalating actions and counteractions that began at a fairly low level. Formerly, the official view had been that strategic strikes would occur from the outset of a general war and thus that the United States should prepare to use nuclear weapons at once.[6] This new assumption further strengthened the position of those who favored maintaining forces capable of dealing with limited actions so that the United States would not be faced with the stark choice between passivity and risking annihilation in the event of aggression. Although neither of these official documents spelled out the obvious implications for U.S. force structure, they were increasingly on the minds of some top policy makers.[7]

FOLLOWING THROUGH WITH THE NEW LOOK

Opposing these arguments for limiting NATO's reliance on nuclear weapons were equally strong pressures for taking the New Look and MC 48 to their logical conclusion. In the United States, these pressures grew out of a persistent desire to reduce substantially the size of the U.S. armed forces, even if doing so meant withdrawing a significant fraction of the units stationed in Europe. This viewpoint achieved its fullest expression in the so-called Radford plan of July 1956.

The possibility of U.S. force reductions in Europe remained an implicit component of the New Look. Although the administration had decided in October 1953 to postpone making any decision on the issue, the initial military guidance drafted on the basis of NSC 162/2 the following December had nevertheless called for a withdrawal of two NATO-assigned divisions by 1957.[8] And influential voices within the administration, especially those of Secretary of Defense Charles Wilson and Secretary of the Treasury George Humphrey, continued to press for such cuts.[9] Indeed, in some government circles, the conviction that the United States stood to gain from putting ever greater emphasis on nuclear weapons was becoming stronger, notwithstanding the growing chorus of concern. This belief was held most firmly in the air force, which had the largest nuclear role among the services, although it was also shared by chairman of the Joint Chiefs of Staff Admiral Arthur Radford and other high officials, such as Humphrey and Wilson, who placed a premium on achieving economies in defense spending and felt that nuclear weapons offered a relatively inexpensive substitute for conventional firepower.[10]

The issue of force reductions came to the fore again in early 1956. As the year began, the Eisenhower administration was still struggling to bring defense spending down sufficiently to enable it to eliminate the federal budget deficit without raising taxes. Hoping to identify possible force reductions, Wilson directed the Joint Chiefs of Staff to prepare "an outline military strategy for the United States which best meets the demands of our national security and which can serve as a basis for guidance for the determination of the size, nature, composition, and deployment of U.S. armed forces for Fiscal Years 1958 and 1959." He stipulated that the study should take into account the importance of a sound economy and the use of atomic weapons from the outset of hostilities in a general war and whenever of military advantage in the case of lesser hostilities.[11]

Wilson's directive triggered a new debate over the proper degree of nuclear reliance and the nature of a future war as the services struggled to draft the strategic concept on which to base military requirements.[12] Radford strongly favored placing greater reliance on nuclear weapons so as to reduce U.S. forces, and both he and the air force hewed to the view that a general war would inevitably begin with an all-out Soviet surprise attack, necessitating an immediate American nuclear riposte.[13] Consistent with NSC 5501 and NSC 5602/1, however, the army insisted that the concept account for the possibility of atomic parity, at which time the Soviets would pursue their objectives through means such as "subversion, infiltration and local aggression" to which general nuclear warfare would not be an appropriate response.[14]

The inability of the services to resolve their differences required Wilson to take the matter to the president, who decided against Radford and the air force.[15] The resulting paper foreshadowed MC 14/2 in several respects. Noting that mutual deterrence would make limited war more likely, it was divided into two parts dealing with general war and "cold war and military conflict short of general war," respectively. The new concept stated that the United States and its allies would need sufficient military means for deterring any resort to local aggression and that if deterrence should fail anyway, the primary U.S. objective would be to avoid a general war if at all possible. Consequently, the paper called for the capability to deal swiftly and decisively with limited military conflict in a way that would prevent hostilities from expanding.[16]

Although Eisenhower's intervention temporarily stilled the squabbling among the services, it failed to set any new limits on U.S. force levels, as Radford had hoped. As a result, the force levels recommended by the Joint Chiefs for 1960 were no lower than before.[17] In response, Radford declared these recommendations "unacceptable" and insisted that the JCS hold daily meetings until they devised a plan that was "in consonance with the agreed strategic concept and within acceptable limits from a manpower and fiscal point of view."[18] Despite the chairman's pressure tactics, however, the disagreement over force requirements persisted.

At this point, Radford decided to take matters into his own hands. In early July, he outlined his personal views of what the future U.S. force structure should be. Disregarding the agreement on the strategic concept that had been so tortuously negotiated, he argued that "a considerable departure should be made from the present force goals" in order that they "be primarily designed to meet the greatest threat, i.e., a general war commencing with an all-out surprise atomic attack."[19] Specifically, Radford proposed a reduction in the overall number of military personnel of 750,000 to 850,000. By far the largest share of the cut was to fall on the army, which would be reduced by nearly 50 percent in size, from just over one million to only 500,000 to 600,000 men.[20]

The potential implications of Radford's views for NATO's conventional force posture were no less significant. He argued that "efforts must be continued to reduce the numbers of military personnel deployed in overseas areas." The United States should reorganize the large units deployed in Europe into small atomic task forces, which would provide not only substantial fire support but also visible evidence of U.S. readiness to back its allies. Only through such drastic restructuring could the active army be significantly reduced.

The Radford plan was extremely short-lived. The proposed cuts were so startling and so threatening to the interests of the services that they were quickly leaked to the press, making front-page news and provoking an international outcry.[21] Because it suggested inconstancy in U.S. policy and a failure to consult the NATO allies, the disclosure caused considerable embarrassment in Washington.[22] U.S. officials all the way up to the president immediately denied that any such plan had been approved or even seriously contemplated.[23] As a result of this controversy, U.S. strategic planning was reportedly put on hold until after the national elections in November.[24]

DIFFICULTIES WITH IMPLEMENTING MC 48

Difficulties associated with the implementation of MC 48 created further pressures to modify NATO's military posture. The alliance's original strategy had suffered from the fact that it required greater conventional forces than the NATO countries were willing or able to provide. Consequently, the adoption of a new strategy that substituted nuclear weapons for conventional firepower had been looked upon since 1952 as a way of reducing these requirements to acceptable levels.[25]

MC 48, however, offered little solace to those who had hoped for relief for national defense budgets. Although the new strategy did make possible some diminution of the alliance's conventional force requirements, this reduction took place primarily in the number of mobilizable reserve divisions, many of which had yet to be formed.

In contrast, it allowed for no reductions in the standing forces that had already been raised. These units, even when supplemented by the long-awaited German divisions, would still be needed to create enough resistance to oblige enemy troops to concentrate in order to break through the NATO line, thereby presenting lucrative targets for tactical nuclear weapons. To make matters worse, the NATO countries would have to pay for the many new measures intended to adapt their forces for the conditions of atomic warfare.[26]

Even before MC 48 was approved, NATO officials had concluded that the development of the forces it recommended would not allow defense budgets to be reduced and would probably require increased expenditures.[27] During the course of the following year, moreover, SACEUR added several expensive requirements, including improved air defense and logistical support, to the already burdensome list of "minimum" measures. Thus by the end of 1955, it was clear that implementation of the new strategy would involve large additional costs.[28]

There was little hope of increasing NATO military budgets, however, and good reason to fear that the alliance's defense effort might actually decline. In many countries, public support for maintaining military spending at existing levels continued to erode, in large part because of a waning sense of threat. The 1955 Geneva summit conference had raised hopes for an end to Cold War tensions—the so-called spirit of Geneva—and Soviet announcements in 1955 and 1956 of unilateral troop reductions suggested that Soviet conventional capabilities in Europe might decrease.[29] The prospect of such cuts made the maintenance of NATO forces at existing levels, let alone increases, seem less necessary. Western publics were thus less willing than ever to defer spending on social programs to maintain a strong military posture.[30]

The lack of resources to implement MC 48 presented NATO leaders with the difficult choice of either not paying for the additional measures it recommended or reducing reserve forces drastically and even cutting existing active forces. In October 1955 alliance defense ministers and ambassadors expressed an interest in eliminating parts of the NATO defense program to ensure the availability of funds to meet the most urgent military needs, and many even called for a comprehensive review of force goals and priorities similar to that conducted by the Temporary Council Committee in 1951 to reconcile military planning with political and economic realities.[31] Indeed, before MC 48 was a year old, some allied officials had concluded that the conflict between force requirements and financial resources could be resolved only by recasting NATO strategy once more.[32]

ALLIED DEMANDS FOR TACTICAL NUCLEAR WEAPONS

Ironically, the new strategy itself undermined efforts to provide the conventional forces needed to implement it. MC 48's emphasis on the deterrent effect of nuclear weapons made it increasingly difficult for allied governments to justify their non-nuclear military contributions, especially given the growing pressures on them to increase social welfare spending. U.S. officials came to perceive a widespread belief in Europe that only atomic air power would play a decisive role in the initial phases of a general war with the Soviet Union and that ground forces would make little difference. From the European perspective, there was little point in developing and maintaining large conventional capabilities.[33]

Instead, the allies began to press the United States to accord them a nuclear role in the defense of Western Europe. After all, such a role was implicit in MC 48, which envisioned the widespread use of tactical nuclear weapons in the event of a major Soviet attack and called for an integrated atomic capability.[34] There had been no formal decision that the United States alone would provide this crucial component of NATO's military posture. In practice, however, only the U.S. forces in Europe were being outfitted with a nuclear capability, and U.S. officials gave virtually no thought to providing nuclear weapons to the allies.

One reason for the initial failure to provide for allied participation may have been uncertainty about the ultimate size of the U.S. nuclear stockpile.[35] Nuclear sharing was also hampered by the U.S. Atomic Energy Act (AEA), which severely restricted the information that could be provided to the allies. Although the AEA was modified in 1954 to permit the sharing of data on the external characteristics of nuclear weapons, including their effects, the subsequent NATO agreement for the exchange of atomic information did not enter into force until March 1956, and the AEA continued to preclude the transfer of atomic weapons to foreign countries in peacetime.[36]

Notwithstanding these obstacles, by October 1955 the Europeans had begun to express an interest in equipping their forces with "new weapons" and to seek advice on how to reorganize their forces for the nuclear battlefield.[37] In response, the United States agreed to participate in a series of discussions with the allies early the following year on nuclear weapons and the U.S. Army's proposed new divisional structure for atomic warfare. The Eisenhower administration remained cool toward the idea of providing the Europeans with nuclear weapons, however, partly out of fear that the allies might consequently decide not to increase their conventional forces. Thus in the ensuing discussions, the U.S. representatives emphasized the slow and evolutionary nature of weapons development and the continuing value of modern conventional equipment, and they stressed that nuclear warfare would require at least as many forces as would the conventional variety.[38]

BRITISH INTEREST IN REVISING NATO STRATEGY

While the other European allies were pressing for lower conventional force goals and an equal role in NATO's defense arrangements through access to nuclear weapons, British leaders decided that their interests required a further revision of the alliance's military strategy. They hoped that such a change would allow Britain to withdraw a substantial portion of the forces then stationed on the continent and, by affording Britain a greater role in performing NATO's nuclear deterrent functions, could be used to justify the country's still nascent nuclear program.

The British were anxious to reduce their forces in Europe for several reasons. In 1955, Britian had suffered both inflation and a balance-of-payments deficit. Consequently, the cabinet had resolved to reduce domestic consumption and to stimulate exports.[39] These twin objectives would be difficult if not impossible to achieve, however, without reducing the size of the British defense effort. The military budget was absorbing approximately 9 percent of Britain's GDP and was expected to rise substantially over the next four years. In addition, the defense program employed 7 percent of the work force, and it consumed 12 percent of the output of the metal-using industries, which supplied half of Britain's exports, thereby restricting the country's export potential. Finally, maintaining British troops abroad required a large amount of foreign exchange, and the situation was expected to worsen as West Germany phased out its annual support payments of 70 million pounds.[40]

The British also hoped to free military resources for defense programs that they deemed to be more important. In 1955, the government had assigned first priority to the development of an independent thermonuclear capability that would serve both to deter the Soviet Union and to maintain British influence with the United States. Like many Americans, British officials believed that the further development of their country's nuclear capability was the best contribution they could make to the collective NATO effort.[41] At the same time, they felt that they could reduce the forces then stationed in the Middle East and the Far East no further without seriously risking British prestige and interests. They concluded, therefore, that the British non-nuclear contribution to NATO presented the least damaging field for cuts.[42]

The development of an alternative defense program was quickly set in motion. In late 1955, the cabinet agreed to reduce military manpower from 800,000 to 700,000, and in mid-1956, Prime Minister Anthony Eden initiated an even farther-reaching review of Britain's long-term defense needs.[43] By that time, the British government had decided to reduce the British Second Tactical Air Force in Germany (2TAF) by nearly one-third and was actively considering further cuts in 2TAF and a withdrawal of as many as two of the four divisions in the British Army of the Rhine (BAOR).[44] Reductions of this magnitude, however, would have been difficult to reconcile with Britain's 1954 commitment to station the equivalent fighting capacity of four divisions and a tactical air force on the continent. A military rationale would be needed to justify the desired withdrawals to the allies, and the most direct way to devise such a rationale was to revise NATO strategy.

The Development of MC 14/2

MC 14/2 grew out of the British effort to reduce their forces on the continent. In mid-1956, the British government approached the United States with a proposal to reappraise NATO strategy. And at Britain's urging, the NATO foreign ministers approved a new Political Directive at their December 1956 meeting, which was promptly translated by the NATO military authorities into a new strategic concept, MC 14/2, early the following year. Because of allied misgivings, however, neither document bore much resemblance to the original British proposal. By calling attention to the problem of limited aggression and creating a place for limited responses by the NATO forces in Europe, both placed new restrictions on the degree to which the alliance should rely on nuclear weapons. Moreover, subsequent interpretations of MC 14/2 placed great stress on the role of the local or "Shield" forces[45] in deterring and, if necessary, dealing with less than all-out aggression. As a result, there could be no reduction in NATO's force requirements.

THE BRITISH PROPOSAL FOR A REAPPRAISAL
OF NATO STRATEGY

The British first broached the subject of a review of NATO strategy
with the United States in June 1956. Specifically, they sought an
early meeting of the NATO foreign ministers to prepare a new Political Directive to the NATO military authorities as the basis for re-
writing the strategic concept. The resulting strategy, they hoped,
would place even greater reliance on the threat of nuclear retaliation
to deter aggression, allowing a further reduction in the alliance's
conventional force requirements. Such a reappraisal, Eden confided
to Eisenhower, was both "necessary and urgent."[46]

The British proposal was based on the argument that the existence
of thermonuclear weapons in large numbers had made war much less
likely and consequently, the most immediate threats now faced by
the West were nonmilitary. In the present state of relaxed tension,
moreover, the NATO force goals were unlikely to be achieved. Indeed, even the maintenance of NATO forces at their existing levels
was imposing an increasingly intolerable economic burden.[47]

In these altered circumstances, it was necessary to accord "full
weight to the deterrent effects of thermonuclear weapons," which
would enable NATO to reduce its conventional forces to the lowest
possible level consistent with security requirements. These forces
would no longer have to be able to prevent the rapid overrunning of
Western Europe. Nor would they have to be capable of conducting
sustained operations, as called for in MC 48, because the intense
initial phase of any war in Europe could not last more than 14 to 21
days and was likely to be much shorter. Rather, the conventional
forces maintained by the alliance should be the minimum necessary
to prevent external intimidation, to deal with any local infiltration,
and to enable aggression to be identified as such by presenting an
obstacle that could be overcome only by force.[48] In this way, the
British hoped, NATO force requirements in the Central Region might
be reduced by as much as one-half.[49]

State Department officials reacted coolly to the British initiative,
which was disparagingly described as "thermonuclear bombs or
nothing." They feared that any public discussion of the proposal
would create the impression that a serious crisis existed within the
alliance. It might also imperil the enactment of the foreign aid bill
that was pending before Congress and the rearmament legislation
then being debated by the German parliament. Consequently, Washington urged that public discussion be strictly avoided and that no

further approaches to other governments be made at that time, suggesting instead that the British government await the outcome of several military studies then in progress.[50]

Eventually, Dulles offered to hold detailed discussions with the British in mid-August, by which time he hoped that his government would have completed its own assessment of the subject, although U.S. officials insisted that the two sides reach a preliminary agreement before raising the issue in NATO. The Suez crisis arose in late July, however, causing the discussions to be postponed for at least one month. The talks were then further delayed by divisions within the administration over the advisability of force reductions, which prevented the determination of a final U.S. position until early October.[51]

THE DEVELOPMENT OF THE POLITICAL DIRECTIVE

By then, however, the British had decided to try a different tack. In August, they began to prepare a new Political Directive on their own. Although the resulting draft was originally intended to serve as the basis for the long-awaited discussions with the United States, the British government presented it in the NATO Council in mid-October without first securing U.S. agreement.[52]

The British government acted unilaterally for several reasons. First, it was becoming clear that the U.S. and British positions on the matter were far apart and likely to prove irreconcilable. The British had provided the United States with their draft directive in early October but had proven unwilling to make any substantive changes in line with the comments subsequently offered by the Americans.[53] Second, time was running out. To assure prompt budgetary relief, the British felt they needed to gain NATO ministerial approval of the new Political Directive in December at the latest.[54]

Third, parallel developments in NATO military channels threatened to undermine the British effort. In early 1956, the alliance military authorities had begun to update and consolidate the various documents concerning NATO strategy to take account of recent political developments, especially West Germany's accession to NATO, and to eliminate the considerable repetition and duplication among them.[55] The goal was to produce two new documents—MC 14/2, the strategic concept, and MC 48/2, measures to implement the strategic concept—for consideration by the Military Committee in October. Although there had been no intention of introducing new doctrine, the latest draft of MC 14/2 acknowledged the possibility of a limited conventional war in the NATO area, which was clearly contrary to the British position.[56] Thus at the same time that they submitted their proposed directive, the British requested that the NATO military authorities halt work on the new strategic concept until the issue of political guidance had been resolved.

The British draft Political Directive reiterated many of the arguments that had been made to the United States the previous summer. Because of the West's ability to devastate their country, Soviet leaders would seek to avoid global war. Nevertheless, they would continue to attempt to disrupt the alliance, chiefly by increasing their political and economic efforts to undermine the Western position outside the NATO area while resorting to indirect military action wherever this was to their advantage. It followed that the NATO countries would have to devote a greater share of their resources to countering these alternative threats without, however, endangering their economic stability, all in the face of rising weapons costs. Thus apart from providing strategic nuclear forces, the alliance should maintain the minimum land, sea, and air forces needed to meet the limited set of military requirements that the British had proposed in June.[57]

The British proposal was strongly opposed by the United States, which had finally adopted a formal position on the question of NATO strategy in response to the original British demarche. The Eisenhower administration now flatly rejected any strategy that implied total reliance on nuclear retaliation. Although such a capability would remain a principal element of the deterrent, it could not solve all of the alliance's military problems. The NATO area might be subject to a variety of types of aggression, including local attack, and the United States could not commit itself in advance to respond with nuclear weapons in all cases. To avoid diplomatic and military inflexibility, then, the alliance required diverse military capabilities. In particular, it needed sufficient conventional forces to be able to meet limited non-nuclear aggression in Europe with a non-nuclear response.[58]

The United States was not alone in its desire to build greater flexibility into NATO strategy. Many of the other European allies had become equally concerned about the limitations of MC 48 and its heavy nuclear emphasis. They had begun to doubt whether the United States, itself now vulnerable to Soviet retaliation, would be willing to launch a nuclear counterattack in response to a small conventional attack in Europe. At the same time, they feared that in the absence of alternative courses of action, the United States would do just that, resulting in widespread devastation.[59]

Such concerns were especially acute in West Germany. Because of their country's exposed geographical position, the Germans were particularly worried about the possibility of limited forms of aggression. They had good grounds to fear, moreover, that the use of tactical nuclear weapons would lay waste to their country.[60] Consequently, German leaders believed that any conventional attack should be met by conventional forces alone, and they rejected the automatic use of nuclear weapons in such situations. A nuclear response, they felt, should be made only in the event of a nuclear attack. Thus the German government was strongly opposed to any proposal that would place even more reliance on nuclear forces.[61]

Even as the British proposal was being debated, the Soviet invasion of Hungary further heightened these concerns. The incident cast doubt on the British argument that the danger of thermonuclear war had significantly dampened the willingness of the Soviet Union to use force. To the contrary, some Europeans concluded that the risk of aggression, including the possibility of a limited war, had increased.[62]

Even intra-alliance considerations worked against the British. They had hoped that the prospect of force reductions would give their proposed Political Directive broad appeal.[63] But most European countries were more concerned to prevent any diminution in the U.S. military presence on the continent, especially after the leak of the Radford plan, because of the likely negative consequences for the credibility of American nuclear threats on their behalf. Assigning conventional forces a greater role in NATO strategy would provide a further rationale for maintaining the U.S. contribution at its present level.[64]

As a result of this opposition, the final version of the Political Directive adopted by the NATO Council in December bore little resemblance to the original British proposal. As the British had wished, it called for a review of NATO defense planning "to determine how, within the resources likely to be available, the defense efforts of the Alliance and of each individual member can best achieve the most effective pattern of forces." As in the past, the new Political Directive was primarily concerned with deterring and, if necessary, prevailing in a general war with the Soviet Union. And like MC 48, the basic instruments of this strategy would be "a fully effective nuclear retaliatory force" able to absorb an attack and to carry out "an instant and devastating nuclear counteroffensive" and Shield forces for the forward defense of NATO territory "able to sustain operations, without any intention to make a major withdrawal, until the strategic counter-offensive has achieved its objective." The Shield forces, moreover, required "the ability to respond quickly, should the situation so require, with nuclear weapons to any type of aggression."[65]

In contrast to MC 48, however, the Political Directive called attention to the problem of limited forms of aggression, and it created a place for limited responses in such circumstances by the Shield forces. In particular, it required that they have the ability "to deal with incidents such as infiltration, incursions or hostile local actions by the Soviets, or by Satellites with or without overt or covert Soviet support" and that they be able to do so "without necessarily having recourse to nuclear weapons." For the first time in the alliance's history, NATO's ground and air forces in Europe might be used independently of the strategic deterrent.[66] Thus the adoption of the Political Directive represented a victory for those who sought to back away from massive retaliation and to place limits on NATO's reliance on nuclear weapons.

THE PREPARATION OF THE STRATEGIC CONCEPT

The struggle over NATO strategy did not end with the adoption of the Political Directive, however. Instead, it entered a new phase when the Standing Group and the Military Committee sought to revise MC 14/2 in conformity with the new guidance. For Britain, it was now more urgent than ever to secure a favorable interpretation. In December, the British government had concluded that deep reductions in the British forces stationed on the continent could be put off no longer. The expected negative reaction of the allies to these cuts made it imperative to press "with all possible firmness" for acceptance by the NATO military authorities of the British view on alliance strategy and conventional force requirements so as to mitigate the worst political and military consequences of a unilateral withdrawal.[67]

Britain and its allies remained deeply divided on several questions, all of which would have a bearing on NATO's future force requirements. One important issue concerning the likely duration of the period of large-scale, organized fighting involving an intense nuclear exchange that was expected to characterize the initial phase of a general war. The Political Directive was silent on the question, and although the first draft of MC 14/2 suggested that this period could last as long as 30 days, the British believed it would be much shorter.[68] The United States and France, however, estimated that the initial phase might go on for weeks.[69]

A second issue concerned the nature of any subsequent hostilities and whether additional forces would be required. Predictably, the

British argued that the Political Directive did not justify the provision of any special forces for the second phase of a general war. All NATO preparations should be geared exclusively for the brief opening phase, and any subsequent fighting would be conducted by those forces that survived.[70] The United States and other countries, in contrast, sought to recognize the possibility of protracted military operations of a largely conventional character, which would require an additional military capability to exploit any advantages achieved in the first phase.[71]

As before, however, the central issue was the relative weight to be placed on the deterrent and defensive functions of the NATO Shield forces. The British felt that the first draft of MC 14/2 placed too little emphasis on deterrence and too much on contingent measures should it fail. Consequently, they proposed an amendment that stressed the former.[72] The United States and Germany resisted this effort, and although the British amendment was eventually approved, it was modified to give equal weight to the objectives of preventing war and, should war nevertheless be forced upon NATO, of having "the capability to bring it to a successful conclusion."[73]

These differences grew out of equally fundamental disagreements over the nature and significance of the possible limited forms of aggression identified in the Political Directive. The British played down this threat, arguing that NATO was not prepared to countenance limited war.[74] U.S. officials sought to place more emphasis on military operations short of general war and, in particular, wanted MC 14/2 to recognize explicitly the possibility of "limited war" with the Soviet satellites. Several countries, moreover, felt that the strategic concept should address the possibility of a limited conflict that could gradually grow into a general war.[75]

The final version of MC 14/2, which was approved in early April by the Military Committee and formally adopted by the NATO Council a month later, followed the Political Directive in its general outline, although it was considerably more detailed.[76] Consistent with the new guidance, the strategic concept addressed two basic contingencies: general war and more limited alternative threats. Although it regarded the latter as more likely than the former, it nevertheless gave first priority to preparing for a general nuclear war.[77]

The document's assumptions regarding general war were virtually identical to those contained in MC 48. Such a war would probably begin with a massive Soviet nuclear offensive together with a campaign to seize Western Europe. Thus the alliance would be unable to prevent the rapid loss of NATO territory unless it immediately employed both strategic and tactical nuclear weapons, whether or not the Soviet Union had used nuclear weapons. The opening phase would be critical so priority would have to be given to maintaining ready forces, which could contribute to the success of the initial operations.

As in the Political Directive, however, the section on alternative threats explicitly recognized the possibility of Soviet-initiated operations with limited objectives such as infiltrations, incursions, or hostile local actions. It stated that the alliance should be prepared to deal immediately and in appropriate strength with such incidents "without necessarily having recourse to nuclear weapons." MC 14/2 nevertheless added that "NATO must also be prepared to respond quickly with nuclear weapons should the situation require it."[78]

SUBSEQUENT INTERPRETATIONS OF MC 14/2

The new strategic concept was the product of substantial compromise among a wide range of views. To facilitate agreement, it had been progressively watered down during the successive stages of drafting, and differences of opinion were often submerged in vague language. As a result, the wording of MC 14/2 was subject to conflicting interpretations.[79]

Thus before NATO force planning could proceed, several issues required clarification. Of particular significance were the related questions of how to define incidents short of general war and how NATO should respond to them, the answers to which would greatly influence the alliance's force requirements. MC 14/2 had been extremely equivocal on these points. While creating a place for limited responses to Soviet-initiated and supported operations with limited objectives, it simultaneously denied that there was any NATO concept of limited war with the Soviets and noted that if the Soviet Union sought to broaden or prolong a hostile local action, "the situation would call for the utilization of all weapons and forces at NATO's disposal."[80]

The critical interpretation of the new strategic concept was provided by General Lauris Norstad, who had replaced Gruenther as SACEUR in late 1956, in the context of a review of the alliance's minimum force requirements in Europe. Following the adoption of MC 14/2, each of the major NATO commanders was instructed to initiate such a study, and in the fall of 1957 their reports were combined into an overall estimate, labeled MC 70, which was to be considered by the foreign ministers in December. This exercise allowed

Norstad to put his own construction on the vague language of MC 14/2, and in the process he significantly expanded the functions of the Shield forces.

Previously, these forces had been assigned two basic roles in NATO strategy. The first was to contribute to deterrence by convincing the Soviet Union that NATO would respond promptly with nuclear weapons, both strategic and tactical, in the event of a major attack. If NATO held the forward line with sufficient strength, the Soviet Union would have to use substantial force to break through, thereby raising the stakes and increasing the risk of nuclear retaliation. The second role was a defensive one. In the event of a general war, the Shield forces would protect important elements of NATO's retaliatory forces and would defend the people and territory of the alliance by holding the forward line until the nuclear counteroffensive took effect.[81]

To these traditional functions, however, Norstad now added a third: to provide NATO with essential political and military flexibility. He argued that because both sides possessed thermonuclear weapons, it was imperative to avoid precipitating a general war unnecessarily. The alliance must be able to deal with any incident short of an all-out attack using conventional forces alone or, if necessary, low-yield nuclear weapons. Indeed, such a capability would deter limited attacks from being undertaken in the first place. Thus the Shield forces would provide an alternative to massive retaliation by allowing the alliance to meet less-than-ultimate threats with decisive but less-than-ultimate responses.[82]

Norstad's interpretation of MC 14/2 implied the possibility of a limited war in Europe.[83] This view was subsequently incorporated into MC 70, which further broadened the range of contingencies to be handled by the Shield forces alone. In particular, it suggested the possibility that even direct Soviet military action would not trigger a full-scale retaliation by NATO, provided that Soviet objectives and the weapons they used appeared to be limited. The cumulative effect was to blur completely the distinction between "incidents" and "limited war" in Europe.[84]

Strategic and technological developments subsequent to the adoption of MC 14/2 only reinforced the tendency to interpret the new strategic concept in the most flexible manner. The Soviet demonstration of a potential intercontinental delivery capability in the fall of 1957 in particular was widely perceived as increasing the probability of limited aggression, thereby enhancing the importance of the

Shield forces, because the Soviet Union might be even more likely to conclude that the United States would not respond to a limited attack in Europe with strategic nuclear weapons. Thus U.S. policy toward NATO provided for "the continued development of military capabilities to counter infiltrations, incursions, and hostile local actions" to allow "a real choice as to the appropriate response to Soviet aggression."[85]

During the following years, Norstad constantly reiterated his expansive view of the role of the Shield forces in NATO strategy. In 1959, he told Congress that NATO plans were based on the principle of using no more force than was necessary and that the alliance would attempt to deal with hostile situations with conventional weapons alone whenever possible.[86] And in 1960, he stated that nuclear weapons would be introduced only at a high threshold, requiring that NATO forces have a substantial conventional capability. Indeed, the Shield should enable the alliance to respond flexibly and appropriately to any situation up to the level of general war.[87] Clearly, this interpretation of the alliance's formal military strategy bore little resemblance to massive retaliation.

NATO CONVENTIONAL FORCE REQUIREMENTS

MC 14/2 and its subsequent interpretations did not allow for the reduction in NATO's conventional force requirements the British had hoped for. Indeed, U.S. and British officials had recognized as early as late 1956 that the Political Directive would not provide the relief the British had been seeking.[88] The authoritative statement of the alliance's minimum force requirements under MC 14/2 was contained in MC 70, which the NATO Council approved "for planning purposes" in May 1958. MC 70 called for a minimum of 30 ready divisions in the Central Region, no change from the previous requirement under MC 48. By that time, however, the maximum number that could be expected in practice was only 28⅓ divisions, which would include 12 German, 5 U.S., 4 French, 3 British, 2 Belgian, and 2 Dutch divisions, and a Canadian brigade.[89]

Ironically, the new role assigned to the Shield forces under the strategic concept affected the overall size of the MC 70 requirements only marginally. These requirements were based primarily on the general war functions of the Shield, especially the defense of NATO territory during the critical opening phase. Specifically, the new requirements followed from the assumption that each division could be expected to hold a 30-kilometer-wide section of the front during the first few days of such a conflict. No more forces were needed to deal with limited aggression than for a general nuclear war.[90] Nevertheless, the conventional capabilities desired by Norstad ensured that the Shield was intended to be much more than a mere trip-wire.

NATO Conventional Force Levels and Capabilities

In mid-1956, the NATO countries still fielded only some fourteen ready divisions in the Central Region. Thus the adoption of the Political Directive and MC 14/2, which placed greater emphasis on the Shield forces in NATO strategy, and the subsequent development of MC 70, which reaffirmed the alliance's long-standing force requirements, called new attention to the need for a significant increase in the forces deployed there. These important strategic developments, however, had little impact on the alliance's conventional force levels. Although efforts to withdraw substantial numbers of U.S. troops from Europe were defeated, France's NATO contribution remained well below what had been promised in previous years, and the size of the British military presence on the continent declined significantly. In addition, the long-anticipated German buildup failed to materialize as quickly as had been hoped, and the nuclearization of NATO's force structure, which was extended to the forces of the European allies during this period, further eroded the alliance's conventional capabilities. Thus the gap between NATO force requirements and the alliance's actual military strength narrowed only slightly.

THE GRADUAL WEAKENING OF THE U.S. FORCES IN EUROPE

The basic structure of U.S. military presence in Europe remained approximately the same during this period at six division-equivalents. The conventional capabilities of these forces declined somewhat, however, as a result of their progressive reorganization for the atomic battlefield and an October 1956 decision to reduce the number of military personnel on the continent. Nevertheless, proposals to withdraw entire U.S. combat units, both that year and again in 1957, were defeated, after which no further serious challenges to the status quo were mounted.

Before the leak of the Radford plan, several key members of the Eisenhower administration had wanted to reduce U.S. forces in Europe. The leak and the subsequent outcry, however, had made it extremely difficult to explore the subject in any detail. Indeed, by demonstrating the continuing sensitivity of the European allies to any change in the level of U.S. forces, which had come to be regarded as a barometer of the American commitment to the region, the episode had strengthened the hand of those in the government who opposed troop withdrawals. Top administration officials were especially concerned about possible adverse German reactions.[91] Consequently, they concluded in mid-August that although a reduction in the size of the American military presence in Europe was possible and should be effected, it would be a mistake to withdraw any U.S. divisions at that time. Instead, the United States should consider ways of reducing the troop strength of each of the divisions in the theater by 5,000 to 10,000 men.[92]

Despite this apparent consensus, deep differences of opinion persisted within the administration. The Defense Department took the position that budgetary considerations necessitated reducing the number of divisions in Europe within eighteen months and that planning should commence forthwith.[93] State Department officials, however, judged that the political situation did not permit the United States even to propose a withdrawal at that time. Such a move, they argued, would jeopardize the achievement of the planned European defense effort, especially the German contribution, and might even lead to the collapse of NATO and to allied accommodation with the Soviet Union. Thus they saw no alternative to maintaining U.S. forces in Europe at their present levels, although they perceived no objection to some "streamlining."[94]

The issue was finally resolved in early October, at the same time the U.S. position on the British proposal for a review of NATO strategy was decided, when Eisenhower sided with the State Department in a meeting of his top advisers. The president "felt very definitely" that the United States could not take any divisions out of Europe at that time, even though he had always regarded the U.S. deployment on the continent as a temporary expedient. He did, however, agree to a reduction in manpower through streamlining the existing divisions, which he thought would yield substantial savings.[95]

As a result of this decision, U.S. ground forces in Europe were reduced by approximately 16,000 men over the following two years.[96] During the same period, their conventional capabilities were further weakened by the gradual reorganization and reequipment of army units for atomic war. In 1956, the U.S. Army began to test a new divisional structure designed for the unprecedented rigors of a nuclear battlefield. To facilitate dispersal, each unit would be divided into five subunits, instead of the previous three, hence the name "Pentomic." That year, an airborne division was reorganized along the new lines, and in early 1957, the army announced that all U.S.

divisions, including those stationed in Europe, would be converted during the following year.[97]

The Pentomic versions of the infantry and airborne divisions were considerably smaller than their predecessors. The infantry division was reduced from 17,500 to 13,700 men while the size of the new airborne division was slashed by more than a third, from 17,500 to 11,500. Only the armored division variant was approximately the same size as before. In addition, Pentomic divisions were somewhat more lightly armed conventionally.[98]

The preparation of U.S. tactical aircraft for nuclear war also cut into the alliance's conventional capabilities. By September 1958, all of the tactical bombers and fighters assigned to NATO were equipped to deliver nuclear weapons.[99] As a result, many of these aircraft were not able to carry conventional ordnance or would have been withheld for nuclear missions and thus would not be available for conventional operations in the event of a conflict.

The 1956 decision to streamline the U.S. divisions in Europe, while ruling out a reduction in the number of combat units in the near future, nevertheless left open the question of the appropriate magnitude of the American presence on the continent over the long run. Consequently, budgetary pressures the following year provided an opportunity for economy-minded members of the administration to press for further cuts. In mid-July, Secretary of Defense Wilson ordered a reduction in military manpower of 100,000, from 2.8 to 2.7 million, to be implemented by the end of the calendar year, half of which was to come out of the army. And in September, he ordered the elimination of a further 92,000 positions to be carried out in the second half of FY 1958, with the army losing another 50,000 men.[100]

These cuts were not specifically intended to result in the withdrawal of U.S. forces from Europe, which the JCS had explicitly opposed as recently as April.[101] Indeed, in announcing the first set of reductions, Wilson declared that U.S. forces overseas would not be weakened, while the second cut was "to be achieved without materially affecting our deployments of major combat units abroad."[102] When army leaders were told to plan on a second reduction of 50,000 men, however, they calculated that they would have to withdraw approximately one-half of an infantry division from Europe by mid-1959.[103]

Norstad was alarmed by this suggestion and put up fierce resistance. Pointing out how a disastrous chain reaction had only been narrowly averted earlier in the year when the British had announced unilateral cuts, he argued that this was the worst possible time for the United States to reduce its combat forces in Europe.[104] As a result of such opposition, the size of the army was ultimately reduced by only a further 30,000 men, which allowed the number of combat units in Europe to be kept constant.[105]

The defeat of this effort marked the end of the last serious challenge to the size of the American military presence in Europe during the Eisenhower administration. During the following years, top U.S. officials, including the president himself, occasionally raised the idea of withdrawing more U.S. forces so as to reduce the defense budget or the growing U.S. balance-of-payments deficit. On each occasion, however, the opponents of reductions were able to block these suggestions before they could receive serious consideration.[106]

THE STRETCHOUT OF THE GERMAN MILITARY BUILDUP

Much hope had been placed on German rearmament to fill the wide gap between force requirements and actual conventional capabilities in the Central Region. In 1955, the government of Chancellor Konrad Adenauer had agreed to undertake the rapid creation over three to four years of a 500,000-man military force. As called for in the NATO force goals, West Germany was to field an army of twelve conventionally armed divisions, which would represent the largest national contribution to the alliance's forces in the Central Region. At the same time, German political and military leaders showed little interest in nuclear armament. Through mid-1956, they regarded tactical nuclear weapons merely as a powerful supplement to conventional forces, which would continue to be the mainstay of NATO's force posture.[107]

German leaders expected their country's conventional buildup to serve several essential political and military purposes. Militarily, these forces were needed "to reinforce the primary [strategic] deterrent . . . , to counterbalance Soviet superiority in conventional forces, and to discourage or defeat Korea-like challenges by the Soviet Union or its satellites." Many top German officials put particular emphasis on defending against a large-scale Soviet invasion, which was regarded as the major threat to European security. Until the leak of the Radford plan, they continued to believe that conventional forces—not tactical nuclear weapons—would play the decisive role in determining the outcome of such a conflict. Indeed, the creation of a strong conventional army would allow NATO to reduce its reliance on nuclear weapons. Thus there was little concern that Germany's rearmament program was out of step with the growing nuclear emphasis in NATO policy.[108]

Politically, a military contribution "would mark Germany's full acceptance into the Western community and secure greater consideration of German interests." It would give the country a voice in the development of NATO's military plans. And of equal if not greater importance, it would help to ensure the maintenance of U.S. forces in Europe and thus the continuation of the American security guarantee. German leaders believed that the best way to obtain these benefits was through strict adherence to the rearmament objectives that had been established in previous years.[109]

The leak of the Radford plan, however, severely undercut a number of the assumptions that underpinned the German rearmament program. Most immediately, it suggested that unswerving pursuit of rearmament was no guarantee that the U.S. military presence would not be reduced.[110] Thus the German government's first concern was to head off any cuts in the number of American troops in Germany, which might greatly weaken the nuclear guarantee. In addition, unilateral U.S. reductions might prompt the other NATO countries to follow suit, leaving Western Europe "dangerously exposed."[111] Consequently, Adenauer instructed his ambassadors to the other NATO countries to express German opposition to any such actions, and the top-ranking German military officer, General Adolf Heusinger, was dispatched to Washington to plead against any withdrawals of U.S. forces from Europe.[112] More generally, the German leaders initially sought to reaffirm the need for strong NATO conventional forces, which they still regarded as vital to their country's security. Hence they quickly announced their intention to resist any shift in NATO strategy that would place greater reliance on nuclear weapons and to continue with the planned buildup of German forces.[113]

At a deeper level, however, the Radford plan raised searching questions about the appropriate size and indeed the very nature of the German rearmament program. In particular, it suggested that the exclusive emphasis on conventional rearmament had been severely mistaken and that there was certainly no need for as many as twelve German divisions.[114] Thus the Radford plan appeared to confirm the argument of Adenauer's political opposition that the planned buildup was inconsistent with the realities of modern warfare, especially given the emphasis on nuclear retaliation in NATO strategy, and would contribute little to German security.[115] If pursued, it would condemn Germany to a permanent second-class status within the alliance. The incident had also seemingly demonstrated how little influence the Germans actually enjoyed, and it had destroyed Adenauer's claim to a special relationship with the United States by suggesting that the Americans had not bothered to consult or even to inform him about issues of tremendous concern to the Federal Republic. To the contrary, the chancellor had been caught completely by surprise.[116]

This apparent reversal of U.S. policy seriously weakened the position of Adenauer and his government on defense issues, forcing them to reassess the German rearmament program.[117] As a result, a significant shift took place in German policy during the following three months, notwithstanding repeated American assurances that there would be no change in U.S. deployments on the continent and that a contribution of twelve German divisions remained essential.[118] The news from Washington immediately created problems in the Bundestag, where legislation to establish the period of conscription was still pending, by strengthening those who favored a shorter term of service than the government had proposed. In late September, the government announced that it would ask for only a twelve-month term for all conscripts rather than the eighteen-month term it had previously been seeking, which meant that Germany would probably be unable to raise an armed force of 500,000 men by 1959 as planned.[119]

The main lesson for the German leadership of the Radford plan incident, however, was that access to tactical nuclear weapons would henceforth be necessary to ensure German security and to guarantee Germany a place of equal influence and stature within the alliance. Thus in early September, the German representative to the NATO Military Committee suddenly expressed a desire to be informed of recent U.S. Army tactical and organizational developments for atomic warfare. And toward the end of the month, it was reported that the Germans had already begun to sound out their allies on the political considerations involved in eventually equipping the continental countries with nuclear weapons.[120]

The most dramatic indication of the shift in German attitudes toward nuclear weapons was the sudden replacement of German defense minister Theodor Blank by Franz Josef-Strauss in mid-October. Strauss had long been convinced that nuclear weapons were the key to military and political power, and for some time he had insisted that German forces be equal in armament and status to those of the other NATO countries.[121] Shortly after his appointment, Strauss informed the allies that the Federal Republic would not be able to meet its previous manpower pledges. The Germans now planned to raise a small number of ready divisions as soon as possible rather than to lay the groundwork for a full twelve-division force all at once. He indicated that Germany would field five divisions and two brigades by the end of 1957 but assiduously avoided making any comment on the ultimate size of the German buildup. At the same time, both he and Adenauer declared their government's intention to seek atomic armament for the Bundeswehr.[122]

As a result of these policy shifts, the German military buildup proceeded more slowly than the allies had hoped. The total number of men under arms reached 100,000 in July 1957, when the first three German infantry divisions were formally committed to NATO. By the end of the year, these units were joined by two German armored divisions and two brigades, one mountain and one airborne, as promised, and two more divisions were formed during the following year. Many of these units, however, were understrength, not fully trained, and deficient in heavy equipment. By the end of 1960, the size of the German armed forces had reached only 270,000 men, while the total number of ready divisions remained at seven, little more than half of what had been promised.[123]

In part, the slowdown was the result of a lack of trained junior and noncommissioned officers and a severe shortage of accommodations, which demonstrated that the previous plans had been unrealistically optimistic.[124] Nevertheless, the ultimate size of the German military contribution remained in doubt until late 1957, when Germany informed the allies that it planned to have 303,000 men under arms by April 1961. And although the German government reaffirmed its intention to raise a total of twelve divisions, the fact that the overall size of the armed forces would be well below the original goal of 500,000 ensured these units would be significantly understrength even if they were ever created.[125]

REDUCTIONS IN BRITISH FORCES

Despite these revisions in the German rearmament program, Norstad could still hope that West Germany would eventually fulfill its original force goals. Thus perhaps the greatest setback to his plans during this period was the British government's decision in late 1956 to reduce its NATO contribution, notwithstanding the failure of its effort to revise NATO strategy to bring about a diminution of the alliance's force requirements. This decision resulted in a 30 percent decline in the size of the British Army of the Rhine and an even larger cut in the British air forces stationed on the continent.

After proposing the review of NATO strategy in mid-June, the British government continued to elaborate the force reductions it sought to achieve. Eden was eager to cut the British defense budget to £1,350

million and eventually to £1,300 million, which would require reducing the overall size of the armed forces to as low as 450,000 men,
well below the existing level of approximately 750,000. The achievement of these ambitious objectives in turn meant that the British
would have to disband two of the four divisions then stationed in
Germany, and the necessary planning was set in motion. In addition,
the government tentatively planned to reduce the Second Tactical
Air Force in Germany by at least one-third, from some 450 aircraft to
no more than 300, over three years. Following the outbreak of the
Suez crisis in late July, however, these planning efforts were temporarily put on hold.[126]

By late November, substantial conventional force reductions
seemed more imperative than ever. As a result of the Suez crisis,
defense outlays had risen, while the British economic situation had
deteriorated markedly. At the same time, the military setbacks suffered at Suez had increased the importance assigned by the British to
the nuclear program, which was in high gear in preparation for the
first British H-bomb test in May 1957. Nor had any of the other financial pressures that had prompted the British government to press for a
review of NATO strategy abated. Thus the political-economic costs
of adhering to the status quo now appeared to outweigh the political
and military risks associated with deep cuts.[127]

To make matters worse, the development of a new Political Directive was no longer expected to provide any relief. Although the emerging final version established a new mission for the Shield forces and
thus threatened to result in even greater NATO force requirements,
even the original British draft would probably not have been interpreted as allowing any reductions. Thus the British Chiefs of Staff
concluded that there would be no military justification for recommending cuts in the forces stationed in Europe. Any significant withdrawals would have to be justified on economic grounds.[128]

British leaders thus decided in mid-December to take decisive action to reduce spending on conventional forces. Their forces would
be reduced to a size they could afford to equip adequately with up-
to-date—including nuclear—weapons. As before, the achievement of
this goal was expected to require limiting the total number of military personnel to 450,000. The BAOR would be cut from 78,000 to
50,000 men and 2TAF to about 200 aircraft by April 1958.[129] The
government hoped that these reductions in Britain's NATO contribution could be approved before the next defense White Paper was
issued in mid-February.[130]

Contrary to the advice of the Chiefs of Staff, however, British leaders initially sought to justify the intended reductions on military grounds. This approach, they believed, was both dictated by the 1954 Paris agreements and consistent with the progressive nuclearization of their forces. They planned to equip the light bombers in 2TAF to carry nuclear bombs, which they thought would enable the NATO-assigned air forces to maintain their equivalent fighting capacity even while being considerably reduced in size. Similarly, they argued, if British ground forces were given an atomic capability, even four much smaller divisions might be judged as adequate to meet British obligations under the Paris accords.[131] Thus although Britain would inform its allies of its dire economic situation, the troop cuts would be negotiated primarily on the basis of military considerations.[132] The British put great stress on holding preliminary talks with the United States and the NATO authorities before announcing the planned cuts. Indeed, prior consultation with SACEUR was essential because he would have to certify that the overall fighting capacity of the British forces would not be reduced.[133]

British officials approached their U.S. counterparts and Norstad at the December meeting of the North Atlantic Council, shortly before they made a final decision on the size of the cuts to be sought. Their description of Britain's economic plight was received sympathetically by the Americans. And Norstad suggested that the proposed withdrawals would be accepted, although he asked the British to reduce their NATO contribution in such a way that it would not trigger a "chain reaction" among the other countries.[134]

Norstad later made it clear, however, that he would oppose the proposed reductions if they were based on the military argument that a smaller force equipped with nuclear weapons would be no less capable. The potential atomic capability of the units presently stationed in Germany had been taken into account as early as 1954, he argued, so he was not prepared to certify that a contingent of only 50,000 men would have the equivalent fighting capacity of the present force. He was concerned, moreover, that other countries would use the same argument to try to justify a reduction in their own NATO contributions. But if the British simply stated that they could not afford to provide more than 50,000 troops without endangering Britain's economic stability, he would do all he could to support them. The British accepted this approach, especially because it meant that the size of their NATO contribution could be decoupled from the specific wording of the new strategic concept.[135]

In the meantime, however, even deeper cuts in the BAOR had become necessary. In early 1957, the British government had decided to terminate national service and to put the armed forces on an all-regular basis, which would require reducing the overall manpower ceiling to 375,000 men. To attain this goal, the Chiefs of Staff concluded that the BAOR would have to be limited to only four brigade groups and 44,000 men, although no further cuts in 2TAF would be required.[136]

The biggest hurdle the British now faced was gaining the acceptance of their Western European Union (WEU) allies, a majority of which would have to approve any reductions under the terms of the Paris agreements. The other European countries had been kept in the dark about the details of the British plans, which were not provided to them until mid-February at a meeting of the WEU council. Not surprisingly, they were strongly opposed to the proposed reductions. In addition to weakening the British commitment to NATO, the cuts were viewed as threatening to unbalance the entire Western defense posture, because the British forces, like those of the United States, were regarded as an essential counterweight to the growing West German army, and as being at variance with the new trend in NATO strategy toward reduced reliance on nuclear weapons. The allies feared, moreover, that such a move might encourage the Americans to reduce their own forces in Europe.[137]

During the following weeks, however, the outlines of a compromise emerged. At about the time the British government presented its plans to the WEU, Norstad recommended that Britain phase the proposed cuts over two years instead of one and that it maintain a 5,000-man strategic reserve force on the continent rather than redeploy it to the British Isles. The British felt that a consensual agreement was preferable to a unilateral breach of the Paris treaties, even if it meant somewhat higher outlays, and they soon concluded that the WEU countries would eventually accede to the reductions if they accepted the adjustments suggested by Norstad.[138]

The WEU agreed in mid-March that Britain could withdraw half of 2TAF and 13,500 soldiers during the following year, which would leave approximately 220 aircraft and eight brigade groups on the continent. In October, the allies would discuss a second cut of the same size in the BAOR, which would result in a force of 50,000 men and six brigade groups.[139] This agreement was confirmed and made public in the 1957 British White Paper on defense, which was belatedly released in early April.[140]

The initial WEU agreement, however, only postponed the question of whether Britain would be permitted to make the second round of cuts it had requested. The plan to limit British forces to 375,000 men, moreover, presupposed cutting the BAOR by a further 6,000 men and one brigade group and reducing 2 TAF to only 100 aircraft in the early 1960s. Although Norstad had been informed of these plans, he had asked the British not to disclose them to the allies for fear of possible repercussions.[141]

As expected, the issue flared up again in October. Rather than demanding the entire 13,500-man cut, the British government declared its willingness to maintain a total of 55,000 men and seven brigade groups in Germany—as Norstad had recommended—during the following year if acceptable arrangements were made to cover their foreign exchange costs, and a second agreement was struck on these terms in January.[142] The required financial arrangements were not forthcoming, however, and in any case, the British soon insisted that the BAOR would eventually have to be reduced by a further 10,000 men, to 45,000 men and only five brigade groups. As before, the British plans stirred fears of a chain reaction within NATO, especially because they disregarded Norstad's judgment that the 55,000-man force was the minimum essential U.K. contribution. To avoid controversy, the British decided to defer approaching the WEU again until October 1958, and as a result of the Berlin crisis that erupted at the end of the year, the plans for further reductions were shelved indefinitely. Nevertheless, the level of British ground forces on the continent had fallen by some 30 percent, while 2 TAF was eventually reduced to only a fraction of its original size.[143]

THE FORCES OF THE OTHER ALLIES

The slowdown of the German buildup and the British cuts created widespread concern that additional countries would follow suit. These fears proved to be exaggerated; none of the other European allies altered their NATO contributions in ways that had equally significant implications for the overall strength of the alliance's conventional capabilities. Still, most continued to be delinquent in meeting their force goals and some fell even further behind.

The size of the French conventional force contribution to NATO remained a source of disappointment to those concerned with the defense of Western Europe. Many French units had been dispatched to Algeria in 1954 and 1955, causing a decline in available French forces in the Central Region from some five ready divisions to two nominal divisions, each of which was at only two-thirds of its full strength. In contrast, approximately fourteen French divisions were stationed in North Africa in mid-1956.[144] The gap between French requirements and actual forces was narrowed by a further reduction in the NATO requirements for France, to only four ready divisions under MC 70. Yet even then, France remained unable to meet its commitments, allowing its military contribution to be surpassed by that of West Germany as early as 1957.[145]

During this period, the smaller countries in the Central Region also struggled to maintain their forces at existing levels, in some cases unsuccessfully. In mid-1956, Belgium decided to reorganize its divisions under a smaller "new look" concept, reducing the size of each division by two battalions. Following the German decision to limit national service to a year, moreover, the Belgian government came under pressure to reduce its own period of conscription from eighteen to fifteen months, which it eventually did in July 1957.[146]

The Netherlands also decided in 1956 to reorganize its single ready division "in the light of new developments in atomic warfare" the following year. As in Belgium, the new-style division was significantly smaller, with only 15,000 rather than 18,000 men, and had somewhat less conventional capability. Although the Dutch were persuaded to plan to upgrade a reserve division into a second active division in line with NATO force requirements, this new division would not be available for several years.[147]

NUCLEARIZATION OF NATO FORCES

The gradual nuclearization of allied ground and tactical air forces that began in the late 1950s contributed further to NATO's conventional weakness in the Central Region. In late 1955, the United States had resisted the demands of its allies for a nuclear role in the defense of Western Europe, and the following year saw little change in U.S. attitudes. Although the Eisenhower administration requested funds to provide "advanced weapons" to the allies and reaffirmed at the December 1956 NATO meeting that the United States would make nuclear-capable systems available to the Europeans and would even help them develop and produce such weapons, no arrangements had been made for providing the allies with the actual warheads.[148]

As Soviet atomic capabilities grew, however, continuing to deny the Europeans a nuclear role made increasingly less sense strategically. By the end of 1956, moreover, the allies had the solid backing of SHAPE. Norstad showed himself to be a strong proponent of providing them with a nuclear capability, going so far as to issue new planning guidance that assumed the integration of U.S. atomic-capable weapons into all NATO forces.[149] Consequently, the allies renewed their demand at the December 1956 NATO meeting, launching what appeared to be a coordinated offensive. The defense ministers of Britain, France, Germany, and the Netherlands all called for access to tactical nuclear weapons in some form, stressing the need both to strengthen Europe's defense and to reduce manpower requirements. And once again, the U.S. representatives refused to budge on the basic issue of making nuclear warheads available to the Europeans.[150]

Soon after the NATO meeting, however, the Eisenhower administration did begin to give serious consideration to the matter. By then, U.S. officials may have realized that there was some room for compromise between the Congress, which jealously guarded U.S. atomic secrets, and the allies. The Europeans had not insisted that the United States grant them complete control of the warheads. Indeed, they had offered no specific proposals as to how greater access might be achieved.[151]

By mid-February, the administration had formulated the basic outlines of a plan to establish a NATO atomic stockpile. Under this arrangement, the United States would store nuclear weapons at strategic points in Europe. In the event of a Soviet attack, the warheads would be distributed to allied forces under authorization from Washington.[152] This approach had the benefit of requiring no change in the Atomic Energy Act. It was founded instead on the president's wartime authority as commander in chief to turn warheads over to allied forces that were equipped with nuclear delivery vehicles and trained in their use.[153] Although the plan encountered considerable opposition, especially from the Joint Chiefs of Staff, it gained the backing of Dulles, Norstad, and, most important, the president and was formally adopted as U.S. policy toward the end of the year.[154]

Dulles announced the United States's willingness to participate in the creation of a NATO nuclear stockpile at the December 1957 meeting of the allied heads of government. Under the administration's plan, nuclear warheads would be deployed under U.S. custody in accordance with NATO defense planning and in agreement with the nations directly concerned. In the event of hostilities in Europe, the warheads would be released to SACEUR for employment by nuclear-capable NATO forces. The United States would maintain exclusive positive control of the weapons in the stockpile, although host countries would have a veto over their use.[155] Immediately following Dulles's announcement, the NATO leaders agreed in principle to establish the necessary arrangements, and by early March, SHAPE had drafted a "basic plan" for the NATO atomic stockpile.[156]

Thus three years after the adoption of MC 48, the stage was finally set for the nuclearization of the allied ground and air forces in Europe.

It would still be some time, however, before this nuclear capability would become a reality. Because detailed bilateral arrangements between the United States and each of the participating European countries had to be worked out, several more years passed before the NATO nuclear stockpile system was actually in place. Some allies, moreover, chose not to participate, and France never permitted the stockpiling of U.S. nuclear weapons on its soil. In addition to these legal hurdles, the training of allied forces in the use of nuclear-capable weapons did not commence until late 1957 in many cases, and the delivery of the launchers themselves did not begin until the second half of 1958. Thus nuclear-capable allied units did not become operational until 1959, and many did not receive their nuclear warheads before 1960.[157]

Notwithstanding these setbacks, the gradual nuclearization of allied forces soon began to erode NATO's conventional capabilities in two ways. First, because military budgets were already stretched to the limit, the procurement of nuclear delivery vehicles absorbed defense resources that might otherwise have been devoted to bolstering the alliance's conventional strength.[158] Second, the conversion of existing conventional weapons systems, especially tactical aircraft, into dual-capable platforms meant that they were increasingly likely to be withheld from combat for nuclear missions.

According to the MC 70 force requirements, the European allies with forces in the Central Region were expected to field at least 65 missile units by the end of 1963.[159] Some of these missiles would be provided by the United States under its military assistance program.[160] The bulk of the new equipment, however, would have to be produced or purchased by the allies themselves.

The effects of nuclearization were most apparent in the British and German force contributions to NATO. The 1957 British White Paper, which announced the planned reductions in the British forces in Europe, also revealed that "atomic rocket artillery" would be introduced into the BAOR and that some of the remaining squadrons of 2TAF would be provided with atomic bombs. By 1958, according to one study, the defense planning and tactics of British units in Germany were conceived entirely in nuclear terms.[161]

The impact on West Germany's military posture was even greater. After 1956, the Federal Republic consistently sought to provide its forces with "the most modern weapons," meaning nuclear-capable aircraft, artillery, and missiles. This new direction in policy, however, entailed "expenditures far exceeding those foreseen for conventional equipment." During the five-year period covered by MC

70, more than half of the money spent by the German government on weapons procurement went for aircraft and missiles, most of which were intended to deliver nuclear weapons. The German air force, moreover, was redesigned for nuclear interdiction and counterair missions, while tactical air support was virtually neglected. This shift of emphasis was given substance in 1958 by the decision to build the air force around the American F-104 Starfighter, which was primarily a nuclear strike aircraft.[162]

Ironically, even setbacks to the nuclearization process could have a negative effect on NATO's residual conventional capability. The French refusal to accept stockpiles of nuclear weapons under U.S. control forced Norstad to redeploy nine squadrons of aircraft to Britain and the Federal Republic. As a result, these valuable assets would either be stationed farther from the potential battlefield or concentrated on a smaller number of bases near the front, making them more vulnerable to preemptive strikes.[163]

SUMMARY

Thus despite the increased emphasis on conventional forces in NATO strategy, the alliance's ability to mount a successful defense of Western Europe seemed farther than ever from realization.[164] In early 1958, Norstad could count on only 18⅓ nominal ready divisions in the Central Region.[165] Many of these units, moreover, were greatly understrength, were seriously deficient in essential equipment, and lacked adequate supporting units.

By early 1961, the situation was not much improved. Largely because of the German buildup, NATO now disposed of 22⅔ divisions on paper in Central Europe, still well short of the 30-division requirement. Because of continuing shortages of personnel and equipment, however, these amounted to the equivalent of only 16 fully ready divisions in combat strength.[166] And it is unclear how much capacity for conventional combat these forces retained. Notwithstanding the new strategic concept, NATO's military posture was "totally dependent on early use of nuclear weapons, strategic and tactical."[167]

Analysis

Why was NATO military strategy revised again so soon after the adoption of MC 48? Why did the new strategy place greater emphasis on conventional forces? And why did the alliance's conventional capabilities fail to increase commensurately with their enhanced importance?

As in earlier years, the changes in NATO's conventional force posture during this period were driven substantially by variables emphasized by the balance-of-power perspective, especially the increasing availability of thermonuclear weapons, the continued growth of Soviet nuclear capabilities, and a further amelioration of Western perceptions of Soviet intentions. For the first time, however, intra-alliance factors assumed equal if not greater importance. Because of differences in geography, capabilities, and interests, how the allies interpreted and sought to respond to these developments varied substantially. Institutional factors also had an unprecedented if still subordinate effect, helping to determine the precise form of the new NATO strategy as well as allied force levels.

Three developments captured by the balance-of-power perspective shaped NATO strategy and force levels during this period. One was the introduction of large numbers of hydrogen bombs into the strategic arsenals of the United States and the Soviet Union. Related to this development was the steady growth in all aspects of Soviet nuclear power, as the Soviet Union expanded the number and diversity of its nuclear warheads and acquired an intercontinental nuclear delivery capability. As a result of this increase in estimated Soviet capabilities, many in the West concluded that the Soviet Union would soon achieve nuclear parity with the United States, if it had not done so already. At the same time, however, Soviet intentions seemed to moderate even more. Before the onset of the second Berlin crisis and with the notable exception of its invasion of Hungary, the Soviet Union was regarded as increasingly unlikely to use force as a result of a general improvement in East-West relations as well as specific conciliatory gestures on its part.

These three developments had three important implications for NATO's conventional force posture. First, a military conflict with the Soviet Union, especially an all-out war, seemed even more improbable than at any time since the outbreak of the Korean War. Soviet leaders appeared to have little or no desire to resort to direct aggression in Europe, and even if they had objectives that could be achieved only through the use of force, they were viewed as unwilling to risk a general war because of the potentially disastrous consequences.

Second, the resulting low sense of threat translated into yet greater domestic pressure to reduce defense spending and other burdens imposed by the NATO military effort. Consequently, most countries found it increasingly difficult to maintain their military contributions at existing levels, let alone to augment them. The British and U.S. governments in particular constantly sought ways to trim their defense budgets, and the German government came under pressure to limit the term of conscription to twelve months even before the leak of the Radford plan.[168]

Third, and somewhat at variance with the first two consequences, allied leaders, including U.S. officials themselves, raised serious doubts for the first time about the credibility of the American threat of nuclear retaliation, which had until then been the backbone of NATO strategy. As the United States became increasingly vulnerable to nuclear devastation, it seemed reasonable to conclude that American presidents would become reluctant to use nuclear weapons on behalf of U.S. allies in response to Soviet aggression as long as American territory had been spared direct attack. Consequently, limited forms of aggression in Europe would be more likely.

Although these developments precipitated a further modification of NATO strategy, their implications were not so obvious as to enable the alliance easily to reach a consensus on the form the new strategy should take. The interpretations the allies placed on these developments and the responses they advocated differed much more than they had in previous years. As a result, the allies were forced to resort to a high degree of ambiguity in the formulation of MC 14/2 to reconcile their conflicting strategic preferences and to make an agreement possible.

These differences were most marked in regard to allied views about the implications of the H-bomb and nuclear parity. Britain emphasized that war with the Soviet Union had become extremely unlikely and that NATO therefore had only to maintain the minimum forces necessary to ensure deterrence. Other countries, led by the United States, stressed the greater relative likelihood of limited forms of aggression and thus the need for a wider range of military options than a high degree of reliance on strategic nuclear forces would provide. At the same time, and despite U.S. objections, many European governments concluded that they could no longer afford not to have a nuclear role in NATO strategy.

Allied views of the implications of the Soviet invasion of Hungary also varied. On the one hand, the invasion and the simultaneous disturbances in Poland demonstrated that the Soviet Union could no longer count on the loyalty of the forces of its satellites, assuming it ever had been able to do so.[169] On the other hand, these events suggested that no diminution had occurred in the Soviet Union's willingness to use force when necessary to achieve its objectives. Consequently, one could draw conflicting conclusions about the risk of war and whether the NATO countries could afford to relax their military efforts.

The allies' contrasting perspectives and strategic preferences, and ultimately their uneven efforts to provide conventional forces, followed from differences in their locations, interests, and military resources. During this period, the geography of the alliance assumed heightened importance. Previously, the United States had been largely invulnerable to direct Soviet attack, thanks to its separation from the continent. U.S. leaders could afford to threaten to respond with the full weight of their atomic arsenal to deter Soviet aggression in Europe. As their country became increasingly subject to Soviet retaliation in kind, however, they began to place equal emphasis on the objective of preventing hostilities, should they occur, from escalating to the level of all-out nuclear war. Thus the logic of geography dictated that the United States should develop an interest in making NATO strategy more flexible in response to the growth in Soviet strategic capabilities.

Interestingly, Germany's very different geographical situation resulted in similar strategic preferences during much of this period. Because their country was located on the front line, the Germans were especially worried that, under the strategy embodied in MC 48, even a relatively minor incident would rapidly become a full-scale nuclear war waged on German territory. Consequently, they were also anxious to build greater flexibility into NATO's military posture.

Britain's strategic views were strongly colored by its atypical out-of-area interests and commitments. These absorbed a large share of the British defense budget and thus were a major consideration in the government's decision to seek a revision of NATO strategy that would justify a reduction in Britain's conventional force contribution to the alliance. Indeed, the adverse economic impact of its behavior in the Suez crisis was so great as to require Britain to take emergency action to reduce its forces on the continent, notwithstanding its failure to secure alliance endorsement of its strategic preferences. France's deep and continuing involvement in Algeria had a similar negative impact on the quantity and quality of the French forces remaining in Europe.

Differences in military power and potential, chiefly in the form of varying degrees of access to nuclear weapons, also influenced developments during this period. Britain's possession of an independent nuclear program helped shape that country's approach toward

NATO strategy because a favorable revision of the strategic concept would help to justify the commitment of substantial resources to the development of a thermonuclear capability at the expense of Britain's conventional forces. For many of the other European allies, which had previously lacked access to nuclear weapons, the acquisition of a nuclear role and the necessary delivery capabilities occurred at the cost of the size and effectiveness of their conventional forces.

Finally, institutional factors began to assume an important role in this period. By the mid-1950s, not only had SACEUR become an influential actor in his own right, but NATO conventional force levels were embedded in an increasingly dense web of international commitments, norms, and expectations. These growing institutional constraints on national behavior were most clearly manifested when Britain sought to reduce its forces on the continent.

British behavior was conditioned and ultimately restrained by several institutional considerations. Government officials were acutely aware of the commitment Britain had undertaken in 1954 to maintain forces of a certain size on the continent, and they were anxious to gain the consent of their allies to the proposed reductions rather than act unilaterally. Consequently, they repeatedly sought to justify the withdrawals in terms that were consistent with the commitment. First, the government attempted to engineer a change in NATO strategy that would result in lower force requirements. When that failed, it argued that the nuclearization of the remaining forces would leave them with the equivalent fighting capacity. Under pressure from Norstad, the British consented instead to use the economic escape clause contained in the 1954 Paris agreements. Finally, to secure the approval of the WEU, they agreed to defer discussion of half of the proposed reduction until the end of the year and, subsequently, to scale it back in size by approximately one-third.[170]

Determinants emphasized by the institutional perspective also helped greatly to shape the size of the U.S. conventional force contribution to NATO during this period. In this case, however, formal institutional constraints were less important than less tangible norms, expectations, and beliefs, which nevertheless tended to reinforce the status quo. Unlike Britain, the United States had not committed itself to maintain a specific level of forces on the continent.[171] But by this time, the notion that the U.S. military presence should consist of six division-equivalents had acquired considerable currency, and powerful allied expectations had developed around this norm. These expectations were complemented by strong and widely held beliefs within the U.S. government as to how the allies were likely to react to an American withdrawal. In this view, the mere hint of a reduction risked disrupting the alliance by suggesting a weakening of the U.S. commitment. Thus Eisenhower judged that it was still too soon for the United States to take any combat units out of Europe.

Institutional actors, especially SACEUR, also played a significant role in shaping alliance strategy and conventional force levels. The major NATO commanders introduced the concept of limited war into the deliberations over NATO strategy in 1956. Later, Norstad stamped a broad interpretation on the function of the Shield forces in his development of new force requirements on the basis of MC 14/2 and in subsequent statements. And it was he who convinced the British government to stretch out and to minimize its force reductions in 1957 and helped to dissuade the United States from withdrawing any forces later that year.

CHAPTER 5

To the Adoption of Flexible
Response, 1961–1968

In 1967, the North Atlantic Council approved Military Committee document (MC) 14/3, the last formal revision of NATO's strategic concept before the end of the Cold War. Commonly known as flexible response, the new military strategy represented a further shift away from massive retaliation. It placed even greater emphasis on the possibility of limited forms of conflict and on the need to prepare conventional responses to aggression than had MC 14/2.

As is well known, the adoption of flexible response was preceded by a heated debate within the alliance over the proper role and strength of conventional forces. Not surprisingly, the new strategy was even more ambiguous than its predecessor, allowing a wide range of interpretations of how it should be implemented. Less familiar is the fact that the administration of John F. Kennedy never made a serious attempt to alter NATO's strategic concept. After reviewing the alliance's military posture during their first months in office, Kennedy and his advisers concluded that the existing strategy document was sufficiently flexible for their purposes. Consequently, they decided to pursue the goal of stronger conventional forces within the framework provided by MC 14/2.

U.S. officials did not express a strong interest in revising the strategic concept until the end of 1964, and even then they decided against making an effort to do so. As a result, as late as September 1966, obtaining a new NATO strategy document was not a top U.S. priority. Instead, much of the impetus for the formulation of a new strategic concept was provided by the trilateral negotiations on force levels and financial burden sharing among the United States, Britain, and West Germany, which began that autumn, and the formulation of the new strategic concept was greatly facilitated by the operation of the NATO force planning process, which had only recently been adopted.

In contrast to the greater emphasis on flexibility and conventional options that were incorporated into NATO strategy, the alliance's overall conventional capabilities in the Central Region showed only modest net growth during this period. Virtually all of this increase, moreover, was the result of the completion of German rearmament in the early 1960s and, secondarily, the modernization and reorganization of the U.S. contribution to NATO. Although additional American troops were dispatched to Europe during the Berlin crisis of 1961, they were subsequently withdrawn. In addition, the expansion of the Bundeswehr was offset somewhat by France's withdrawal from the integrated military structure in 1966 and, more important, the removal of some American and British units in 1968 as a consequence of the trilateral negotiations. And as the Johnson administration prepared to leave office, several other countries were actively considering further reductions in their conventional force contributions.

The Kennedy Administration and NATO's Conventional Force Posture

When John F. Kennedy became president in 1961, a large gap remained between NATO's ground force requirements for the Central Region, as established in MC 70, and its actual forces-in-being. Most of the alliance's tactical aircraft, moreover, were postured for nuclear missions and were supplied with little conventional ordnance. As a result, NATO continued to lack the means to implement even the relatively limited non-nuclear responses that MC 14/2 had foreseen.

During the early 1960s, the United States undertook a series of initiatives intended to correct these deficiencies. Although widespread American support for improving NATO's conventional capabilities had emerged during the last year of Eisenhower's presidency, U.S. efforts to this end did not begin in earnest until the new administration took over. Kennedy and his advisers were already predisposed to provide the United States and its allies with a wider range of military options. Their initial interest in obtaining stronger non-nuclear forces was reinforced by the Berlin crisis of 1961, which clearly demonstrated the limitations of NATO's existing military posture, and the growing conviction of many U.S. officials, as a result of detailed assessments of Warsaw Pact capabilities, that an adequate conventional posture was well within the alliance's grasp.

These efforts, however, met with little success. Although the Berlin crisis prompted the United States to strengthen its forces on the continent and caused West Germany to expand and accelerate its rearmament program, the responses of the other Europeans were extremely modest. With few exceptions, the allies proved unable or unwilling to increase their military contributions. By the time of Kennedy's assassination in late 1963, the U.S. campaign to improve the alliance's conventional capabilities had bogged down completely.

THE INITIAL POLICY OF THE KENNEDY ADMINISTRATION

Some of the groundwork for Kennedy's policy toward NATO's conventional force posture was laid during the last year of the Eisenhower administration. Consistent with Norstad's liberal interpretation of MC 14/2, considerable support developed within the government for a significant strengthening of the NATO Shield forces. A midyear report for the State Department prepared by Robert Bowie, who had served as Dulles's assistant secretary for policy planning, concluded that the threat of a strategic nuclear response to less than all-out aggression was no longer credible and, consequently, that the alliance should enhance the conventional component of its Shield forces to the point that they could contain an unreinforced Soviet attack for as long as several weeks without resort to tactical nuclear weapons.[1] A second study, conducted independently in the Pentagon, called for "a non-nuclear capability sufficient to deny the Soviet bloc any choice between all-out assault and abstention from attack."[2]

In November, these parallel positions were brought together in a major report titled "NATO in the 1960's," which was prepared jointly by the State and Defense departments at Eisenhower's behest. Starting from the premise that the Soviet Union would be increasingly willing to initiate limited aggression under the umbrella of nuclear parity, the report concluded that the NATO countries would have to devote additional resources to defense and proposed that the United States press its allies to take vigorous measures toward fulfilling their force requirements.[3] Consequently, at the December ministerial meeting of the alliance, the U.S. representatives called upon the alliance to provide the NATO commanders with "that *flexibility of response* that will enable them to meet any situation with the appropriate response."[4]

Thus much of the U.S. national security community had come to favor the development of a greater conventional capability for the purpose of fighting a limited war in Europe by the time Kennedy took office. This shift in U.S. policy toward NATO came too late in the

Eisenhower administration for it to have any noticeable effect, however. More important, Eisenhower himself continued to oppose the investment of scarce defense resources in forces for limited war, a contingency that he thought was highly unlikely.[5] With the change of government in January 1961, the challenge of modifying the alliance's military posture was left to the new administration.

Kennedy's conception of the nature of conflict in the nuclear era differed sharply from that of his predecessor.[6] Unlike Eisenhower, he was not convinced that a confrontation in Europe would inevitably escalate to the level of all-out nuclear war. To the contrary, he believed that it was possible, indeed imperative, to limit hostilities before they crossed the nuclear threshold. Although he made an issue of the purported "missile gap" in the 1960 presidential campaign, Kennedy had long been critical of the Eisenhower administration's emphasis on strategic nuclear forces. The threat of massive retaliation, he argued, was not sufficient to deter the broad array of much more likely forms of limited aggression. To rectify this imbalance in its force structure, the United States needed to develop a wider range of military options for the deterrence of such lesser contingencies as well.

For Kennedy, simply placing greater reliance on tactical nuclear weapons was not the answer. The use of such weapons would prompt Soviet retaliation, cause tremendous destruction, and possibly turn a limited conflict into an all-out war. The primary role of such weapons, he felt, was to deter their use by the Soviet Union. Thus the principal solution Kennedy proposed to America's lack of military flexibility was stronger conventional forces. Specifically, he promised to increase the number of men under arms, modernize America's conventional armaments, and provide more airlift and sealift capacity to transport U.S. forces quickly to wherever they might be needed.

The president's top advisers—Secretary of State Dean Rusk, Secretary of Defense Robert McNamara, and Special Assistant for National Security Affairs McGeorge Bundy—shared these views,[7] which quickly became part of the administration's initial comprehensive policy statement on NATO. The new policy was based on a report prepared by a small group under the direction of former secretary of state Dean Acheson during the first months of 1961.[8] The Acheson report started with the assumption that a local conflict in Europe was more likely than a general war. Although this view had been expressed in MC 14/2, the strategic concept had nevertheless placed primary emphasis on acquiring the nuclear capabilities needed to deter or to deal with more extreme eventualities, which contributed to the neglect of NATO's non-nuclear forces. In contrast, the Acheson report recommended that the alliance's military preparations should henceforth be geared toward contingencies short of a nuclear or a massive non-nuclear attack because these were precisely the contingencies that NATO forces in Europe could do the most about. In particular, it stated, top priority should be assigned to strengthening NATO's conventional forces to give them the flexibility to meet a wide range of more limited military challenges.[9]

The principal military objective set forth in the report, however, was no more ambitious than any of the proposals that had circulated within the Eisenhower administration the previous year.[10] It called for improving the alliance's conventional capabilities merely to the point at which NATO would have the option of halting Soviet forces then in or rapidly deployable to Central Europe long enough to allow the Soviets to appreciate the wider risks of their course of action and to ensure that any allied decision to use nuclear weapons could be based on careful deliberation. In other words, the Acheson report forswore developing an ability to defeat every conceivable conventional aggression in Europe without using nuclear weapons. Although the NATO countries would have to raise the manning levels, modernize the equipment, and improve the mobility of their conventional forces as a matter of highest priority, requiring a significant increase in the present rates of European and U.S. defense spending, the report did not call for any overall growth in the size of NATO forces.

Moreover, the Acheson report explicitly cautioned against seeking to revise NATO's strategic concept, arguing that it should be possible to pursue the desired conventional improvement program "by a constructive re-interpretation of existing doctrine." Indeed, American officials had determined that the achievement of the existing MC 70 force requirements would be sufficient to accomplish U.S. objectives. An attempt to rewrite MC 14/2 should be made, the report concluded, only "if needed European energies and resources cannot be mobilized in any other way, and if it is clear that NATO agreement can be reached on a revision."[11]

The conclusions of the Acheson report were approved in only slightly modified form by Kennedy as official U.S. policy on April 21 and circulated as National Security Memorandum (NSAM) 40.[12] Just days later, the U.S. ambassador to NATO, Thomas Finletter, spelled out the details of the new policy to the allies. Summarizing the arguments and conclusions of the Acheson report, Finletter emphasized the importance of strengthening NATO's conventional forces—as a matter of the highest priority—so that they would "be able to enforce a pause in the event of substantial Soviet conventional aggression." He assured the allies that there was no need to revise the strategic concept, nor would it be necessary to exceed the requirements established in MC 70. Nevertheless, although the United States would maintain its forces in Europe and increase their conventional capabilities, the allies would be expected to make a vigorous effort to provide the balance of the forces needed to achieve this goal.[13]

Simultaneously, U.S. officials approached the more important allies on a bilateral basis to smooth the way for NATO agreement. For example, Acheson met personally with French president Charles de Gaulle to explain the American position, assuring the general that the United States had no intention of fighting a long conventional war in Europe, and to make a special appeal to France to play a greater role in European defense. Once French troops returned from North Africa, he noted, achievement of the MC 70 goals would be well within sight.[14]

Initially, the allies seemed receptive to the U.S. proposals. De Gaulle agreed that the West's reliance on nuclear weapons could result in inaction in the event of a crisis over Berlin and that it would be desirable to raise the nuclear threshold, and the French ambassador to the United States indicated his country's support for strengthening NATO's conventional forces after Finletter's presentation.[15] As time passed, however, the French leader's attitude became increasingly critical. He was already on record as doubting the willingness of the United States to use its nuclear forces except in response to a direct Soviet attack on American territory, and he subsequently remarked that the U.S. insistence on increasing NATO's conventional capabilities had strengthened this conviction.[16] French doubts about the U.S. nuclear guarantee and the wisdom of strengthening NATO's conventional forces were reflected in de Gaulle's military program for the period 1960–65, which contained no provisions for improving France's non-nuclear capabilities. Indeed, the size of the French army was scheduled to be reduced from nineteen to only five or six divisions, while the French nuclear program shifted into high gear, with nearly half of the defense research budget earmarked for the development and production of nuclear weapons and delivery vehicles.[17]

The Germans were even less enthusiastic about the U.S. proposals. Like the French, they feared that a conventional buildup could undermine deterrence, which in their view depended primarily on the threat of escalation to general nuclear war. A potential aggressor, they argued, must be convinced from the outset of the West's readiness and willingness to use nuclear weapons. Strengthening NATO's conventional capabilities would suggest a weakening of the West's resolve and might actually tempt the Soviets to attempt limited aggression.[18] The most vocal German critic of U.S. policy was Defense Minister Franz-Josef Strauss, who strongly opposed any differentiation between the alliance's nuclear and non-nuclear forces. In addition, Strauss worried that the new emphasis on conventional forces might obstruct his efforts to provide the Bundeswehr with dual-capable weapons.[19]

To make matters worse, the American initiative clashed with Norstad's proposed force requirements for 1966. U.S. policy effectively called for a moratorium on investment in NATO's nuclear capability, whereas Norstad had requested a significant increase in the number of nuclear delivery vehicles over the MC 70 force levels and even more over existing levels.[20] The NATO civilian bureaucracy was skeptical of the new U.S. position as well.[21]

The Kennedy administration responded to allied doubts and criticisms by emphasizing the continuing importance of nuclear weapons in U.S. policy.[22] Indeed, U.S. officials sought to portray the proposed conventional buildup as a natural extension of previous policy that would enhance rather than detract from the credibility of the nuclear deterrent. Specifically, they argued that the threat of nuclear war was more credible if it appeared NATO would first create a substantial level of non-nuclear violence, over Berlin or any other issue, than if NATO merely threatened to initiate general war *ab initio*.[23] Thus on every possible occasion, administration spokesmen reaffirmed the U.S. pledge to use nuclear weapons in response to a nuclear attack on Western Europe or in the event that NATO's conventional forces were about to be overwhelmed.[24]

Such hedging was possible because of the vagueness of the initial U.S. position. The administration had yet to put forward any specific proposals for NATO force improvements. Although McNamara had asked the U.S. Joint Chiefs of Staff to study NATO's force requirements in early April, the study quickly became bogged down in serious differences among the services and was not expected to be completed before the middle of the summer.[25]

THE IMPACT OF THE BERLIN CRISIS

Before matters could come to a head, the Kennedy administration's initial attempt to obtain allied agreement to a sustained program for increasing NATO's conventional capabilities was overtaken by the Berlin crisis of 1961.[26] The disposition of Berlin had been a source of East-West tension since the end of World War II. The beginning of the 1961 crisis, however, is commonly associated with the early June summit meeting between Kennedy and Nikita Khrushchev in Vienna, at which the Soviet leader revived his threat, first issued in 1958, to hand over the Soviet Union's responsibility for the city to East Germany. Such a move would have effectively terminated the legal basis for American, British, and French access to and presence in West Berlin, thus presenting the three Western powers with a difficult choice. Either they would have to negotiate with the East German government, in violation of their long-standing refusal to do so, or they would have to insist on the continuing validity of their occupation rights, by force if necessary, at the risk of precipitating a military confrontation with the Soviet Union.[27]

Although Khrushchev's strident behavior at the meeting disturbed Kennedy deeply, the president left Vienna determined to preserve the West's rights in Berlin. He reportedly viewed the matter as a test of U.S. credibility. A firm stand on the issue would preserve the allies' confidence in the United States, while a failure to meet its commitments could result in the dissolution of the alliance.[28]

Kennedy and his advisers had become aware, however, that the West might have few military options should the Soviets make a move in Berlin. The existing contingency plans called for NATO to attempt a series of military probes down the Autobahn connecting the city to West Germany if access were blocked. Given the West's limited conventional capabilities, however, "these probes [would be] too small to convey serious intent and would surely be quickly contained by the Soviets or even by the East Germans alone."[29] Consequently, NATO would have no choice but to use nuclear weapons. The president feared, however, that Khrushchev might be able to cut off access to Berlin so gradually that the West could never justify such a forceful reaction.

To Kennedy, the required response was clear: a rapid increase in NATO's conventional capabilities in Central Europe. He desired enough forces to demonstrate to the Soviet Union that Berlin was of vital importance to the United States, to prevent an easy seizure of the city by the East Germans, and to permit a true pause before the West would have to resort to nuclear weapons in the event of hostilities. These views were reinforced by the conclusions of a new study undertaken by Acheson concerning the Berlin problem, which called for a "prompt, serious and quiet" buildup of conventional and nuclear forces to convince Khrushchev that the United States would go to war rather than back down.[30]

The details of the U.S. response were formulated during July.[31] Despite a general consensus within the administration on the need to increase U.S. conventional strength, considerable differences existed over the precise form such an effort should take. In the middle of the month, the president was presented with four broad military alternatives. These ranged from the substantial reinforcement of U.S. forces in Europe, requiring a large-scale call-up of reserve and National Guard units, the declaration of a national emergency, and the prompt request of up to an additional $4.3 billion for the fiscal year 1962 military budget, to combinations of more modest measures.[32]

Although there was initially considerable support for a dramatic gesture, Kennedy and an increasing number of his closest advisers opposed the declaration of a national emergency. They believed that such a move was not absolutely necessary and would have negative repercussions.[33] On July 19, the president decided to take all possible steps necessary, short of an immediate mobilization of major reserve and National Guard units, to enable the United States to deploy six additional divisions and supporting tactical air units in Europe by the beginning of the following year. To pay for these measures, he authorized a request of an additional $3.2 billion from the Congress. Finally, he directed that negotiations be promptly undertaken with the allies to elicit parallel efforts to achieve a higher level of military readiness.[34]

In line with this last point, Kennedy promptly dispatched letters to British prime minister Harold Macmillan, German chancellor Konrad Adenauer, and de Gaulle, declaring that the NATO countries should make a major effort to create a strong military posture in view of the Berlin situation. It was essential, he argued, that they all "initiate immediately a significant buildup of military strength suitable for a peak crisis as well as prolonged tension" to deter the Soviets "by impressing them with our unqualified resoluteness of purpose." He described the steps the United States was planning to take and appealed for comparable efforts from them.[35] This preliminary approach was immediately followed up with a detailed proposal concerning how the allies might increase their forces on the continent.[36] And several days later, McNamara and his top military advisers made a brief trip to Paris and London, where they discussed the U.S. proposal with Norstad and allied leaders.[37]

The allies' initial response to the U.S. proposal was dismaying. They refused to take any concrete measures until the foreign ministers of the United States, Britain, and France conferred as previously scheduled early the following month and could work out a common program.[38] Although Rusk did meet with his counterparts to discuss the proposed buildup and to approve certain measures for increasing military readiness, which he reported to the NATO Council on August 8, European efforts to strengthen their conventional forces gained momentum only after the crisis peaked on August 13, when East Germany began to erect the Berlin Wall.[39] They were apparently shaped less by the U.S. proposal than by an initiative undertaken within the NATO organizational framework.

During the previous days, Norstad had drawn up his own plan spelling out the actions he felt each country should take. This proposal, which was completed by August 18, called for raising the manning and equipment levels of the existing forces, making available additional combat units, increasing the number and capability of support units, and improving the readiness of reserve formations. These measures were expected to result in a significant increase in NATO's combat capability in Central Europe. Most notably, the number of divisions would rise by the end of the year from 21⅔, many of which were understrength, to 24⅓, most of which would have a high combat potential.[40]

The events of August 13 also triggered a noticeable acceleration of U.S. military preparations. The next day, Kennedy told McNamara that the United States should hasten to the continent all the reinforcements needed to make the forces there completely combat ready.[41] By the end of the month, the first reservists had been called up, and on September 9, McNamara ordered that 40,000 troops be sent to Europe.[42]

As a result of these efforts, the strength of NATO's ground forces in the Central Region grew by approximately 25 percent, or the equivalent of four combat-ready divisions, and the number of allied squadrons of tactical aircraft increased by 19 (see Table 5.1).[43] Most notably, the U.S. Army in Europe grew from 228,000 to 273,000 men and gained an armored cavalry regiment as well as several smaller combat units, while a further 11 U.S. fighter squadrons were deployed to the continent, for a total of 32. In addition, the United States began to preposition stocks of equipment in Europe for two complete divisions, whose personnel would be airlifted to the continent in a crisis.[44]

By contrast, U.S. officials regarded the allies' efforts as less than satisfactory. Although the number of Europeans under arms had increased by 100,000, most of the allies still fell short of meeting their obligations under MC 14/2.[45] Among the Europeans, West Germany made by far the greatest contribution, notwithstanding German officials' previous reservations about strengthening NATO's conven-

TABLE 5.1
Central Region Ground Strength, 1961–1962
(Divisions)

	April 1961	April 1962	MC 26/4 requirements
Belgium	2	2	2
Canada	⅓	⅓	⅓
France	2⅓	2⅓	4
Germany	7	8	11⅓
Netherlands	2	2	2
United Kingdom	3	3	3
United States	5⅔	5⅓	5⅔
Total	22⅔	24	29⅔
Combat equivalent	16	20	29⅔

SOURCE: DOD, "Remarks by Secretary McNamara."

tional forces. Following the national elections in September, which returned the ruling coalition to power, the Adenauer government decided to extend the term of military service from twelve to eighteen months, as Norstad had recommended, and increased the defense budget substantially. Although they agreed to add only one ready division in 1961, the Germans increased the number of men under arms by more than a quarter, from 300,000 to 380,000, by the end of the year and affirmed their original rearmament goals of 500,000 men and twelve divisions.[46]

The French also took steps to strengthen their forces in Europe. De Gaulle dispatched 10,000 troops to reinforce the two and one-third French divisions in Germany and eventually redeployed two divisions from North Africa to northwest France. The arrival of the second division was delayed, however, by a renewal of hostilities in mid-August, and the two new divisions were never formally reassigned to the alliance. Thus France did not fully comply with Norstad's recommendation that it meet its MC 70 requirement of four divisions under NATO command.[47]

U.S. officials were most disappointed by Britain's performance.[48] Norstad had recommended that the British raise the manning level of the seven brigades in Germany to 90 percent and be prepared to increase the number of brigades to nine, for a total of three fully combat-ready divisions in the Central Region. The British had replied, however, that they were unlikely to be able to deploy an additional two brigades, and by November it was apparent that there would be no increase in the size of the British Army on the Rhine.[49] Britain's failure to make a significant effort was regarded as particularly unfortunate because "many allies would take a sterner attitude on the part of the UK as indicative of the seriousness of the situation."[50]

SUBSEQUENT U.S. EFFORTS TO STRENGTHEN NATO'S
CONVENTIONAL FORCES

Even as the tension spawned by the Berlin crisis dissipated in late 1961, the Kennedy administration renewed its campaign to induce the European allies to increase their conventional force contributions to NATO. The crisis had reinforced the administration's belief in the need for a wider range of military options, giving this effort an even greater sense of urgency. As careful reassessments of the threat resulted in lower estimates of Soviet military strength, moreover, U.S. officials increasingly believed that a fully adequate conventional capability was not beyond the alliance's resources.

These further attempts at persuasion, however, were no more successful than the previous ones. Apart from continued progress toward the completion of the German buildup, allied force levels failed to increase and even declined somewhat. The U.S. effort was hampered primarily by differing European views of how to interpret and implement NATO's existing strategy. To varying degrees, each of the major allies disagreed with the American belief that a fully adequate conventional capability was both desirable and feasible. At the same time, U.S. actions suggested a less than full commitment to achieving the declared objective. Although the Kennedy administration lowered its sights in 1963, emphasizing instead a reform of NATO planning procedures that might eventually result in stronger conventional forces, even this seemingly uncontroversial approach was stymied by French fears that it would validate a more flexible military strategy than they were prepared to accept.

Lessons of the Berlin Crisis. The Berlin crisis reinforced the conviction of many Kennedy administration officials that NATO needed stronger conventional capabilities in order to have a wider range of military options in Central Europe.[51] First, it illuminated critical deficiencies in the alliance's existing forces, underscoring the limited number of available alternatives. In addition to problems with the ground forces, the crisis had revealed grave weaknesses in NATO's tactical air capabilities. Indeed, there had been almost no prior planning for conventional air operations.[52]

Second, the crisis showed that NATO decision makers would consider even the limited use of nuclear weapons only with the greatest misgivings. The allies had been reluctant to discuss the use of nuclear weapons, and top U.S. officials had demonstrated extremely cautious behavior in a simulated war game played in September.[53] In short, it appeared that NATO's relatively well-developed nuclear posture offered no real options in a crisis. Finally, the peaceful resolution of the Berlin crisis suggested that the U.S. approach had been effective and that NATO's military efforts had already begun to bear fruit. Kennedy in particular believed that the alliance's conventional force buildup had helped to prevent a confrontation that might otherwise have reached the nuclear level.[54]

U.S. Reassessments of the Conventional Balance in Europe. In 1961, the objective of being able to respond with conventional forces alone to a broad spectrum of non-nuclear aggression in Europe had been merely desirable. During the following years, however, Kennedy administration officials became increasingly convinced that it was feasible as well. This conclusion was the result of a downward revision in U.S. assessments of Soviet and Warsaw Pact military capabilities in the Central Region.

Throughout the 1950s, NATO planners had had to work with relatively crude estimates of the threat. Western observers had generally credited the Soviet Union with 175 ready divisions and the ability to expand this force to some 300 to 320 divisions after 30 days of mobilization. Twenty-eight to 30 ready divisions were believed to be stationed in Eastern Europe. Another 60 or so were deployed in the western military districts of the USSR and could be quickly brought to bear on the central front.[55]

The Berlin crisis prompted the Kennedy administration to reexamine these figures. In the summer of 1961, McNamara set in motion a detailed review of Warsaw Pact military capabilities.[56] Preliminary results were available by early the following year. In February, McNamara announced that many Warsaw Pact divisions were a good deal smaller at full strength than their NATO counterparts and that many were understrength. Further analysis suggested that NATO might face a maximum of 60 Soviet divisions on the central front and that to generate even this reduced force would require 30 to 60 days of mobilization.[57] Finally, Soviet forces might be able to count on little support from their East European counterparts, whose reliability was regarded as questionable.[58]

This revised assessment was based on several factors. First, the total number of combat-ready Soviet divisions was now estimated at less than 80. In addition, it had become evident that some Soviet forces were likely to be tied down along the USSR's long frontier with China. Given the possibility of nuclear attack, moreover, the Soviets were thought to be reluctant to commit more than 60 divisions at a time, and a successful NATO air interdiction campaign would reduce the ability of the Soviet logistics system to support any more.

As the threat was cut down to size, a rough balance of forces emerged, and the recently adopted MC 26/4 force requirements appeared to be an adequate basis for achieving the administration's objective of sufficient military flexibility. McNamara concluded that the NATO countries had all the resources they needed to match the Soviets at any level of aggression in Europe. With relatively small increases in the efforts of the Europeans and especially with greater efficiency in the use of the defense resources that were already available, NATO's conventional forces could be made adequate to deal with a wide range of contingencies.[59]

Renewed U.S. Efforts. The Kennedy administration renewed its effort to convince the allies to improve their conventional contributions at the December 1961 meeting of the North Atlantic Council, where McNamara argued once again that the threat of general war would not deter lesser Soviet provocations and that NATO should bring its conventional and nuclear forces into better balance.[60] The secretary of defense presented the U.S. position perhaps most fully, however, at the NAC ministerial meeting in Athens the following May. Although the so-called Athens speech is better known for his unprecedentedly candid remarks about U.S. strategic nuclear forces and plans for their use, McNamara put equal emphasis on the need to strengthen the alliance's conventional forces.[61]

McNamara began his remarks by reaffirming the U.S. pledge to use nuclear weapons in response to any nuclear attack on the European allies or any conventional assault so strong that it could not be dealt with by non-nuclear means alone. In his view, however, Soviet aggression of this magnitude was unlikely. Rather, the Berlin crisis exemplified the type of challenge that the West should expect to face in the NATO area in the future.

Because of the changing nature of the military threat, McNamara argued, the most likely contingencies would not justify a response with strategic forces, which in any case were no longer adequate to deter limited provocations. The threat to use tactical nuclear weapons, moreover, though perhaps somewhat more credible, was itself fraught with risk. Given the likelihood of Soviet nuclear retaliation, their use would probably guarantee the devastation of Europe and lead to general nuclear war, while yielding no marked military advantage. Because of the inadequate state of NATO's conventional forces, however, the West might have no choice but to initiate the use of nuclear weapons in a crisis.

To remedy this situation, McNamara called for conventional forces that were "ready and able to deal with any Soviet non-nuclear attack less than all-out." Citing the successful outcome of the Berlin buildup as proof that creating more conventional strength would reinforce deterrence, he appealed to the allies to provide the forces recommended for the Central Region in MC 26/4, which still amounted to some ready 30 divisions.[62] As evidence of American earnestness, McNamara enumerated the new U.S. military programs. At the same time, he took pains to assure the allies that the goal was not for NATO to be able to defeat any conceivable act of non-nuclear aggression with conventional forces alone, reiterating that U.S. nuclear forces "would rapidly come into play if an all-out attack developed."

These themes were repeated in a steady stream of speeches and statements by American officials during the following year, and McNamara redoubled his efforts at the meeting of the NAC in December 1962, by which time the U.S. belief in the need to strengthen NATO's conventional forces to widen the range of available military options had been further reinforced by the Cuban missile crisis.[63] Through the end of 1963, however, the European response remained disappointing. Apart from the continuing growth of the Bundeswehr, which had little to do with the renewed U.S. pressure, allied conventional strength in the Central Region showed no net increase. Although France reconstituted a fifth division in Europe in early 1962 and made plans to add a sixth, all but the two divisions in Germany remained uncommitted to NATO. The overall size of the French army, moreover, steadily declined, from 700,000 to 450,000 men in 1963, as the Algerian conflict wound down. That same year, de Gaulle shortened the term of military service from two years to eighteen months.[64] Canada withdrew four of twelve aircraft squadrons stationed in Europe, and the Netherlands began to reduce the size of its 100,000-man army.[65]

Sources of Allied Opposition. One reason for the failure of the Kennedy administration's efforts to elicit greater allied contributions was the fundamental difference in strategic outlook that divided the United States and the European countries.[66] Although each of the principal allies had its own qualms about the U.S. campaign, their criticisms had some points in common. Above all, they argued that efforts to strengthen NATO's conventional forces would diminish deterrence rather than enhance it, as the United States contended, by suggesting a lack of resolve to use nuclear weapons if necessary. Thus a conventional buildup would make limited forms of aggression more likely, not less so.[67]

By implying that the Europeans should devote the bulk of their resources to fielding the necessary conventional forces while the United States alone would provide NATO's nuclear backing, the U.S. proposals also raised the specter of an unequal division of labor. Such an arrangement would reverse the hard-won gains of the previous half-decade in the field of nuclear sharing and would once again relegate the Europeans to second-class status within the alliance. Indeed, the proposals were viewed in cynical circles as being intended to deprive the Europeans of all access to tactical nuclear weapons.[68]

Third, the Kennedy administration's efforts prompted resistance simply because they represented a departure from previous U.S. policy. Why should the allies be expected suddenly to change course, especially when the Americans might acquire yet another strategic preference only a few years hence? It was not easy or inexpensive to modify military force structure, training, doctrine, and equipment.

Fourth, Europeans questioned many of the assumptions on which the U.S. proposals were based, especially the feasibility of a successful conventional defense. They were skeptical of the revised U.S. figures for Soviet capabilities, regarding them as politically motivated. Consequently, the allies maintained that much larger Western forces would be needed than the Kennedy administration claimed and thus continued to feel that greater investment in conventional forces would be pointless.[69]

Finally, an easing of East-West tensions made conventional force improvements seem unnecessary, even if they could make a difference in the military balance. Just as the Berlin crisis had strengthened the case for increasing NATO's conventional capabilities, the conciliatory moves following the Cuban missile crisis combined with mounting evidence of a major split within the communist bloc suggested that war in Europe was increasingly remote. The achievement in 1963 of the Limited Test Ban Treaty in particular, by contributing to the atmosphere of detente, was seen in the United States as impeding efforts to strengthen NATO's forces in Europe.[70]

Initially, such criticisms were voiced most strongly in West Germany, the country that would be most directly affected by the American proposals. For the Germans, who lived in the middle of the potential battlefield, there was little to distinguish the destruction that would result from extended conventional conflict or nuclear warfare, both of which would be disastrous. Consequently, they now talked exclusively of strengthening deterrence and, in the event that it nevertheless failed, conducting a forward defense to protect as much German territory as possible.[71]

The Germans believed that successful deterrence depended on

making the risks of aggression incalculable to a potential aggressor. If NATO were to suggest that certain provocations would meet with only a conventional response, the Soviets would be able to estimate the likely costs and benefits and perhaps be tempted to engage in limited aggression. To ensure that the risks remained incalculable, NATO had to be prepared to initiate the use of tactical nuclear weapons under all circumstances.[72]

The Germans also regarded greater emphasis on conventional forces as incompatible with forward defense because they believed that any Soviet attack would presage a full-scale effort to overrun their country and that NATO could not withstand such an attack employing conventional means alone.[73] At a minimum, NATO forces would have to give ground if nuclear weapons were withheld. For this reason as well, the Germans insisted on the immediate use of tactical nuclear weapons in all but the most limited contingencies.

Nevertheless, there were clear limits to how far the Germans were willing to press their criticisms because they did not want to precipitate a U.S. disengagement from Europe. No such inhibitions deterred France, however, which was ultimately the source of the most virulent attacks on the U.S. proposals. French opposition was compelled in part by de Gaulle's commitment to the development of an independent nuclear force, which he regarded as essential for restoring France's status as a world power. To justify an increasingly expensive program, French strategists argued that the only effective deterrent was the threat of immediate nuclear retaliation and that the approach of nuclear parity meant that the Europeans could not count on the United States to make a strategic response on their behalf unless its own territory was under attack. This extreme formulation of the requirements of deterrence did not admit any potential role for stronger conventional forces.[74]

French criticism served the additional purpose of loosening the ties that bound Western Europe to the United States. De Gaulle had long felt that NATO was too dominated by the Americans. From the first days of Kennedy's presidency, he had sought to drive a wedge between the United States and its allies so that France could assume what he viewed as its natural place as the leader of Western Europe. Hence the French suggested that the U.S. proposals represented a desire on the part of the Americans to renege on their security commitments to the region.[75]

The British reaction to the U.S. proposals was the most muted, though no less opposed. Although the 1962 British Defence White Paper called for a balance between conventional and nuclear forces,

British officials continued to believe that a major Soviet conventional attack would quickly require the use of nuclear weapons.[76] As a practical matter, moreover, there was no money in the British treasury for a conventional buildup on the continent. The Conservative government was firmly wedded to the further development and maintenance of an independent national nuclear force, and the country continued to be burdened with expensive military commitments in distant parts of the world. Finally, stationing more troops in Germany would further aggravate Britain's chronic balance-of-payments deficits.[77]

Mixed Signals from the United States. The obstacles raised by these differences of perspective between the United States and its allies were reinforced by a series of contradictory signals emanating from Washington, which raised doubts about the Kennedy administration's commitment to strengthening NATO's conventional forces. One was the continuing deployment of tactical nuclear weapons and dual-capable delivery systems in Europe. During 1962 and 1963 alone, the number of U.S. tactical nuclear weapons stationed on the continent increased by 60 percent.[78] At the same time, the United States agreed to sell its most modern nuclear-capable missiles to the allies to bring in more foreign exchange. Although the increase in warheads took place entirely within the context of programs established during the Eisenhower administration and much of the government's behavior could be attributed to its desire to reassure the allies of the firmness of the U.S. commitment to their security, these actions suggested that the Americans did not take entirely seriously their own warnings about relying too heavily on nuclear weapons.[79] Similarly, U.S. efforts to promote the idea of a NATO multilateral nuclear force (MLF) in 1963 further undercut U.S. attempts to strengthen NATO's conventional forces by validating European beliefs in the importance of possessing a nuclear capability.[80]

Even more counterproductive was the gradual withdrawal of the American forces that had been dispatched to Europe during the Berlin crisis and the suggestion that further reductions would follow. After all, if the Americans were so concerned about the state of Western Europe's defenses, why were they not willing to make additional sacrifices themselves?

Initially, the Kennedy administration had sought to set an example for the allies by continuing to strengthen U.S. conventional forces, both at home and in Europe. The three infantry divisions on the continent were converted to mechanized status, while all sixteen divisions of the expanded U.S. Army were reorganized and reequipped with more modern equipment for greater flexibility, mobility, and conventional firepower in accordance with the new "ROAD" concept.[81] The size and conventional capabilities of the U.S. tactical air force were also increased.[82] To improve the ability of the United States rapidly to reinforce its forces in Europe, the administration completed the prepositioning of equipment for two divisions that had begun in 1961 and launched a massive buildup of U.S. airlift capacity.[83] Furthermore, it steadily increased U.S. stocks of spare parts and combat consumables to lengthen the time existing units could fight before the economy was mobilized and wartime levels of production were achieved.[84]

By the end of 1962, however, U.S. officials had decided that the balance of the required conventional capabilities should be provided by the other NATO countries alone.[85] This conclusion stemmed from several considerations. First, it appeared that much of the necessary increase in strength could be obtained simply by correcting deficiencies in existing allied forces.[86] Second, the tremendous growth of European economic power that had occurred since the formation of the alliance meant that the allies could now make a larger contribution to their own defense. Indeed, members of Congress began to agitate for greater burden sharing and the termination of all U.S. military assistance to Europe.[87] Third, simply maintaining U.S. forces in Europe at their existing levels—let alone expanding them—was becoming increasingly difficult because of a persistent U.S. balance-of-payments deficit. Since 1958, the United States had experienced a net currency outflow of between $3 billion and $4 billion per year, contributing to a steady dwindling of the country's gold reserves. Although the cause of these chronic deficits was complicated, a large percentage could plausibly be attributed to the substantial American military presence on the continent.[88]

Initially, the Kennedy administration sought to neutralize the deficits by arranging for the allies to purchase offsetting amounts of American military equipment.[89] The persistence of the deficits at roughly constant levels in 1962 and 1963, however, created pressure to reduce U.S. forces in Europe, at least to the levels that had existed before the Berlin crisis. The departure of Berlin-related reinforcements began as early as mid-1962, although it was initially limited to noncombat formations.[90] The pace quickened the following year, however, when many of the remaining round-out units were withdrawn.[91] And the administration's frequent pronouncements on the need to improve the U.S. balance of payments suggested that it might seek even deeper cuts.[92]

Allied concern about U.S. troop reductions reached a crescendo in October 1963, when the United States staged a major airlift exercise in which the personnel of an entire armored division of 15,000 men were deployed to Europe in a matter of days.[93] Although intended to demonstrate the capability rapidly to reinforce U.S. forces on the continent, exercise Big Lift was widely regarded by the Europeans as a prelude to the withdrawal of as many as five of the six American divisions on the continent. To make matters worse, Big Lift was followed by a major Defense Department policy statement and persistent rumors suggesting that the administration was indeed contemplating a large reduction.[94]

The rumors were not without foundation. In late October, the administration had begun to lay the groundwork for the most significant redeployment yet of U.S. forces from Europe.[95] Out of concern for possible allied reactions, however, Kennedy had firmly insisted that there be no discussion of the withdrawals until a final decision had been made and a plan of action approved. Despite such precautions, news of the pending action quickly leaked to the press, and the administration was forced to put its plans on hold.[96]

The NATO Force Planning Exercise. In 1963, when it became clear that the allies were unlikely to increase their conventional contributions significantly, the Kennedy administration began to moderate its objectives. U.S. officials explicitly denied that they sought the ability to defeat even an all-out conventional surprise attack without nuclear weapons.[97] Indeed, the contingencies for which they suggested NATO should make the greatest effort to prepare did not require particularly strong conventional forces.[98]

The failure of direct attempts to convince the allies to do more prompted the Kennedy administration to shift tactics, emphasizing instead changes in NATO's planning procedures that, U.S. officials hoped, would facilitate the development of stronger conventional forces. Under the existing process, alliance force requirements were devised by the major NATO military commanders with little consideration for member countries' available resources and strategic views. As a result, the force levels they recommended were never approved as formal national commitments and thus were not binding.[99]

In early 1963, the administration endorsed an initiative proposed by NATO secretary general Dirk Stikker that was intended to close the gap between the alliance's force requirements, as set forth in MC 26/4, and the forces the countries were actually willing to provide. This Force Planning Exercise was expected to relate strategy, force requirements, and country resources in a rational, systematic way.

After securing alliance approval of the proposal in May, Stikker devised a two-stage procedure that would result in force goals that were both feasible and likely to attract "general endorsement."[100]

Even this modest, multilateral approach to strengthening NATO's conventional forces soon ran into trouble. In July, France objected that force planning could proceed only on the basis of an agreed strategy. In response, the procedure proposed by Stikker was modified to take into account a military appreciation then being developed independently by the NATO Military Committee. This broad strategic appraisal, once completed, was intended to serve as a basis for either revalidating or revising MC 14/2. Before much progress could be made, however, France voiced objections to the most recent draft of the document, labeled MC 100/1, thereby threatening to derail the Force Planning Exercise as well as the work of the Military Committee.[101]

French criticism of MC 100/1 centered on two issues: the level of aggression at which NATO should use nuclear weapons and the appropriate form of a nuclear response should one become necessary. The document endorsed the principle of having military options at the conventional, tactical nuclear, and strategic nuclear levels, incorporating in particular the ideas of "flexibility of response" and "limited tactical nuclear warfare." Although they had participated in the development of MC 100/1 up to that point, the French were now unwilling to entertain any meaningful flexibility or any use of nuclear weapons except in the context of a general nuclear war. To justify their nascent national nuclear doctrine, they insisted instead on a trip-wire concept, whereby any border crossing would automatically trigger a "total nuclear response." Thus in mid-November, the French refused to have anything more to do with MC 100/1, insisting that an agreement on the basic strategic issues be worked out first at the political level, and the Force Planning Exercise ground to a halt.[102]

It was just at this moment that Kennedy was assassinated. The president's death both hastened and symbolized the demise of his administration's efforts to strengthen NATO's conventional forces. To allay presumed allied fears that the change in leadership might mean a shift in U.S. policy, American policy makers were anxious to demonstrate that there would be no wavering in U.S. support for NATO.[103] In particular, plans for additional U.S. troop withdrawals from Europe were shelved and suggestions of further force reductions were carefully avoided.[104] Just as important, the new Johnson administration's initial goals regarding NATO's conventional force posture would be extremely modest: to fill gaps in and to improve the efficiency of the existing forces so that the alliance could confidently meet crises and handle conflicts at the lower end of the spectrum of military contingencies.[105]

The Development of MC 14/3

NATO's strategic concept was not formally revised until 1967, when the alliance adopted MC 14/3, commonly known as the doctrine of flexible response. The origins of the U.S. effort to replace its predecessor, MC 14/2, are frequently traced to McNamara's Athens speech.[106] There he articulated a somewhat more ambitious role for NATO's conventional forces than that set forth in the Acheson report, one that seemed to go beyond what even a fairly liberal interpretation of MC 14/2 might have allowed. Whereas the old strategic concept formally recognized the possibility of limited aggression only on a relatively small scale, the contingencies that the secretary of defense regarded as requiring a purely conventional response could involve a high percentage of the available Soviet conventional forces.

McNamara stopped short, however, of calling for a conventional capability adequate to deal with all-out conventional attacks. Not once in the address, moreover, did he describe the existing strategic concept as inadequate or in need of revision. Nor did he ask the allies to provide more forces than the NATO military authorities then regarded as necessary to implement MC 14/2. Although McNamara did suggest in several subsequent statements that he sought nothing less than conventional forces so strong that NATO would be able to meet any level of non-nuclear aggression without having to resort to nuclear weapons,[107] the Kennedy administration never called for a formal review of the strategic concept.

The initiation of the NATO Force Planning Exercise in 1963 has also been interpreted as an attempt to modify NATO strategy, with only French opposition to MC 100/1 preventing the formal adoption of a revised strategic concept.[108] The original purpose of the initiative, however, was to strengthen the alliance's conventional capabilities, even if it did eventually come to focus on the question of NATO strategy. MC 100/1, moreover, was not a substitute for MC 14/2. And even if the document had been approved, it is by no means clear that the old strategic concept would have subsequently been revised rather than revalidated. To the contrary, the concepts to which France objected were not inconsistent with the broad interpretation that MC 14/2 had been given in the late 1950s. In effect, it was

France that sought to modify NATO strategy, with the intention of putting it back on a massive retaliatory basis.

Several factors contributed to the Kennedy administration's reluctance to tamper with the strategic concept. First, the desirability of placing much greater reliance on conventional forces was not universally shared within the U.S. government.[109] In addition, administration officials may have feared a strongly negative reaction on the part of the European allies, many of which had already refused to heed U.S. appeals for stronger conventional forces. Most important, the available evidence suggests that they believed the United States could attain its objectives of stronger NATO conventional forces and greater military flexibility while working within the relatively broad parameters established by MC 14/2.

U.S. officials did not seriously consider revising the strategic concept until late 1964. Within months, however, they decided against making such an effort because of anticipated French opposition. By mid-1966, when France's withdrawal from NATO's integrated military structure had finally made possible the adoption of a new strategy, it was no longer a leading U.S. policy objective. In order to account for the subsequent formulation of MC 14/3, therefore, it is necessary to consider two other concurrent developments, which played equally crucial roles. The immediate impetus was provided by the trilateral negotiations that began that fall between the United States, Britian, and West Germany over U.S. and British force levels on the continent and measures to offset the foreign exchange losses occasioned by these deployments, which required these three key allies to reach a consensus on strategy. The recently established NATO force planning process then served as an institutional mechanism for translating this agreement into new political guidance to the NATO military authorities, which served as the basis for MC 14/3.[110]

INITIAL U.S. INTEREST IN REVISING THE
STRATEGIC CONCEPT

Changes in German and British doctrine in the direction of that favored by the United States helped to lay the groundwork for a revision of the strategic concept following the impasse over MC 100/1 in late 1963. Most important was the shift in German attitudes under Chancellor Ludwig Erhard, who took office in late 1963, and his minister of defense, Kai-Uwe von Hassel. In early 1964, von Hassel publicly acknowledged the need to be able to deal with "minor local conflicts" using conventional means alone.[111] Late that year, the Germans proposed the adoption of a strategy of "graduated deterrence," agreeing that "all levels of possible aggression must be matched with the appropriate means of defense."[112]

One reason for the changing German stance was the desire of the new government to improve relations with the United States after the departures of Adenauer and Strauss. Another factor was the approval of new SHAPE operational planning guidelines in 1963 that established NATO's first line of defense close to the inter-German border. This decision assured Germany that greater emphasis on conventional forces in NATO strategy would not come at the expense of a forward defense, as they had previously feared. And during the following year, new Bundeswehr staff studies yielded more favorable assessments of the prospects for a successful conventional defense.[113]

In early 1965, the new Labour government of Harold Wilson in Britain began to agitate for a revision of NATO strategy. Concerned about nuclear overreliance, the British maintained that NATO should be able to deal with small-scale conflict without automatic resort to nuclear weapons. Just as in 1956, however, they were equally motivated by a desire to reduce military spending and, if necessary, their forces on the continent. Thus the British also argued that the decline in the Soviet threat made deliberate aggression in Europe, even on a limited scale, highly unlikely. The chief contingency that NATO should plan for was an attack arising from political or military miscalculation. In the unlikely event of a general nuclear war, moreover, an extended land campaign was inconceivable. Consequently, they concluded, the alliance's existing conventional forces were more than adequate, especially if dual-capable weapon systems could be dedicated to nonnuclear missions.[114]

The belated emergence of a U.S. interest in formally revising NATO strategy was triggered by several developments. One was the accumulation of further evidence that the alliance already had generally sufficient forces for a successful forward defense by conventional means alone. On the assumption that one U.S. division equaled two Warsaw Pact divisions and that each non-U.S. NATO division equaled 1.2 Pact divisions, Department of Defense officials estimated that the alliance confronted 34 operational divisions in East Germany and Czechoslovakia with the NATO equivalent of 35 Pact divisions. Within 30 days, moreover, NATO could muster the equivalent of 57 Pact divisions in the Central Region. This compared favorably with an estimated Warsaw Pact capability to deploy up to 60 divisions in three to four weeks.[115]

These findings were paralleled by an important shift in the corporate view of the JCS, who had finally come to support a primarily conventional strategy for NATO. Through the early 1960s, U.S. military planning had assigned a prominent place to massive retaliatory responses.[116] In July 1964, after a lengthy debate, however, the Chiefs finally recognized the full implications of the Soviet strategic nuclear capability, particularly the Soviet Union's ability to absorb a nuclear attack and still launch a devastating counterattack. Consequently, they recommended to McNamara that the United States pursue a non-nuclear option in Europe.[117]

Pentagon officials, moreover, believed that NATO's full conventional potential was not being realized because of the way MC 14/2 was then interpreted at SHAPE. Although the strategic concept had established the principle of limited responses to limited forms of aggression, Norstad and his successor, Lyman Lemnitzer, had continued to give first priority to preparing for the less likely contingency of a general war. Indeed, SACEUR's alerting system was designed to put NATO forces automatically into their posture for an all-out conflict. In addition to unnecessarily limiting NATO flexibility, this bias reduced the alliance's capability to respond to lesser threats. The current Emergency Defense Plan for Western Europe called for a third of all dual-capable attack aircraft to be withheld from conventional combat so as to be ready to implement SACEUR's nuclear strike plan, and similar considerations would have inhibited making full use of the short-range delivery systems organic to NATO ground forces. In short, the alliance would not have been able to exercise a major conventional option even though the total military resources at its disposal should have been adequate to provide one.[118]

The only way to realize this potential, McNamara belatedly concluded, was to modify the strategic concept. In particular, he hoped to lessen its emphasis on preparing for general nuclear war, making conventional defense SACEUR's primary task, while placing greater stress on the need for flexible response capabilities, especially non-nuclear ones. NATO forces that were not specialized for nuclear tasks should be committed to non-nuclear missions. A revision of MC 14/2 along these lines would require SACEUR to modify his war plans and to improve his conventional capabilities accordingly. Thus in late 1964, McNamara recommended—for the first time—that the United States "make a concerted effort to obtain an updated political directive and strategic concept."[119]

U.S. interest in revising MC 14/2, however, declined as abruptly as it had arisen. Despite the more favorable German and British attitudes, the secretary of defense's advisers persuaded him that such an effort would be fruitless because of French opposition to the adoption of a new strategy. They recommended instead that he continue to concentrate on obtaining concrete force improvements. Consequently, at the May 1965 meeting of the NATO defense ministers, McNamara stated that little useful purpose would be served by continuing the debate over strategic concepts. Instead, he simply recommended that further development of NATO force goals be guided by an analysis of the forces required to respond effectively to typical situations representative of the full spectrum of likely contingencies and that the alliance revise its force planning procedure. In the fall, the United States followed up this suggestion with a detailed proposal for the establishment of a formal force planning process that would result in the adoption of an updated five-year force plan every two years, which was approved by the NATO ministers in December.[120]

FRANCE WITHDRAWS FROM THE NATO MILITARY STRUCTURE

Ironically, the prospects for a formal revision of NATO strategy brightened appreciably almost as soon as the United States dismissed the idea as an unrealistic policy objective. The principal obstacle was removed in early 1966, when France withdrew from NATO's integrated military structure. In March, President de Gaulle informed the allies of France's decision to alter its relationship with NATO. Most important, he demanded the termination of the assignment of all French land and air forces to NATO commands and of French participation in the integrated command structure by July 1966 and the removal of NATO military headquarters and all foreign—primarily U.S. and some Canadian—units, bases, and installations not under French authority from French territory by April 1967. These changes were the logical, if not inevitable, result of de Gaulle's basic foreign policy objectives, which were the full restoration of his country's sovereignty and the end of French subordination to the United States. The achievement of these goals required regaining total control of the French armed forces while limiting other forms of foreign influence over French behavior to the greatest extent possible.[121]

De Gaulle's action temporarily threw NATO military activities into disarray. During the following months, however, the organizational structure adjusted to accommodate the anomalous position that France wished to occupy within the alliance. Most important, the other countries decided to broaden the mandate of the Defense Planning Committee, which had been established to supervise the NATO Force Planning Exercise and which consisted of the allied defense ministers and their representatives, to deal with all defense-related subjects to which France did not contribute. Although France continued to sit on the NATO Council, it would no longer obstruct defense planning, making possible the consideration of a new strategic concept.[122]

By the fall of 1966, therefore, conditions were finally propitious for a change in NATO's formal strategy.[123] The lifting of the French veto alone, however, was not sufficient to bring about this result; it could be no more than a permissive cause. Moreover, Pentagon officials had temporarily lost interest in modifying MC 14/2, as reflected in a September 1966 policy paper on NATO strategy and forces that was silent on the issue.[124] Yet in the absence of a strong lead from Washington, the adoption of a new strategic concept remained highly problematic. The formulation of MC 14/3 probably would not have occurred as soon as it did were it not for the initiation of the trilateral negotiations and the operation of the new force planning process.

THE TRILATERAL NEGOTIATIONS

The trilateral negotiations, which lasted from October 1966 through April 1967, were not intended to result in a revision of the strategic concept. Rather, they were the product of a hastily improvised American effort to head off looming unilateral troop reductions by the British while ensuring that the allies addressed the United States's own balance-of-payments problems. By mid-1966, Germany was offsetting little more than half of the $250 million in annual foreign exchange costs incurred by Britain in maintaining its forces in the Federal Republic. This net outflow contributed greatly to Britain's overall balance-of-payments deficit. The situation reached crisis proportions in July, when a run on the pound prompted the government to impose severe deflationary measures and to seek a reduction in overseas expenditures of $280 million. Most notably, the British demanded a full offset from the Germans, threatening to cut their forces on the continent as much as necessary to eliminate the loss of foreign exchange. The following month, they informed NATO of the specific measures they would take, including troop withdrawals and a drastic reduction in the levels of stocks they maintained in Germany, if a satisfactory agreement were not achieved.[125]

By this time, the U.S. government was also under increasing pressure to consider major troop withdrawals. The foreign exchange costs of stationing forces on the continent had been neutralized by arms sales to Germany under a series of offset agreements dating to 1961, but the German government made it clear in mid-1966 that it planned to scale back sharply its military purchases in the United States, both because of a looming budget deficit and because the process of equipping the Bundeswehr was largely complete.[126] At the same time, there was mounting congressional sentiment for a substantial reduction in the number of U.S. forces in Europe, in view of the greater ability of the Europeans to provide for their own defense and to meet the manpower needs of the effort in Vietnam. This sentiment was most strongly expressed in a resolution introduced by Senator Mike Mansfield at the end of August.[127]

The convergence of these multiple pressures prompted the Johnson administration to take action. American officials feared that unilateral British troop withdrawals would prompt comparable cuts by other allies, causing NATO's military posture to unravel. They also hoped to preserve the existing U.S. offset arrangement, which they thought might be jeopardized by a separate British deal with the Germans. Consequently, in late August, the United States proposed that it join the bilateral discussions that were already taking place between these two allies.[128]

Initially, the British and the Germans reacted coolly to the U.S. proposal, although both eventually agreed. The British, who had hoped to achieve quick results through direct talks with the Germans, saw U.S. participation as a recipe for delay. Indeed, a primary U.S. objective was to induce the British to postpone any troop cut announcement until mid-January, when the negotiations were expected to conclude. U.S.-German differences focused on the mandate of the talks, which the Germans sought to limit to the issues of force levels and financial burden sharing. At U.S. insistence, however, the agenda was broadened to encompass the logically prior questions of the nature of the threat and NATO strategy.[129]

The negotiations formally began on October 20. Although the Erhard government fell a week later, the three countries quickly reached a basic consensus on NATO strategy and force levels, which was outlined in a set of minutes adopted on November 10. This document called for "a full spectrum of military capabilities" to deter aggression and defined a somewhat broader role for conventional forces than that contained in MC 14/2. Specifically, the alliance needed sufficient conventional forces to deal with limited non-nuclear attacks and to deter larger attacks by confronting the Soviet Union with the prospect of hostilities on such a scale as to involve a grave risk of escalation to nuclear war. The minutes stopped short, however, of mandating a full conventional capability.[130]

THE FORMULATION OF MC 14/3

This strategic consensus quickly worked its way into the NATO force planning process, which was just getting under way. Many of the countries not taking part in the trilateral negotiations had become concerned that any decisions taken in the new NATO mechanism might be rendered moot by the outcome of the talks. Consequently, the force planning process was accelerated and imbued with unintended significance. The first step was to prepare a political guidance paper as the basis for the development of a military appreciation, which was in turn needed for the preparation of a force plan for the years 1968 to 1972. By early December, NATO planners working around the clock produced a draft guidance paper that incorporated portions of the progress report issued by the trilateral negotiators at the end of November. As a result the new paper was no longer regarded as just the first step in the biannual planning process but also as a possible replacement for the 1956 Political Directive.[131]

After exhausting intra-alliance negotiations, the NATO defense ministers approved a final version of the political guidance the following May. The document described Soviet policy, characterized the threat, outlined in general terms NATO strategy and force requirements, and estimated broadly the resources likely to be available for NATO defense. More important, it officially replaced the old Political Directive, as expected, and set the stage for the revision of MC 14/2.[132] During the summer, the Military Committee used the political guidance to draft a new strategic concept, which was adopted by the NATO Council in December as MC 14/3.

THE CONTENT OF FLEXIBLE RESPONSE

Flexible response embodied a new alliance consensus on the nature of the military threat faced by the West.[133] The new strategy, like its predecessor, provided a blueprint for deterring, or resisting if necessary, a wide spectrum of possible Soviet and Warsaw Pact aggression, ranging from small conventional probing actions to all-out nuclear attacks. Whereas MC 14/2 had been primarily concerned with preventing a general war, however, flexible response was addressed mainly to lesser contingencies. Six years after the most recent Berlin crisis, NATO leaders had come to view the military situation in Europe as fundamentally stable. A deliberate, all-out attack, whether conventional or nuclear, seemed extremely unlikely.[134]

Flexible response also represented a compromise among competing conceptions of how NATO should respond to intermediate levels of aggression, especially a large-scale conventional assault, although the allies did seem to agree that attacks at either end of the spectrum of aggression could be deterred with the threat of a symmetrical response. Although the United States desired the ability to stage a full-scale conventional defense of Western Europe, the European allies continued to be unwilling to prepare for more than a brief period of conventional hostilities before resorting to nuclear weapons.[135]

To bridge these differences, MC 14/3 provided for a hierarchy of possible responses. NATO would initially attempt to meet any attack by direct defense on the level at which the enemy had chosen to fight. Thus the alliance would respond to a non-nuclear assault with conventional forces alone. If defense at the first level was not effective and the aggression could not be contained, however, NATO would be prepared to conduct a deliberate escalation, including the possible use of nuclear weapons. Such a response would raise the scope and intensity of combat, with the aim of making the costs and risks disproportionate to the aggressor's intentions and the threat of a large-scale nuclear riposte more imminent. Finally, in the event of a major nuclear attack, NATO would be prepared to initiate a general nuclear response. To implement this strategy, the alliance required a range of military capabilities, including significant conventional, tactical nuclear, and strategic nuclear forces, which would enable NATO to respond "appropriately" to any level of attack.

Overall, MC 14/3 placed less emphasis on nuclear weapons and more on non-nuclear capabilities than had its predecessor.[136] Conventional forces were assigned three specific roles. First, they were to help deter a deliberate non-nuclear attack by denying the Warsaw Pact any confidence of success unless it used a force so large as clearly to threaten NATO's most vital interests. Second, they were to have the capability to deal successfully with a conflict arising out of some unexpected event or through miscalculation during a period of tension or political crisis. Third, NATO was to be able to build up its conventional capability rapidly in a crisis to show determination and to prevent the Warsaw Pact from substantially altering the balance of forces.[137]

Nevertheless, flexible response stopped well short of requiring a capability to deal successfully with any non-nuclear attack without recourse to nuclear weapons. Indeed, the role of conventional forces in the new strategy was less ambitious than that articulated by McNamara more than five years before in his Athens speech.[138]

There would continue to be situations in which, if deterrence failed, NATO would have to initiate the use of nuclear weapons.[139]

Beyond such generalities, however, the strategic concept offered little guidance to defense planners. The need to reconcile conflicting viewpoints resulted in a high degree of ambiguity, leaving many critical force planning issues unresolved.[140] In particular, MC 14/3 did not specify the size of the attacks the alliance would attempt to defeat with conventional forces alone. Nor in the case of larger attacks did it indicate the point at which NATO should employ nuclear weapons. It was also silent on the size and nature of the nuclear response NATO would make once the decision to cross the nuclear threshold had been reached.[141] Finally, neither flexible response itself nor any subsequent study clearly spelled out the levels of forces that would be needed to implement the strategy.

The difficulty of specifying force requirements was further complicated by the concomitant adoption of the assumption of political warning. At the same time that they approved flexible response, the NATO ministers agreed that any hostilities would likely be preceded by a prolonged period of political tension, which would provide weeks or months, rather than days, of warning. This decision allowed the alliance to place less emphasis on ensuring that the forces required to deter or defeat an attack were ready and in place at all times.[142]

The concept of political warning did not go unchallenged, however. Some officials, especially military authorities, questioned whether the West would receive sufficient warning of an attack. Others doubted whether political leaders would respond promptly to indications of Soviet military preparations. Instead, they might be paralyzed by fears that the measures required to ready NATO forces in a crisis would exacerbate tensions and possibly precipitate the very aggression that such precautions were intended to deter.[143]

Changes in NATO's Conventional Capabilities, 1964–1968

NATO's conventional capabilities in the Central Region increased somewhat during the first years of the Johnson administration, although U.S. pressure on the allies to increase the size of their forces had subsided by that time. American officials had concluded, on the basis of their reassessments of Soviet and Warsaw Pact capabilities, that NATO's existing force structure was generally adequate and that a large increase in defense budgets was not needed. Instead, U.S. efforts now focused on getting the allies to correct the deficiencies in their existing forces through redeployments and some reallocation of the resources that were already available.[144]

The strengthening that occurred was primarily due to the completion of the German military buildup. In 1965, the size of the Bundeswehr reached twelve NATO-assigned divisions, ten wings of tactical aircraft, and more than 450,000 men under arms.[145] Also important was the further modernization of the U.S. military contribution to NATO. The ROAD conversion was completed in 1964, and the U.S. airlift capacity continued to expand. During this period, the shortage in stocks of non-nuclear ordnance for the tactical air force was largely overcome.[146]

At the same time, the U.S. force structure in Europe remained basically unchanged, although several minor downward adjustments in force levels took place. The United States finally completed the withdrawal of the small combat units dispatched during the Berlin crisis, and it modestly streamlined the support structure of the remaining force of six division-equivalents. Nevertheless, proposals for further reductions in the American military presence made little headway.[147]

When U.S. troop deployments to Vietnam increased in 1965, however, concern grew in Europe that the United States would eventually have to draw on its forces on the continent to meet the demands of the escalating ground war in Asia. To allay such fears, McNamara pledged at the December 1965 NATO ministerial meeting that the U.S. effort in Vietnam would not require the withdrawal of any "major combat units."[148] The lingering suspicion that this disavowal did not preclude a hollowing out of the U.S. force structure in Europe was confirmed the following spring, when the Johnson administration revealed that it would withdraw 15,000 "specialists." Although U.S. officials emphasized that these cuts were only temporary and would be restored by the end of the year, many European allies feared that further reductions would follow.[149]

Ironically, the development and approval of MC 14/3 in the later Johnson years, which formalized a somewhat more expansive role for conventional forces in NATO strategy, coincided with a decline in the alliance's conventional capabilities. Perhaps most significant were the withdrawals of U.S. and British forces that resulted from the trilateral talks. Subsequently, several other countries, notably West Germany, Belgium, and Canada, sought to capitalize on these moves as well as a diminishing sense of threat to reduce their own forces.

A further weakening of NATO's conventional strength was only averted by the Soviet intervention in Czechoslovakia in August 1968. In addition, the American war effort in Vietnam reduced the number of reserve divisions in the United States available to reinforce NATO, which eventually fell to four.[150] Harder to assess is the impact of France's withdrawal from NATO's integrated military structure because this involved no redeployment of the French ground forces stationed in Germany. Taken together, these developments offset many of the gains that had been so painstakingly made during the first half of the 1960s.

MILITARY CONSEQUENCES OF FRANCE'S WITHDRAWAL

Concerns about the potential impact of the Vietnam War on the U.S. military presence in Europe were temporarily overshadowed by the announcement that France would withdraw from NATO's integrated military structure. The French decision raised the prospect of a sudden unraveling of the alliance. The actual impact on NATO's conventional capabilities, however, was less than many had feared.

In terminating the formal French military contribution in the Central Region, de Gaulle merely completed a pattern of disengagement that had begun more than a decade before, when the majority of the NATO-assigned French ground forces were redeployed to North Africa. The process of de-integrating the remaining forces began soon after de Gaulle took power. In addition to withdrawing all French naval forces and personnel from NATO commands during the following years, de Gaulle imposed strict limits on France's participation in the defense of the Central Region. Beyond his refusal to reassign the divisions that returned to the continent during the Berlin crisis to NATO, he minimized French involvement in the alliance's integrated air defense arrangements and saw that all French forces in Germany remained stationed near the French border, even as NATO's initial line of defense moved forward, ensuring that they would be limited to a reserve role. In 1965, France refused to participate in NATO's annual fall staff exercise on the grounds that some of the scheduled activities were intended to simulate a conventional defense, in violation of the French interpretation of NATO strategy.[151]

Thus when de Gaulle informed the allies of France's intention to sever all links with the integrated military structure, relatively little work was left to be done. The principal unresolved issue was whether the French forces stationed in Germany would remain there or would be withdrawn.[152] Although the small air force contingent was redeployed to France, de Gaulle was willing to leave the ground forces in place under the terms of the convention on the stationing of foreign forces in Germany that was part of the package of agreements signed in October 1954 to clear the way for German rearmament. The other allies, and especially the Germans, sought to link the continued presence of French forces to the development of satisfactory arrangements for the cooperation of those forces with those of NATO. Although France had an interest in keeping its troops in Germany for reasons of prestige, de Gaulle was unwilling to compromise, refusing to accept any NATO authority over French forces. The West German government, in contrast, was divided. Many officials were eager to preserve good relations with France and thus willing to keep French troops in Germany on de Gaulle's terms as a symbol of a continuing French commitment to German security. Eventually, because of the asymmetries of interest and bargaining power inherent in the situation, West Germany effectively agreed to the French conditions, clearing the way for a continued French military presence.[153]

Subsequently, French and NATO military authorities developed military contingency plans that assumed French participation in NATO responses to a variety of acts of aggression. Because French involvement could not be guaranteed, however, plans that excluded France were given priority. Nevertheless, U.S. officials assumed that French forces would be available in the event of a deliberate, large-scale attack, precisely the contingency in which they would be most needed.[154]

The other principal consequence for NATO's force structure of the French withdrawal was the departure of all foreign troops from French soil. U.S. forces constituted the largest contingent, yet by 1966 these amounted to no more than 30,000 military personnel and only eight squadrons of aircraft. Five of the air units were redeployed to Britain, and although the remaining squadrons (and some 18,000 military personnel) were withdrawn to the United States, two of these were classified as dual-based, with a second home in Germany, and remained assigned to NATO. Thus the relocation of the U.S. forces from France was thought to have very little adverse effect on NATO's military strength.[155]

Nor did the loss of access to French territory create serious problems. By the mid-1960s, the strategic depth afforded by French territory had lost much of its significance because NATO now planned a truly forward defense of the Central Region. Similarly, the need for bases and other installations in France had declined, although greater reliance would have to be placed on the more vulnerable lines of communication running through the low countries and northern Germany. Perhaps the most important question concerned NATO overflight rights during a major crisis, although France eventually made these available on the same terms as before in peacetime.[156]

THE IMPACT OF THE TRILATERAL TALKS

Much more consequential for NATO's conventional force structure were the trilateral talks, which resulted in the first formal downward adjustment of alliance force levels in the Central Region since the late 1950s. The agreement reached in late April 1967 and subsequently ratified by the alliance as a whole authorized the United States to withdraw two-thirds of a division and four squadrons of aircraft, comprising approximately 35,000 military personnel, while Britain would remove one brigade group of 5,000 men and one squadron.[157] The U.S. reductions in particular were justified on the grounds that there would be substantial political warning of an attack and that the United States now possessed sufficient strategic airlift capability to redeploy the withdrawn forces to Europe during the time available.[158] Still, it was difficult not to conclude that NATO's conventional capabilities in the region had diminished.

When the talks began in October 1966, the positions of the three parties were far apart. Britain insisted that Germany fully offset the additional costs of some $250 million that were incurred annually by stationing forces on the continent and threatened to reduce those forces unless a satisfactory agreement was reached by the end of the year.[159] Similarly, the United States demanded a full offset of its foreign exchange costs, which amounted to more than $700 million per year, although there were differences within the government as to what form the offset might take. Germany, however, was prepared to purchase only up to $88 million in military equipment from Britain and $350 million from the United States during the next year, far less than was necessary to meet its allies' demands.[160]

Because of the collapse of the Erhard government in late October,[161] little could be accomplished until a new ruling coalition was in place in Germany and had developed its position. Consequently, U.S. officials proposed extending the deadline for reaching an agreement until May. To forestall unilateral British reductions in the meantime, the United States offered to make a onetime purchase of $35 million of military equipment from Britain.[162]

The delay also provided the United States with additional time to determine its own position. The Johnson administration was deeply divided on the issue of whether to seek a reduction in the number of U.S. forces stationed in Europe. McNamara recommended dual-basing two divisions and six air wings, which would allow the United States to withdraw as many as 72,000 troops. In view of the substantial increase in U.S. airlift capability that had taken place since the early 1960s and the likelihood that a serious crisis would be preceded by a substantial period of political warning, he argued, the United States would be able to fly these forces back to Europe "rapidly enough to meet the likely threat which would require their employment."[163]

In contrast, the Joint Chiefs of Staff opposed any U.S. troop withdrawals. NATO's military posture, they argued, should be based on the adversary's military capabilities, not estimates of intentions, especially because alliance leaders might find it politically undesirable to take the actions necessary to reinforce their forces in a crisis. And in light of current Warsaw Pact capabilities and the ability of the Soviet Union to augment rapidly its forces in Europe, there was no military justification for reducing the strength of the U.S. forces in Europe—or those of any other NATO country.[164]

The JCS position was supported by the principal U.S. negotiator in the trilateral talks, John McCloy, though more on political than military grounds. In particular, McCloy argued, any U.S. troop withdrawals would probably trigger allied force cuts, resulting in a substantial reduction in NATO's overall conventional capability and possibly the disintegration of the alliance and a concomitant loss of U.S. influence in Europe. Both he and the Joint Chiefs of Staff felt that McNamara underestimated the amount of time required to return any withdrawn forces to the continent.[165]

President Lyndon Johnson was reluctant to make any cuts in U.S. forces because of possible disruptive effects on the alliance. Nevertheless, he concluded that he would have to withdraw some troops because of renewed pressure to do so in the Senate.[166] In January, Mansfield, frustrated by the slow pace of the trilateral negotiations, had reintroduced the resolution he had first submitted the previous August, obtaining the support of 44 cosponsors.[167] The president's decision was facilitated by McCloy's willingness to back a compromise advanced by the State Department, whereby the United States would dual-base one division and three air wings.[168]

The final U.S. position for the negotiations was determined in early March.[169] In addition to limiting the withdrawal of U.S. forces as suggested by the State Department, the United States would count German purchases of U.S. Treasury bonds toward its offset needs, no longer insisting on military procurement. In return, the United States asked the Germans to pledge not to convert their dollar holdings for gold and the British to limit their reduction to one brigade. Although Johnson hoped that the other allies could reach an agreement on their outstanding financial differences, he was prepared to provide some further assistance to bring the two sides together if necessary.

In fact, the offset gap dividing Britain and West Germany had narrowed considerably. The British were now prepared to accept less than 100 percent compensation and sought only $154 million in German purchases. The Germans, realizing that a substantial British withdrawal might prompt the United States to make even larger cuts than those proposed, had increased their offer from $88 million to $114 million, leaving a difference of only $40 million. To close the remaining gap, the United States agreed to purchase an additional $20 million in arms from the United Kingdom.[170]

The final sticking point concerned the size of the planned U.S. redeployment, the details of which were revealed to the allies only in early April. The U.S. plan was to withdraw two brigades, one armored cavalry regiment, and an appropriate share of support units from one division (approximately 30,000 men) and 144 of the 216 tactical fighter aircraft in three wings (approximately 6,000 men). Nevertheless, the ground units would be provided with duplicate sets of equipment in Germany, and all the withdrawn forces would remain committed to NATO, be maintained at a high degree of readiness, and be flown back to Europe for annual field exercises. This arrangement was expected to reduce U.S. foreign exchange expenses by $75 million per year.[171]

Although the Germans were prepared to accept the dual-basing of one division, they were shocked by the number of aircraft involved—two-thirds of all U.S. strike aircraft in Germany—which in their view would significantly reduce the NATO deterrent. In response, the Germans proposed that the redeployment be limited to only 72 aircraft. To conclude the talks on time, both sides agreed to a further compromise suggested by McCloy involving a withdrawal of 96 aircraft, or four squadrons.[172]

FURTHER FORCE REDUCTIONS, 1967–1968

The trilateral accord on U.S. and British redeployments was endorsed by the NATO ministers in December 1967 and carried out the following year. This agreement did not end the threat of a serious contraction of the alliance's conventional capabilities, however. To the contrary, the negotiations triggered a series of attempts by several of the other allies to obtain budgetary relief through cuts of their own. The Belgian government began to seek a reduction in its NATO commitment from six to four brigades as early as the fall of 1966 and subsequently adopted a reorganization plan to that effect.[173] The ink was hardly dry on the trilateral agreement, moreover, when the German government proposed a reduction in its armed forces from 460,000 to 400,000 men over several years. Germany immediately came under strong pressure from the United States to reverse its position and eventually decided against making the threatened cuts.[174] Nevertheless, the size of the German armed forces remained below the long-standing goal of 500,000 military personnel. And in the spring of 1968, it was reported that the Canadian government was reexamining its commitment of one brigade and six air squadrons in Germany.[175]

In the United States, meanwhile, a resurgence of congressional pressure threatened to result in yet further cuts in the American forces in Europe. The reasons, though familiar, had become even more compelling: to reduce the U.S. balance-of-payments deficit, which had reached $3.5 billion in 1967, the highest level since 1960, and to free trained manpower for use in Vietnam, especially following the Tet offensive in early 1968.[176] In January 1968, Mansfield once again revived his troop cut proposal, and in April, Senator Stuart Symington went even further, introducing an amendment that would have prohibited the use of funds to support more than 50,000 troops in Europe after the end of the year.[177] That summer, Mansfield delivered a blistering attack on the latest U.S.-German offset arrangement and reiterated his call for substantial force reductions.[178]

This seemingly inexorable movement toward a further weakening of the alliance's conventional capabilities was averted by the August 1968 Soviet intervention in Czechoslovakia, which caused NATO leaders once again to revise their assessments of the threat. The manifest willingness of the Soviet Union to use force argued for a more sanguine view of Soviet intentions.[179] The resulting deployment of Soviet forces in the country, moreover, seemed to shift the balance of forces to NATO's disadvantage. Consequently, many observers concluded that the threat to the West had increased.[180] Finally, the alliance's reluctance to enhance its military readiness in order to deny the Soviet Union a pretext for invading Czechoslovakia confirmed doubts about the wisdom of the assumption that NATO would respond in a timely manner to political warning.[181]

Thus the Soviet action temporarily arrested the trend toward further force reductions. Mansfield immediately retreated from his position,[182] and alliance members pledged to maintain their military capabilities.[183] Subsequently, Belgium and Canada agreed to defer the cuts they had planned to make.[184] The intervention did not, however, result in any significant strengthening of NATO's conventional forces. U.S. attempts to induce the allies to increase their military efforts bore little fruit, and the alliance settled for a modest program of qualitative improvements.[185]

Analysis

Why was NATO's formal military strategy revised once more during the 1960s, yet not until nearly seven years after the Kennedy administration had assumed office committed to increasing the alliance's military options? How are we to explain the even greater degree of ambiguity embodied in the new strategy? And why, apart from the completion of German rearmament, did so little change occur in the alliance's conventional capabilities, notwithstanding the increased emphasis placed on conventional forces by flexible response, on the one hand, and the relaxation of Cold War tensions that characterized much of this period, on the other?

The content of MC 14/3 reflected shifts in the military balance of power. NATO was losing its advantage in nuclear weapons while arguably gaining the ability to defend itself conventionally. At the same time, flexible response was predicated on a different view of the threat than that contained in MC 14/2. A war in Europe was now considered even less probable than it had been in the mid- to late 1950s. If hostilities were to occur, moreover, they were as likely to be the result of accident and miscalculation as of premeditated aggression. Consequently, any conflict would probably be preceded by a substantial period of warning and limited in scope at the outset. It was more important than ever that the alliance be able to respond militarily in a manner tailored to fit the circumstances.

Thus the change in NATO strategy was also driven by a further evolution in Western perceptions of Soviet intentions. The steady improvement in East-West relations that occurred after the Berlin and Cuban crises early in the decade enabled the Soviet Union to shed much of its aggressive image. The country's emerging conflict with China, moreover, was seen as placing an important constraint on its willingness to use force in Europe.

In addition to its impact on NATO strategy, this increasingly benign view of Soviet intentions also helped to precipitate the modest conventional force reductions that occurred in the Central Region in the late 1960s. The alliance was likely to receive substantial warning of an attack so it was no longer imperative to match Warsaw Pact forces man for man or division for division in peacetime, although an adequate mobilization capability was still regarded as essential. Conversely, much of the increase that took place in NATO's conventional capabilities early in the decade could be attributed to the temporary perception of a heightened risk of war that accompanied the Berlin crisis, while the invasion of Czechoslovakia in 1968, by raising new doubts about Soviet intentions, served to arrest the cascade of force reductions that had begun the previous year and threatened to weaken NATO's military posture substantially.

An exclusive focus on the balance-of-power perspective nevertheless leaves several important questions unanswered. One concerns the considerable amount of time that transpired before the strategic concept was formally revised. Previously, NATO strategy had been modified every three to four years. Of course, one reason for this delay was the ambiguity of MC 14/2, which was amenable to a wide range of interpretations. Consequently, even the Kennedy administration concluded that the old strategy offered sufficient flexibility and thus did not immediately press for its modification. Nevertheless, the emergence of unprecedentedly cordial East-West relations soon undermined the assumptions on which it had been based.

Perhaps more puzzling from this perspective was the alliance's continued failure to provide the conventional forces called for in MC 14/2 in the early 1960s. The unwillingness of most of the European countries to increase their contributions to NATO during the Berlin crisis may be viewed as especially problematic. By the same token, it may be reasonable to question why the Soviet invasion of Czechoslovakia, rather than simply ending the process of reductions that had been set in motion, did not result in more corrective action on NATO's part.

Conversely, one might have expected a greater decline in alliance force levels before mid-1968, when NATO's requirements for ready units on the continent became decoupled from estimated Warsaw Pact capabilities under the assumption of political warning. Instead, the redeployments negotiated by the United States and Britain in the trilateral talks were considerably smaller than the largest cuts that had been contemplated, and Germany decided against making reductions of its own even before the Czech crisis.

Some of these puzzles can be explained in terms of determinants emphasized by the intra-alliance perspective. By the 1960s, the United States and its European allies had become deeply divided over two related issues.[186] First, the two sides disagreed over the requirements of deterrence and thus the desirability of placing greater reliance on conventional forces. The United States emphasized the importance of possessing credible responses to aggression. From this standpoint, the availability of a wider array of conventional options would enhance deterrence by ensuring that NATO would not fail to respond to any level of attack. The West Europeans, however, emphasized the importance of ensuring that the costs of aggression would outweigh any possible gains. Thus they regarded the risk of escalation to all-out nuclear war inherent in the use of tactical nuclear weapons as the most effective deterrent. In contrast, strengthening the alliance's conventional forces, by suggesting that NATO might seek to limit its response, would undermine deterrence and make aggression more likely.

Second, the United States and its European allies differed over the balance of conventional forces in Europe and thus the feasibility of the American proposals. The U.S. position was that an adequate conventional capability was well within NATO's grasp. The Europeans, who took a much more pessimistic view of the balance, believed that any attempt to create a strong conventional defense would be futile.

These contrasting viewpoints can in turn be understood in terms of the different positions that the allies occupied within the alliance and the international system more generally. First, differences in location ensured that countries would be unevenly exposed to the threat. From the perspective of the Europeans, and especially the West Germans, who were on the front line, any military conflict would assume near total proportions. Consequently, they were anxious to minimize the possibility that any hostilities would occur. For the United States, which was far from the potential battlefield, not all wars would have the same consequences. Hence it was only natural for American leaders to place equal weight on the goal of controlling escalation should a conflict nevertheless break out.

Geography also implied different degrees of responsibility for ensuring the success of the conventional force buildup. Because of their proximity to the Warsaw Pact nations, the European allies would inevitably bear the greatest share of the burden. For the United States, the potential economic costs of strengthening the alliance's conventional forces were much lower, making it easier to advocate such a policy.

The views of the allies were also shaped by differences in their international status and their roles within the alliance. In general, the European countries sought to increase their influence within NATO—or their independence from it—chiefly by enlarging their nuclear roles. France, for example, assigned top priority to its program to develop a national nuclear force. This effort, however, absorbed a high percentage of France's defense resources, which might otherwise have been devoted to strengthening the country's conventional forces. It also required France to oppose any interpretation of NATO strategy that would call into question the need for such a capability. By contrast, West Germany which had forsworn the development of an independent nuclear capability, viewed a substantial nuclear role in NATO as a prerequisite for being treated as a full partner. Accepting greater responsibility for conventional defense would effectively downgrade its status within the alliance.

In view of these differences, it is not surprising that, apart from the Germans, who had yet to complete their rearmament program, the European allies repeatedly failed to increase their conventional contributions despite external pressures. Similarly, the delay in the adoption of flexible response seems much more understandable. It took time—and France's withdrawal from the debate—to overcome disagreements among the allies and to fashion a new strategic consensus. Even then, considerable compromise was required to reconcile their conflicting preferences, resulting in the high degree of ambiguity that characterized flexible response.

Institutional factors also contributed to the resistance that greeted U.S. proposals for strengthening the alliance's conventional forces. The NATO bureaucracy showed little enthusiasm, and by the early 1960s, national military organizations had become strongly wedded to their new nuclear roles. Thus both the U.S. and German air forces opposed attempts to reconfigure their nuclear-armed aircraft for conventional missions.[187]

Institutional constraints played an even greater role, however, in limiting proposed force reductions during this period. Although the NATO military authorities last established alliance conventional force requirements in 1961, there was a strong presumption that NATO members would maintain their force contributions at existing levels. Any retreat from this norm would be viewed not only as diminishing the security of one's alliance partners but as a violation of one's international commitments. Thus no country could afford to underestimate the possible negative repercussions of troop reductions, for it risked incurring the wrath of its allies.

Such considerations were apparent in the cautious behavior of the United States in its handling of the withdrawal of the Berlin round-out units and especially during the trilateral negotiations, when the U.S. Air Force reductions in Europe were cut back sharply in response to German criticism. Nor were the European allies immune to the pressure created by such expectations, as when West Germany decided to scrap a proposed reduction of the Bundeswehr the same year.

Finally, this period offers evidence of how norms concerning alliance force levels can acquire strong domestic and cognitive roots. For example, the U.S. military presented a solid front in opposition to McNamara's force reduction proposals in 1966 and 1967. Despite its steadily increasing engagement in the war in Vietnam, the army continued to see its principal long-term mission as lying in the European theater, and although the air force had previously resisted denuclearization, its interests were presumably threatened at least as much by the proposed withdrawal of six air wings. McCloy's views, which had almost certainly been formed in the 1950s, about the potentially disastrous consequences of any U.S. troop withdrawals, were also influential.

CHAPTER 6

Implementing Flexible Response,
1969–1989

The two decades following the adoption of flexible response witnessed no further modification of NATO's formal strategic concept and the role it assigned to conventional forces or any serious attempt to revise it. Although the allies often disagreed on how precisely MC 14/3 should be interpreted and implemented, they were generally content to work within its broad parameters.[1] The only explicit appeal for a revision of the strategy—the proposal in the early 1980s that NATO adopt a doctrine of no first use of nuclear weapons—originated outside the alliance's organizational structure and member governments and consequently had little impact.

During the same period, NATO's conventional force levels in the Central Region grew only modestly. The number of regular combat units rose from approximately 26⅓ divisions and 70 brigades in 1969 to 28⅔ divisions and 77 brigades in 1989, while the number of active military personnel remained constant. These changes largely reflected the restructuring of U.S. and German forces in the 1970s to produce greater combat power and the reorganization of the French and British forces in Germany, which resulted in a larger number of divisions. The alliance's conventional capabilities were enhanced by the introduction of more capable weapons systems, the prepositioning of equipment for three additional reinforcing divisions from the United States, and the creation of six well-armed German territorial army brigades that were maintained at high degrees of readiness. But much of the increase in force levels simply compensated for the reductions of the late 1960s.

These relatively static in-place force levels over an extended period contrast with the greater fluctuations and more abrupt changes of earlier periods. They are especially striking in view of the strong pressures, first for further decreases, then for increases in NATO's conventional capabilities, that racked the alliance during these years. In 1969 and the early 1970s, renewed fiscal constraints, improving East-West relations, and continued U.S. involvement in the Vietnam War threatened to culminate in a second, and possibly much deeper, round of allied force reductions. Then, in the mid-1970s, the formalization of strategic nuclear parity between the superpowers and steady improvements in the conventional capabilities of the Warsaw Pact led to warnings that the European military balance was shifting decisively to NATO's disadvantage.

In response to these pressures, the alliance undertook several initiatives intended to strengthen—or at least to prevent the further weakening of—its conventional forces. Most notable among these were Alliance Defense in the Seventies (AD-70) in the early 1970s, the Long Term Defense Program (LTDP) in the late 1970s, and the Conventional Defense Improvement (CDI) initiative in the mid-1980s. In most respects, however, these initiatives were modest. They rarely entailed efforts that went far beyond those alliance members were already planning to undertake. Because they emphasized low-cost and qualitative rather than more expensive, quantitative measures, moreover, they resulted in few easily measurable improvements in NATO's conventional capabilities, even when successfully implemented, and many important proposed actions were not carried out. Each initiative soon lost momentum and disappeared from the NATO agenda within a few years. Thus perhaps the most significant gains were the result of primarily unilateral U.S. and German efforts, and even these were offset by the expansion and the steady modernization of the forces of the Warsaw Pact.

Internal Challenges and Resistance, 1969–1973

During the first several years after the adoption of flexible response, NATO's conventional force posture in the Central Region faced several internal challenges because pressures to reduce national force contributions continued in the United States and several allied countries. In response to these pressures NATO officials devised a comprehensive program, Alliance Defense in the Seventies, designed to improve NATO's conventional forces over the following decade. As a practical matter, however, both AD-70 and the parallel European Defense Improvement Program (EDIP) were primarily intended to prevent the further weakening of NATO's existing forces and, in particular, to stave off unilateral American reductions by demonstrating that the European allies were bearing an equitable share of the defense burden. Consequently, neither initiative was particularly ambitious. They resulted in few concrete force improvements, and AD-70 was subsumed in the regular NATO force planning process almost as soon as the threat of reductions had subsided.

These internal challenges culminated in the early 1970s, when Senator Mike Mansfield introduced successive legislative amendments that would have mandated significant U.S. troop withdrawals from Europe. Although the Mansfield amendments garnered substantial support in the Senate, all were eventually defeated. Nevertheless, the Nixon administration was forced to go to great lengths to block these troop reduction efforts, and even then its ultimate success was due in no small part to the assistance of unexpected allies.

INITIAL PRESSURES FOR REDUCTIONS IN NATO FORCE LEVELS

The Soviet intervention in Czechoslovakia in 1968 arrested a series of unilateral conventional force reductions by various NATO countries that threatened to unravel the fabric of the alliance. The crisis had hardly subsided, however, when pressure for cuts reemerged. In 1969, Belgium resumed the contraction of its armed forces, from six to only four active brigades, which had begun two years before. And that spring, Canada announced its intention to slash its military contribution in Europe by two-thirds the following year.[2] Allied leaders feared that such action by Canada would not only weaken NATO's conventional capabilities directly but trigger a much more serious chain reaction of force reductions.[3]

These fears were not without foundation. The situation remained unsettled in the United States, where there was talk of further troop withdrawals. In fact, the incoming administration of Richard M. Nixon was immediately confronted with a plan devised by its predecessor to withdraw 30,000 support and administrative personnel from the continent in fiscal year 1970. Thus one of Nixon's first tasks was to formulate a position on the issue of American force levels in Europe. Despite some differences of opinion, the administration eventually decided against making any substantial troop reductions for the time being, even if this meant breaking the previous link between the maintenance of the U.S. military presence at its existing size and the achievement of satisfactory offset agreements with Germany.[4] In December, Secretary of State William Rogers declared that U.S. combat units in Europe would be maintained at "essentially present levels" until at least mid-1971.[5]

This affirmation of the status quo did nothing to mollify those senators favoring a further reduction in the U.S. commitment to NATO, and they steadily increased their pressure on the administration. On December 1, 1969, Mansfield reintroduced his troop withdrawal resolution, which quickly gathered 51 cosponsors. And during hearings on the subject of U.S. forces in Europe the following year, Senator Stuart Symington peppered witnesses with questions about the consequences of a 50 percent reduction.[6]

Senate interest in further troop reductions during this period stemmed from a variety of considerations, some new and some familiar. The need to reduce the U.S. balance-of-payments deficit, which reached a record $7 billion in 1969, by limiting foreign exchange outlays seemed more pressing than ever. Nor had there been any diminution of American resentment of the allies for their perceived unwillingness to bear a fair share of the NATO defense burden, a factor that was now aggravated by increasing European economic competition. By this time, moreover, the Vietnam War had fostered a general discontent with military commitments abroad. Finally, the renewed atmosphere of detente with the Soviet Union raised further questions about the need to maintain such a large force in Europe, especially when it diverted scarce resources from the social programs initiated in the 1960s to which the Democratic-controlled body remained strongly committed.[7]

EUROPEAN RESPONSES

The debate in the United States over U.S. force levels was not lost on the West Europeans, many of whom feared that major reductions were in the offing. Even the administration's commitment to maintain the status quo was couched in such a way as to suggest the possibility of cuts after mid-1971. Thus the allies gradually concluded that they would have to take positive action to influence the debate if further American troop withdrawals were to be forestalled.[8]

The first to act were the British. In March 1970, Britain agreed to move the infantry brigade withdrawn as a result of the trilateral talks back to the continent later that year. British officials felt that NATO forces could no longer be safely reduced and, indeed, that some increase was desirable. The redeployment would also help to close the gap created by the Canadian pullback. Most important, it might help to convince skeptical Americans that the Europeans were determined to take their defense seriously.[9]

The principal collective European response to the threat of U.S. force reductions was the European Defense Improvement Program, which was proposed during the second half of 1970 by an association of European defense ministers from ten countries known as the Eurogroup. Formed in 1968, the Eurogroup offered the perfect forum for such an initiative. Its declared purpose was "to strengthen the alliance by increasing the effectiveness and cohesiveness of the European defense contribution to NATO." Politically, however, it represented "an effort to demonstrate the significance and magnitude of [this] contribution and thus to counter arguments that NATO burdens [were] carried disproportionately by the United States and that Europe [was] insufficiently committed to its own defense."[10]

In June 1970, the Eurogroup concluded that a special European effort was needed to avert American force reductions, and it began to elaborate a program that would help to neutralize the pressures in the United States for cuts. Initially, this effort focused on the development of a direct European financial contribution to the United States that would relieve some of the budgetary and foreign exchange costs associated with its NATO commitments. During a visit to Europe at the end of September, however, Nixon expressed a preference for European force improvements over subsidies to the United States. Consequently, the Eurogroup shifted gears and spent the following two months hastily working out the details of such a program so that they could be announced at the NATO ministerial meeting scheduled for early December.[11]

The result of these labors was the European Defense Improvement Program, which provided for nearly $1 billion in additional spending over five years by the Eurogroup members. The EDIP consisted of three elements: a collective contribution of $420 million to the NATO common infrastructure fund (to build more aircraft shelters and to accelerate the development of a NATO integrated communication system); $450 million in specific national force improvements not previously planned; and the provision of $80 million in transport aircraft by West Germany to Turkey.[12]

Consistent with the true purpose of the exercise, however, these commitments were explicitly linked to the continued presence of U.S. forces in Europe at their existing levels.[13] The Nixon administration wasted no time in reassuring the Europeans on that score. In November, the National Security Council had considered yet another study of the question of U.S. force levels in Europe, and once again, the president rejected the idea of making any cuts. Consequently, Secretary of Defense Melvin Laird announced that the United States would extend its commitment to the status quo through mid-1972, and, going a step further, Nixon sent a statement to the North Atlantic Council in early December indicating that "given a similar approach by the other allies, the United States would maintain and improve its own forces in Europe and would not reduce them unless there is reciprocal action from our adversaries."[14]

The EDIP was announced with much fanfare. Nevertheless, it was widely—and properly—viewed as little more than cosmetic. It represented little additional effort on the part of the Eurogroup members, only $200 million per year, or an increase of merely three-quarters of 1 percent in the combined defense spending of the European NATO countries. Moreover, only four countries—West Germany, Britain, Belgium, and Norway—committed themselves to improving their national forces.[15] To the degree that it influenced the Nixon administration's decision to reaffirm the U.S. troop commitment in Europe, however, the EDIP was at least modestly successful as a political maneuver.

ALLIANCE DEFENSE IN THE SEVENTIES

The main NATO response to the threat of further troop reductions was a comprehensive conventional force improvement program that was also adopted at the end of 1970. This effort was based on a detailed study of the defense problems the alliance would face over the next decade known as Alliance Defense in the Seventies, or AD-70, that originated within the NATO organizational structure.[16] Rather than simply try to resist the pressures for further troop cuts that permeated the alliance in 1969 and early 1970, the new SACEUR, General Andrew Goodpaster, decided to counterattack. First, he instructed his staff to generate a set of compensatory measures to be taken in the event that Canada went through with its proposed force reductions.[17] Subsequently, he held informal discussions with NATO secretary general Manlio Brosio and the other NATO military authorities in which they decided to take the initiative away from the proposals for cutbacks by making a thorough survey of how to strengthen the alliance's military posture.[18]

In February, Nixon provided the occasion to launch just such an initiative. In his foreign policy report to Congress, the president called for a thorough reexamination of alliance strategy and forces. Seizing the opportunity, Brosio proposed the following month that the NATO organization itself undertake such a study, which was set in motion by the allied ambassadors in May. This undertaking was then given formal blessing by the NATO defense ministers, who asked that the ambassadors prepare a report in time for consideration at their next meeting at the end of the year.[19]

The AD-70 study was completed in October 1970, and the final report was approved by the defense ministers in December. Consistent with MC 14/3, the report affirmed that NATO forces should be able to deter and counter a wide variety of possible aggressive actions, and it concluded that there were critical imbalances in conventional forces between NATO and the Warsaw Pact. To remedy these deficiencies, the report identified eight key problem areas in which the alliance's conventional capabilities should be strengthened over the next decade.[20]

Most of the AD-70 recommendations were too general to be useful for detailed planning, however. Although the report had described NATO's most important military deficiencies, it had not suggested specific remedies.[21] Consequently, two more years were spent refining the program. In May 1971, the NATO defense ministers requested a further report that would recommend specific force improvement measures and establish relative priorities among them. In December, they endorsed the priority areas that had been recommended and identified certain measures within these areas for early action, some of which were included in the five-year force plan adopted at that time for the period 1972–76 as part of the regular force planning process.[22] Additional AD-70 measures were incorporated into the new NATO force goals adopted by the defense ministers in May 1972, and the alliance instituted an annual review of progress made toward their implementation that September.

Greater specification and prioritization alone, however, did not guarantee implementation. The defense ministers soon concluded that some overall increase in defense spending would be needed to improve NATO's conventional forces in the problem areas identified in AD-70. In response, the Eurogroup announced in December 1971 that its members would increase their defense budgets by a total of $1 billion in 1972, ostensibly for the purpose of implementing the AD-70 measures. This so-called Europackage represented an increase in European defense spending of approximately 3 to 4 percent in real terms.[23]

Even this extra effort, however, was insufficient to have much impact. The cost of the force improvements recommended by AD-70 had been estimated at $3 billion per year over five years, whereas the Europackage largely represented actions to replace equipment that had been planned before the report was adopted the previous year.[24]

The reluctance of NATO countries to implement the program is not surprising because, like the EDIP, it was as much intended to stanch the pressure for troop withdrawals as to result in significant force improvements. Thus although the stated intention had been to fashion a decade-long effort, enthusiasm quickly waned, and by late 1973, there were no more references to AD-70 in NATO communiques. As a result, most of the problems the study had identified remained unresolved. Yet another initiative would be needed if NATO were to improve its conventional force posture.

THE MANSFIELD AMENDMENTS

Because of their modest natures, AD-70 and the efforts of the Eurogroup did little to mollify congressional advocates of U.S. troop reductions. To the contrary, the Nixon administration's parallel pledge to maintain and even to improve U.S. forces in Europe triggered a series of unprecedentedly serious challenges in the Senate to the NATO status quo. In both 1971 and 1973, Senator Mike Mansfield introduced binding legislation that would have mandated substantial American withdrawals. These challenges were ultimately turned back, however, and by the mid-1970s, the arguments for troop cuts had lost much of their force as a result of changes in the international environment.

The 1971 Mansfield Amendment. In December 1970, Mansfield announced that he would no longer limit his efforts to seeking the passage of nonbinding legislation regarding troop withdrawals. True to his word, on May 11, 1971, the majority leader introduced an amendment to the draft extension bill then working its way through the Senate that would have required a 50 percent reduction in the U.S. military presence in Europe, to 150,000 servicemen, by the end of the year. Unlike his previous resolutions, this measure would have had the force of law if approved.

Predictably, although not without conviction, Nixon administration officials roundly condemned Mansfield's action. They stressed the tremendous damage that would be caused both to the alliance and to U.S. interests if unilateral reductions were imposed. From the administration's perspective, however, the amendment could not have come at a less propitious moment. The U.S. balance-of-payments deficit was mounting at a record clip—$5 billion in the first three months of 1971 alone—and resentment in Congress toward the Europeans was high. Consequently, the amendment "created almost total panic in the administration."[25]

Although initially caught off guard, Nixon and his advisers quickly regained the initiative. They managed to get the vote on the amendment postponed for nearly a week while they undertook a major lobbying effort on Capitol Hill. Most notable was their success in enlisting the assistance of an impressive array of elder statesmen whose ties with NATO went back to the beginning of the alliance. The administration obtained statements of support signed by former presidents Truman and Johnson, nearly all living former secretaries of state and defense, and numerous other former high officials. Culminating this carefully orchestrated campaign, Nixon called the proposal for unilateral cuts in U.S. forces "an error of historic dimensions."[26]

At this critical juncture, the administration received timely assistance from an unlikely source: Soviet leader Leonid Brezhnev. Although NATO had pressed for negotiations with the Warsaw Pact on mutual force reductions since June 1968, little progress had been made. Instead, the Soviet Union and its allies had promoted a European security conference that would codify the postwar territorial status quo.[27] During a major speech on May 14, just three days after the amendment was introduced, however, Brezhnev declared that his government was now ready to participate in force reduction talks. The administration immediately seized upon this apparent Soviet shift to argue that unilateral cuts by the United States would undermine the West's bargaining position and thus remove any Soviet incentive to negotiate.[28]

Largely because of Brezhnev's intervention, the Mansfield amendment was voted down on May 19 by a vote of 61 to 36. In November, a second, much more modest amendment, which would have limited the number of U.S. military personnel in Europe to 250,000 after mid-1972, was also thwarted.[29] During the following election year, Senate pressure for troop reductions abated temporarily.

The 1973 Debate. The issue reemerged in 1973 with even greater intensity. The arguments that had traditionally been advanced in support of reducing the number of U.S. forces in Europe seemed more compelling than ever. The general mood of retrenchment that accompanied the American withdrawal from Vietnam in combination with the atmosphere of detente in U.S.-Soviet relations cast doubt on both the wisdom of and the need for a large military presence in Europe. Continuing balance-of-payments problems, moreover, raised questions about the U.S. ability to finance it.[30] Finally, considerable antagonism persisted in the United States toward Western Europe as a result of both the region's increasingly assertive economic policies and its perceived failure to do more for its for defense.[31]

Senate attempts to legislate troop withdrawals in 1973 must also be viewed in the context of the larger domestic power struggle then taking place between the Republican administration and the Democratic-controlled Congress. The perennial dispute over the role of the Congress in foreign policy making was at a peak in a year that saw the passage of the War Powers Resolution. There was also substantial disagreement over national priorities, with the Democrats wanting to devote greater resources to domestic programs even if doing so pinched foreign commitments. These tensions were further aggravated by a series of administration challenges to congressional authority that seemed almost designed to provoke the legislative branch.[32]

The 1973 debate began in mid-March, when the Senate Democratic caucus approved a resolution calling on the president to reduce "substantially" the size of the American military presence in Europe and Asia by mid-1974. The Nixon administration quickly responded with the familiar argument that unilateral cuts would destroy the prospects for mutual force reductions by removing any incentive the Soviets might have to seek a negotiated withdrawal.[33] Administration officials nevertheless took the threat of congressionally imposed troop cuts very seriously.[34] In June, Secretary of Defense-designate James Schlesinger asked his NATO colleagues to consider the development of a new multilateral arrangement for sharing the additional budgetary and foreign exchange costs attributable to the U.S. forces in Europe.[35]

In response to this pressure, the alliance established a group to study "financial problems arising from the stationing of forces on the territory of other NATO countries." The prospects of financial relief, however, were not bright. The study group agreed that measures to alleviate the U.S. burden should not be undertaken at the expense of force improvements the allies had already proposed. Yet because real increases in defense spending seemed out of the question in most NATO countries, they could not assume a greater share of U.S. costs without making such cuts.[36]

Thus the administration found itself in a quandary when Mansfield prepared to offer another troop reduction amendment. Once again, however, it received critical assistance from an unexpected source, this time conservative Democratic senators who favored U.S. withdrawals only if the additional financial costs of the American presence could not be eliminated in any other way. In September, Senators Henry Jackson and Sam Nunn introduced an amendment to the military procurement bill requiring the president to seek compensation from the allies sufficient to offset fully any balance-of-payment deficit incurred in FY 1974 as a result of U.S. deployments in Europe in fulfillment of NATO obligations. Only if a full offset proved to be unobtainable would the United States reduce its forces by a percentage equal to that of the shortfall.[37]

The Jackson-Nunn amendment passed overwhelmingly in late

September. It was strong medicine for the administration to swallow, but it eroded support for unconditional, unilateral reductions, and it held out the hope that cuts could be avoided. As a result, when Mansfield and Senator Alan Cranston subsequently introduced amendments to reduce U.S. forces overseas by 40 to 50 percent over three years, the measures were rejected.[38]

The Demise of Senate Pressure for Troop Reductions. Although the Jackson-Nunn amendment restored the link between the maintenance of U.S. forces in Europe at existing levels and a resolution of the balance-of-payments problem, which Nixon had severed in 1969, its passage indicated that the Senate was much more concerned with obtaining financial relief from overseas commitments than with reducing the number of American forces abroad per se. Consequently, the immediate threat to the U.S. military presence in Europe was largely surmounted when the balance of payments improved and actually registered a surplus for the whole of 1973.

Several other developments late that year ensured that there would not soon be another serious congressional challenge. One was the formal commencement of the long-awaited East-West talks on mutual and balanced force reductions (MBFR) in October. Henceforth the administration's argument that unilateral cuts would undermine the West's negotiating position would carry even more weight.[39] Another was the Yom Kippur War in the Middle East, which stimulated an American reappraisal of detente and, more specifically, of Soviet intentions toward Europe. After the war, "the burden of proof shifted to those who wanted to reduce the US military presence in Europe." In addition, the subsequent sharp rise in world oil prices provoked a crisis in Europe that made it inopportune for Americans to raise the issue of withdrawals. It was one thing to propose a change in NATO's defense burden-sharing arrangements when the allies seemed economically strong and quite another to do so when they were not.[40]

In June 1974, not long before Nixon resigned as president, Mansfield made one last attempt to legislate a troop reduction. Compared with his previous amendments, this one was modest. It proposed to cut overseas ground forces by a total of only 76,000, and the reduction did not have to be made in Europe. Nevertheless, as a result of the altered international situation, the amendment was easily defeated, by a margin of 54 to 35.[41]

External Challenges and NATO Responses, 1973–1981

Several long-term trends further undermined sentiment for U.S. troop withdrawals. By this time, the United States had fully extricated itself from Vietnam, and the American public's disenchantment with foreign military commitments had begun to wane. There was also renewed concern about the military threat to Western Europe. The Soviet Union was rapidly narrowing the United States's lead in strategic nuclear forces, and since the mid-1960s, it had steadily strengthened its conventional capabilities in and readily deployable to the Central Region. By the early 1970s, the Warsaw Pact appeared capable of overrunning NATO in a short, intense war, much like the blitzkrieg of World War II, and by the middle of the decade, some well-informed observers had concluded that the Soviets and their allies could launch such an attack with little or no warning.

Concern about nuclear parity and the "new Soviet threat" in Europe generated proposals in the mid- and late 1970s for strengthening the alliance's conventional capabilities. These included qualitative improvements such as modernizing equipment, restructuring NATO's existing forces to augment their combat power, increasing the number of reserve and reinforcing units that could be made quickly available, and using more efficiently the alliance's available defense resources.

NATO's primary collective response to the adverse trend in the conventional balance was the Long Term Defense Program, which was launched soon after President Jimmy Carter took office in 1977. This comprehensive initiative embraced a wide range of largely qualitative force improvement measures while stressing greater alliance cooperation and long-term planning and was accompanied by a pledge to seek real annual increases in defense spending on the order of 3 percent. Because of its broad scope, however, definition of the LTDP continued well after its formal adoption by NATO in 1978, and implementation did not begin in earnest until the following year. Because of its extraordinary nature, moreover, the program was resisted by other countries and much of the NATO bureaucracy. Thus by the time Carter left office, the LTDP had resulted in few significant achievements. Instead, the most important improvements in NATO's conventional capabilities during this period were primarily the result of largely unilateral U.S. and German actions.

RENEWED CONCERN ABOUT THE SOVIET THREAT

During the decade following the adoption of flexible response in 1967, the West's overall military situation vis-à-vis the Warsaw Pact seemed to deteriorate markedly. This shift was perhaps most obvious in the strategic realm, where the Soviet Union achieved formal parity with the United States. By the early 1970s, the two superpowers were roughly evenly matched in numbers of strategic nuclear delivery vehicles. This equivalence was codified in the first agreements of the Strategic Arms Limitation Talks (SALT), which were signed in May 1972.[42]

Many American officials thought strategic parity called into question NATO's reliance on nuclear weapons. They believed that U.S. strategic forces would reliably deter in a narrowing range of contingencies. A major part of the burden of deterrence would fall increasingly on other forces. Under these circumstances, they argued, greater reliance should be placed on conventional forces, which required strengthening NATO's non-nuclear capabilities.[43]

Improvements in the Warsaw Pact's Conventional Capabilities. The task of strengthening the alliance's conventional capabilities to the point that they could compensate for the seemingly declining ability of nuclear weapons to deter Soviet aggression was complicated, however, by simultaneous improvements in the Warsaw Pact ground and tactical air forces facing NATO in the Central Region. The Soviet Union now stationed five divisions and approximately 70,000 military personnel in Czechoslovakia in addition to the 26 divisions already deployed in East Germany, Poland, and Hungary. At the same time, the average size of the Soviet divisions in Eastern Europe had grown significantly, and the Soviet Union had added to its combat capability at the army and front levels. Thus whereas there had been only some 400,000 Soviet ground troops in the region in 1967, they numbered approximately half a million by 1977, an increase of 25 percent. These forces, moreover, could be augmented by some 30 East European divisions and another 40 or so Soviet divisions located in the western military districts of the Soviet Union, of which 12 were combat ready and the rest could be mobilized in one month.[44]

This expansion in the number and size of the Soviet divisions facing NATO was paralleled by a substantial growth in their mobility and firepower through the addition of large amounts of modern equipment. The Soviet Union introduced new main battle tanks, the T-64 and T-72, and increased the amount of armor in each of its motorized rifle divisions by 40 percent. Consequently, the Soviet tank inventory in East Germany rose from 5,000 to 6,500, while the holdings of its Warsaw Pact allies also grew substantially. The Soviet Union also converted from vulnerable towed to armored, self-propelled artillery and increased the number of artillery pieces per division in Eastern Europe by 50 to 100 percent. Finally, new infantry combat vehicles, antitank guided missiles, attack helicopters, and mobile air defense systems were deployed among the ground forces.

As a result of these and other improvement measures, the overall strength of each Soviet division was estimated to have grown by 25 percent.[45]

The strengthening of the Soviet ground forces was accompanied by a major transformation of the air threat to NATO. Until 1968, the Soviet tactical air forces had been oriented almost exclusively toward air defense. They had been composed of short-range, low-payload aircraft with little capability for attacking targets on the ground.[46] In the early and mid-1970s, however, large numbers of new aircraft with longer ranges and larger payloads were introduced. In addition to narrowing NATO's traditional qualitative advantage in the air, this modernization effort gave the Soviet Union a substantial capability for close air support of the ground forces and deep interdiction of NATO airfields and theater nuclear capabilities. As a result, NATO's tactical air assets, upon which the alliance depended heavily to rectify the balance on the ground, would henceforth have to ensure their own survival before they could influence the land battle.[47]

Implications for NATO. The implications of these developments for NATO were sobering. By 1973, many analysts concluded that the Soviets were preparing to wage a short, intense war. Rather than try to wear down NATO forces in a long war of attrition, they would attempt to overrun Europe in blitzkrieg fashion.[48] This conclusion was supported by changes in Soviet military doctrine, which had come to accept the possibility that war in Europe would not be nuclear and called for rapid, armored thrusts. Further evidence was found in the high tooth-to-tail ratio of Soviet ground forces—only 25 percent of Soviet manpower was in support units—which made it difficult for them to sustain operations for an extended period of time.[49]

These improvements in Warsaw Pact capabilities seemed specifically designed to take advantage of certain characteristics of NATO's conventional force posture. Many NATO forces were structured for a long war of attrition rather than a short war of high intensity. They were heavy in support troops at the expense of combat troops, sacrificing firepower in favor of the ability to wage sustained combat. Nowhere was this more evident than in the U.S. contingent in Europe, which had a built-in logistics system designed for indefinite support.[50] Another NATO weakness that the Soviet Union appeared to be preparing to exploit was the uneven north-south distribution of its forces. The majority of the Soviet divisions were situated astride the North German plain, which was defended by NATO's Northern Army Group (NORTHAG). The bulk of the alliance's most capable forces, those of the United States and the Federal Republic, however, were located in the southern half of the Central Region, from where they could provide little immediate assistance in the event of an attack in the north. Such considerations led many analysts to the grim conclusion that NATO could be overrun in a few days, depending on the intensity of the conflict.[51]

As the danger of a Soviet blitzkrieg began to receive attention, yet another dimension of the new threat emerged: the possibility of a surprise attack launched with little or no warning. Previously, NATO's conventional force posture had been largely predicated on the belief that any hostilities would be preceded by a substantial period of mobilization and deployment by the Warsaw Pact, giving the alliance time to bolster its defenses. For planning purposes, U.S. officials assumed that the Soviet Union would attack only after 30 days of active preparation and that NATO mobilization would lag by only 7 days.[52]

By 1973, however, serious doubts had arisen in the United States about the validity of this assumption. Because of the improvements in the Warsaw Pact's conventional capabilities, the Soviets would no longer have to augment substantially their forces in the Central Region before an attack. Increasingly, it appeared that they might be able to launch an offensive with little or no mobilization, perhaps using only the forces available in Eastern Europe. On the basis of such considerations, U.S. officials sharply reduced their estimates of the amount of time that NATO would have to prepare for hostilities.[53] And by the mid-1970s, more pessimistic observers concluded that a massive Warsaw Pact attack might be preceded by only a few days of warning. In any case, the alliance would have little or no time to mobilize and to position its forces in Europe, let alone to deploy reinforcements from the United States and Britain.[54]

INITIAL NATO RESPONSES

Initially, the allies responded to the Soviet achievement of strategic nuclear parity and the growth in Warsaw Pact conventional capabilities with a series of modest and largely uncoordinated initiatives. These efforts included correcting critical qualitative deficiencies and restructuring U.S. and German forces to increase their immediately available combat capabilities. Throughout this period, however, NATO countries labored under severe financial constraints. Competing domestic programs, continuing hopes for detente, and the repercussions of the oil price shock exerted downward pressure on defense budgets, while the steadily rising costs of new weapons and manpower meant that real increases in spending would be needed just to keep force levels constant. Consequently, alliance leaders pinned their hopes for strengthening NATO's conventional forces largely on measures that could be achieved at little or no additional cost, such as greater cooperation and long-term planning.

Basic Issues Initiative. The first noteworthy effort at force improvement occurred concurrently with Senator Mansfield's last serious attempt to legislate U.S. troop reductions. In June 1973, Secretary of Defense-designate James Schlesinger proposed that NATO assign top priority to correcting deficiencies in several less visible but nevertheless critical items needed in the early phases of a conflict. Specifically, he called for increases in antitank weapons, aircraft shelters, and war reserve stocks.[55]

Although Schlesinger mentioned the threat of congressionally mandated restrictions on U.S. force levels in Europe, his proposal was motivated primarily by American concerns about the implications for deterrence of strategic nuclear parity. In contrast, he took a relatively optimistic view of the conventional balance in Europe. Schlesinger argued that there was no inherent reason why the Warsaw Pact should enjoy conventional superiority. To the contrary, according to his calculations, NATO would not be at a serious disadvantage either at the beginning or after a month of Soviet mobilization. In short, the alliance was already providing the resources needed to erect a powerful non-nuclear defense, and the existing deficiencies could be remedied at relatively modest additional cost.[56]

The "basic issues" proposed by Schlesinger for special attention were approved by the NATO defense ministers in December 1973, although the number of measures was subsequently broadened to six.[57] This initiative made little progress, however, and was largely abandoned by 1976. Declining gross domestic products in both Western Europe and the United States following the 1973–74 oil crisis kept military budgets static.[58] Equally important in limiting the program's success was a lack of allied support. Initially, some Europeans believed that the real U.S. objective was to reduce further the role of nuclear weapons in NATO strategy, which in their view would dangerously weaken the American nuclear guarantee. Others feared that the Nixon administration sought allied force improvements so as to prepare the groundwork for U.S. troop withdrawals.[59]

Above all, the allies as well as the NATO staff were extremely skeptical of the U.S. claim that an adequate conventional defense was feasible. They thought the alliance was far from enjoying parity with the Warsaw Pact. In fact, the balance seemed to be shifting against NATO as the Soviet Union modernized its forces. Consequently, allied analysts believed that NATO should plan for a major attack after a mobilization period of two weeks at the most, and they expected that any conflict would become nuclear within 30 days.[60]

U.S. and German Force Restructuring. A second and somewhat more successful response to the growing conventional threat was the restructuring of the U.S. forces in Europe to increase their combat capability. This response was originally proposed by independent defense analysts outside of the U.S. government, who argued that NATO—and the United States in particular—was emphasizing sustainability at the expense of combat power.[61] To correct this imbalance, these analysts proposed that NATO reverse its investment priorities. Reducing support forces would enable the alliance to release resources for increasing the firepower of existing divisions or increasing the total number of combat units at no additional net cost. Restructuring might also create additional reserve divisions and increase the rate at which U.S. reinforcements could be deployed to Europe.[62]

These ideas soon caught the attention of Senator Nunn, who was particularly concerned that the United States would not respond to warning in a timely manner and thus desired to increase in-place U.S. combat capability. In 1974, Nunn proposed an amendment to the defense authorization bill requiring that the United States reduce the number of army support personnel in Europe by 18,000. These troops could be replaced, however, by an equivalent number of men in combat formations, enabling the United States to generate additional combat power at little or no extra cost.[63] The administration of Gerald R. Ford responded positively to Nunn's initiative, drawing up plans to deploy two new brigades in Europe over the next two years and generally giving priority to combat forces over sustainability.[64]

A related set of suggestions involved relocating combat units to correct the maldeployment of NATO forces. Independent analysts proposed that as much as half of the American contingent be transferred to NORTHAG. And in 1976, the United States decided to move one of the two new brigades to that area.[65]

West Germany took similar steps to restructure its forces within existing personnel ceilings to increase their immediately available combat power. Even before the Americans, the Germans had come to fear a short, intense war launched with little warning and thus to emphasize forces-in-being and rapid mobilization capabilities over the ability to wage prolonged combat.[66] In 1975, the Federal Republic introduced a new army structure involving the creation of three additional armored brigades, for a total of 36. By placing greater reliance on reserves to perform support functions, the new structure was also expected to enhance the operational readiness and strength of the fighting units. At the same time, Germany announced plans to form six Home Defense Brigades. Unlike the other territorial forces, these units would be heavily armed with tanks, armored personnel carriers, artillery, and antitank weapons and would be maintained at fairly high levels of readiness in peacetime. As a result, they would be capable of supporting forward defense operations if necessary.[67]

Rationalization, Standardization, and Interoperability. The squeeze created by limited financial resources on the one hand and spiraling costs on the other spawned other proposals for strengthening NATO's conventional forces at little or no additional expense, which became collectively known as rationalization, standardization, and interoperability (RSI). These proposals, which included a wide range of possible measures, all involved improving the coordination of national defense programs in the service of overall alliance military requirements through greater cooperation and long-term planning. RSI seemed to have tremendous potential benefits. According to one estimate, the lack of equipment standardization reduced NATO's conventional capability by 30 to 50 percent; another expert calculated that the alliance was wasting at least $11 billion per year at a time when the combined annual defense expenditures of the European allies was only $30 billion.[68] In short, NATO's weaknesses were apparently due as much to the inefficient use of the available resources as they were to resource constraints.

NATO bodies began to explore the potential of RSI as early as 1973. The defense ministers urged increased cooperation, especially in the areas of logistics and armaments research, development, and production, with the goal of greater standardization of equipment. They sponsored a study, proposed by the Netherlands, of the possibility of defense role specialization in the Central Region. And they commissioned reports on other areas of potential rationalization and specialization. This activity culminated in May 1975, when the defense ministers adopted a long-range defense concept that placed increased emphasis on cooperative measures and the establishment of rigorous priorities "in order to obtain maximum efficiency from the force levels and resources which the Alliance can reasonably expect to have at its disposal."[69]

That same year, the newly appointed SACEUR, General Alexander Haig, undertook an examination of ways to improve the capabilities of the forces under his command. This so-called flexibility study focused on operational constraints on the use of these forces rather than force structure. Completed in December 1975, the study identified some 235 possible remedial actions. The following year, Haig initiated a program to correct the most serious deficiencies in the areas of readiness, reinforcement, and the harmonization of doctrine, tactics, and procedures.[70]

RSI also became a major issue in the United States in 1974, when Senator Nunn proposed another amendment aimed at improving the commonality and standardization of NATO weapons and equipment. Specifically, the amendment directed the secretary of defense to work to make standardization in research, development, procurement, and support an integral part of the NATO planning process.[71] At the same time, the Department of Defense was completing its own study on NATO's potential for rationalization, and it subsequently commissioned the Rand Corporation to assess further the possibility of large-scale rationalization of the alliance's defense posture. The resulting report recommended assigning first priority to correcting deficiencies in NATO's ground forces, altering the alliance's planning procedures to include a ten-year development program for major items of equipment, increasing the size of the NATO planning staff, and making rationalization an explicit item on the NATO ministerial agenda.

THE CARTER ADMINISTRATION AND THE LONG TERM DEFENSE PROGRAM

Even though many of NATO's weaknesses had been identified and reasonable solutions had been suggested, progress toward improving the conventional military balance in Europe was slow. Indeed, the alliance seemed to be falling even further behind the Warsaw Pact. The very number and diversity of proposals fostered indecision and prevented the concentrated application of effort. What was needed, proponents of strengthening NATO's conventional forces asserted, was "a comprehensive, coordinated, and concerted program undertaken by the Alliance as a whole."[72]

An attempt to remedy this situation began in 1977 with the development of the Long Term Defense Program, which became NATO's primary collective response to the new Soviet threat. The LTDP sought to provide a "coherent management framework" within which the many existing proposals could be integrated and related to one another.[73] Like previous studies of rationalization, it emphasized greater cooperation and long-term planning to make optimal use of the available defense resources. Going farther, however, the LTDP attempted to create new bodies and procedures for the purpose of selecting and implementing a limited number of high-priority force improvement measures. And it secured high-level political attention for these efforts to increase their chances of success.

The LTDP originated in the Carter administration, which was strongly committed to strengthening NATO's conventional forces and proposed the initiative at the May 1977 NATO summit meeting in London. Specific action programs in the force improvement areas embraced by the LTDP were hastily devised over the course of the following year and approved at a second NATO summit in May 1978. At that time, however, many details remained to be worked out and other proposed measures were accepted subject to reservations. Implementation of the program began even while specific objectives and timetables continued to be developed. By the end of the Carter administration, the LTDP had resulted in few concrete achievements, and interest in the program, which had been intended to last through the decade, was waning. Thus the most significant improvements in NATO's conventional capabilities during this period followed instead from largely independent efforts on the part of the United States to modernize its forces and to increase its ability to deploy reinforcements to Europe rapidly in a crisis.

The Origins of the LTDP. The LTDP grew out of a second Rand Corporation study, *Alliance Defense in the Eighties*, or AD-80, which was completed in November 1976.[74] The report began with the proposition that NATO would have to undertake a major conventional force improvement effort over the next decade to address the growing Soviet ability to launch a massive attack with little warning and proceeded to outline the elements that such an effort should include. Although the study was loosely modeled on AD-70, its recommendations were drawn largely from the long-range defense concept that had been adopted by the NATO defense ministers as part of their biannual ministerial guidance in May 1975 and the major NATO commanders' flexibility studies. Like the 1975 ministerial guidance, AD-80 assumed that allied defense spending would increase only modestly in real terms and thus emphasized making better use of the available defense resources through greater alliance cooperation. An adequate conventional capability could be obtained at an affordable cost, but it required overhauling NATO's structures and processes, something AD-70 had not attempted to do.

Although AD-80 proposed concrete measures for improving NATO's conventional force posture, giving top priority to rectifying deficiencies in the ground forces in the Central Region,[75] many of these were familiar. More novel were its recommendations for institutional and procedural reform, which were based on a blistering critique of the NATO force planning system. Specifically, the report called for broadening NATO defense planning to cover overall defense resource allocation, not just national force goals; integrating functional and cooperative programs that cut across national boundaries into the planning process; establishing rigorous priorities and identifying trade-offs;[76] and putting NATO planning on a longer-term basis to gain greater influence over national programs and thereby increase standardization and interoperability.[77] Finally, AD-80 advocated the creation of a special high-level task force with a broad mandate to propose the institutional and procedural changes needed to hasten the development and implementation of the required force improvement program.

The principal author of the Rand study, Robert Komer, soon received the opportunity to put AD-80's conclusions into practice. During the 1976 presidential campaign, he had prepared background papers for the eventual victor, Jimmy Carter. Upon taking office, the new secretary of defense, Harold Brown, appointed Komer as his special assistant for NATO affairs, granting him considerable power and access within the government.[78] Making full use of his new position, Komer immediately set out to convince the administration that the United States should propose a major force improvement initiative at the NATO summit meeting scheduled for May and quickly secured broad support for the idea. Once a consensus had been achieved in Washington, the next step was to lay the groundwork for alliance approval, a difficult task in view of the short time remaining before the summit. Nevertheless, U.S. officials held informal discussions with their European and NATO counterparts and were able to gain at least tacit agreement before the LTDP was formally proposed to the allies.[79]

Thanks in part to these careful preparations, the initiative got off to a strong start. In his remarks to the other heads of government, Carter called for the development of "a long-term program to strengthen the Alliance's deterrence and defense in the 1980's" that would emphasize greater cooperation to ensure that the combined resources of the alliance were used most effectively.[80] The NATO leaders immediately endorsed the U.S. proposal and asked the defense ministers to prepare such a program in time for a summit meeting the following spring.[81]

The warm reception the U.S. initiative received is also partly explained by its timing, which could not have been more propitious. The NATO defense ministers had first acknowledged the growing danger of a short-warning attack only the previous December, and they reiterated their concern in the biannual ministerial guidance issued following the May summit, which noted that "the Warsaw Pact ground forces have the capabilities to stage a major offensive in Europe without reinforcement."[82] To meet this new threat, the guidance called for a conventional force improvement program that paid particular attention to NATO's ability to respond to an attack after very little warning so as to avoid the need to use nuclear weapons early in a conflict.[83] One way to achieve the necessary improvements, it noted, was better allocation of defense resources, especially through greater alliance cooperation, which "would be greatly facilitated by the establishment of a more comprehensive framework for defence planning incorporating a longer term approach." The document also called for real annual increases in defense spending of about 3 percent, suggesting that further rationalization alone would not reverse the adverse trends in the NATO–Warsaw Pact military balance.[84]

The 1977 ministerial guidance did not explicitly refer to the LTDP. Preparation of the document had begun long before the program was proposed. Nevertheless, the U.S. initiative fit the bill and was immediately seized upon as a way to implement the guidance. The defense ministers identified nine areas in which the need for conventional force improvements was most pressing and directed the NATO ambassadors to prepare time-phased action programs for each of them.[85]

Development and Adoption. To conduct the detailed work of developing the LTDP action programs, the ambassadors established nine separate task forces, which were given considerable independence. Although directed by NATO civilian and military authorities and composed of experts drawn from both NATO and national staffs, the task forces were kept separate from the alliance's formal organizational structure. This arrangement was intended to prevent them from being unduly influenced or inhibited by existing bodies, standard operating procedures, and national instructions. The principal planning constraint was that their proposals reflect the financial resources likely to be available.[86]

The preparation of the task force reports was only the first step in the process of launching the LTDP. Before they could become the bases for the LTDP action programs, the reports had to receive the blessing of each of the NATO countries. This requirement created problems. First, the reports were not completed until March, leaving little time for national consideration. To make matters worse, they tended to be voluminous, packed with detail and substantive proposals and containing many new ideas. Consequently, most national staffs were hard-pressed to absorb and evaluate fully the proposed programs before the planned meetings of the NATO defense ministers and heads of government in May.[87]

The difficulties posed by the limited time available and the great amount of work to be done were compounded by the controversy provoked by the content of the reports. Many proposed measures conflicted with national plans and priorities. As a result, it appeared that the process of evaluating and approving the action programs would have to extend beyond the May summit. To keep the LTDP on track and to maintain its momentum, its proponents agreed to dilute some of the measures in order to obtain high-level approval at the NATO meetings, with the intention of making them more specific later.[88]

Because of these last-minute compromises, the NATO defense ministers were able to approve action programs in all nine areas, which were in turn endorsed by the heads of government in late May. At the same time, the ministers adopted new force goals for the 1979–84 period, which they described as interrelated and complementary with the LTDP.[89] The similarities between the action programs and the force goals stopped there, however. As intended, the LTDP seemed to be much more highly prioritized, containing only 123 measures, whereas there were some 1,300 force goals. The LTDP, moreover, extended planning farther into the future by including actions that would not be fully implemented until the late 1980s and by projecting defense needs into the 1990s. Finally, its nine functional areas cut across national lines whereas the force goals continued to be issued on a country-specific basis.[90]

Refinement, Implementation, and Monitoring. Nevertheless, much work remained to be done before implementation could begin. The price of gaining high-level backing at the May summit had been high. In giving their consent, several countries had inserted reservations about certain provisions or indicated only general agreement, with final approval pending further study and refinement. In some cases, the task forces themselves had called for additional analysis to establish the need for improvements and to define proposed actions.[91] Most of the LTDP measures still lacked detailed implementation plans specifying the quantities required, the precise steps to be taken, timetables, and costs.[92]

To impose some semblance of order, three categories of action were established: agreed defense improvement measures; measures requiring further elaboration to facilitate implementation; and measures in need of further study before any action could be taken.[93] Simply deciding how the action programs would be further refined and how implementation plans would be developed, however, was the subject of yet another intra-alliance struggle following the summit. The United States desired to maintain the LTDP as a separate exercise and to establish additional ad hoc mechanisms for carrying it out. Some of the allies, including Britain, however, sought to redirect the LTDP into the normal force planning process.[94]

The U.S. view prevailed, and the LTDP continued outside the regular NATO channels. In July 1978, special action bodies were formed to spell out each approved measure in greater detail and to develop implementation plans. Later that year, high-level NATO officials were designated as program monitors, who would issue annual reports assessing progress in each of the nine program areas, identifying problems, and suggesting remedial action.[95]

Notwithstanding these notable institutional developments, refinement and elaboration of the original LTDP measures continued into 1979 and beyond.[96] In addition, no formal provision existed for updating the action programs and introducing new measures as further deficiencies in NATO's conventional force posture were identified. Komer had originally envisioned an annual review of the LTDP, much like the regular NATO defense planning system, but such a process was never instituted. The program monitors were able to recommend some additional measures, although not many.[97]

Despite the slowness with which the LTDP was spelled out, the alliance duly began to implement it. Each country was required to report on its progress toward meeting the LTDP goals, highlighting the actions it had programmed or planned to take to implement program measures in its annual response to the NATO Defense Planning Questionnaire. Beginning in the fall of 1979, these country reports were evaluated by the nine program monitors, who submitted comprehensive reports of national efforts in each area to the defense ministers.[98] These assessments were hampered, however, because the program monitors lacked a data baseline against which to measure progress in implementation. In addition, nations often did not describe their efforts in sufficient detail to enable the monitors to render a complete evaluation.[99]

Achievements, Failures, and Shortcomings. The LTDP reached its high-water mark during the last two years of the Carter administration. By December 1979, the majority of its medium-term measures had reportedly been taken into account by national plans. All applicable measures were incorporated into the new force goals adopted in May 1980.[100] At the same time, the defense ministers approved a formal procedure that extended NATO defense planning farther into the future.[101]

Even as the LTDP was gaining momentum, however, other events and developments began to divert NATO's attention and collective energies. The first of these was the debate over long-range theater nuclear forces (LRTNF) in Europe, which culminated in the December 1979 decision to pursue a two-track policy of new deployments and arms control talks.[102] The LRTNF decision was immediately followed by the Soviet invasion of Afghanistan. Although this event prompted the NATO defense ministers to designate a small number of especially urgent LTDP measures for accelerated implementation at their May 1980 meeting,[103] the allies remained deeply divided on the issues of how to respond to the Soviet action and how to compensate for the possible diversion of NATO-committed U.S. reinforcements to the Middle East.[104] These political impediments were compounded by the sharp rise in oil prices in 1979–80, which plunged many Western economies into recession. Subsequently, few countries were able to achieve the goal of 3 percent real increases in defense spending established just two years before.[105]

Thus by the end of the Carter administration, the LTDP's achievements left much to be desired. In December 1980, the NATO ministers acknowledged that "the rate at which improvements were being made was not commensurate with the sustained growth in the Soviet and other Warsaw Pact forces" and that "continued and increased efforts would be needed to maintain the necessary capabilities for deterrence and defense." And a May 1981 congressional study concluded that "overall progress in implementing the LTDP has been slow and disappointing." Indeed, many measures were still written in general terms, requiring further refinement. Even Secretary of Defense Brown conceded just before leaving office that the LTDP and related efforts to rationalize NATO's conventional force posture had yielded few positive results and that "enormous inefficiencies and duplication of effort" among the allies persisted.[106]

The defeat of Carter's reelection bid dealt a further blow to the LTDP. The initiative had been sustained largely by the high-level support of his administration. The president himself had proposed that NATO adopt such a program and had called for the follow-up summit at which the alliance leaders gave it their blessing. Komer's departure from the government, moreover, removed the driving force behind the LTDP at the operational level.

For all of its apparent virtues, especially its emphasis on greater prioritization, cooperation, and extending NATO planning farther into the future, the LTDP suffered from serious shortcomings, which guaranteed that its life would be short and its accomplishments limited. Perhaps most notable is what the LTDP did not try to do, which was to increase the size of NATO's conventional force structure. Instead, its emphasis was almost entirely on force modernization and other qualitative improvements.[107] U.S. planners had hoped that the LTDP would call for the creation of six more allied division-equivalents. Yet the allies indicated plans only to form an additional six reserve brigades, and not before the late 1980s, and the LTDP entailed no commitment to buy enough equipment for the new units.[108]

In other respects, the scope of the LTDP was too broad. Notwithstanding the rhetoric about setting rigorous priorities that surrounded its inception, the program excluded few desired force improvement measures. Thus, although the concern that had prompted the program was the growing Soviet ability to launch an overwhelming conventional blitzkrieg in the Central Region with little warning, the LTDP also included measures intended to strengthen NATO's maritime posture and to improve the air defense of peripheral areas, such as Britain and even Portugal. In short, it seemed to contain something to satisfy virtually every national and service interest.

Even within the program areas that addressed the threat of a massive surprise attack most directly, such as readiness, reinforcement, and reserve mobilization, the LTDP may have attempted to accomplish too much. Although initially it included only 123 measures, many of these were quite broad, encompassing potentially large numbers of specific actions. Building up war reserve stocks of antitank missiles, improving defenses against chemical attack, and modernizing and increasing the number of tactical aircraft, to name but three measures, each implied a variety of often expensive tasks for many of the NATO countries. The number of measures, moreover, continued to expand as the original nine program areas were further refined and totaled more than 150 by 1981.

Consequently, full implementation of the LTDP would have required significant increases in defense spending, notwithstanding its deemphasis of changes in force structure and its underlying philosophy of achieving the necessary force improvements within the level of resources likely to be available. To be sure, many of the proposed measures were of a procedural or organizational nature and could be carried out at little or no additional expense.[109] But the overall package, when costed out by the Pentagon, was found to be extremely expensive.[110]

A final shortcoming was that the LTDP itself provided few mechanisms for institutionalizing long-term planning, cooperation, and prioritization. In most cases, the task of devising new means of enhancing alliance cooperation was left to existing bodies, which diluted the focus that was the initiative's strength. Although medium-term LTDP measures were incorporated into the NATO force goals wherever possible beginning in 1980, the time horizon of the regular force planning process was not extended beyond six years to create a place for longer-term measures. Meanwhile, such innovations in the force planning process as NATO did achieve were not explicitly linked to the LTDP.

U.S. Contributions to Strengthening NATO. Perhaps the most significant improvement in NATO's conventional force posture during this period resulted from a Carter administration initiative to increase the U.S. ability to send reinforcements to Europe rapidly. In 1977, the United States already maintained duplicate sets of equipment on the continent for the bulk of three U.S.-based divisions whose personnel would be flown to Europe in an emergency.[111] Because of shortages of airlift and the shipment of some equipment to Israel during and after the 1973 war, however, the United States would have been able to deploy little more than one reinforcing division within the first ten days after the order to do so was given.[112]

The Carter administration immediately began to correct these deficiencies. And while the LTDP was being developed in early 1978, the administration announced its decision to preposition the equipment for three more heavy divisions during the following several years. When completed—the target date was 1982—the rapid reinforcement program would enable the United States to double its ground forces in Europe to ten full divisions and supporting units in as little as ten days. Plans were also made to triple the number of squadrons of tactical aircraft on the continent, from 28 to 80, in the same period of time.[113]

The buildup of the U.S. reinforcement capability suffered setbacks, however. The Soviet invasion of Afghanistan and the resulting concern about threats to Western access to Persian Gulf oil raised questions about the availability of U.S. reinforcements in a crisis. Although the first additional divisional set of equipment was put in place on schedule in 1980, Congress did not appropriate money for the establishment of the last two division sets until three years later, and even then, funding was made conditional on the achievement of adequate equipment levels in active army units. As late as 1987, the stocks of prepositioned equipment remained incomplete. Thus a decade after the inception of the program, and despite considerable progress, the United States still could not meet the objective of placing ten divisions in Europe within ten days.[114]

The Carter administration also made significant efforts to modernize U.S. conventional forces, especially those stationed in Europe. In 1977 and 1978, the United States added 47,000 modern antitank guided missiles to its NATO inventory and accelerated efforts to upgrade nearly all ground force equipment. American officials estimated that this modernization would increase the capabilities of the U.S. ground forces in Europe by a full one-third between 1978 and 1984.[115] In addition, the United States began to deploy a new generation of combat aircraft, including the F-15, A-10, and F-16, which was expected to reestablish NATO's qualitative lead in the air by the mid-1980s.[116]

These optimistic assessments had to be tempered by the fact that the European allies did not plan to modernize their forces nearly as much. Thus NATO's overall ground force capabilities would increase by only 19 percent by 1984, according to U.S. projections in 1979. This gain was only slightly better than the anticipated 18 percent improvement in Warsaw Pact forces during the same period.[117]

The Reagan Years, 1981–1989

NATO's conventional force posture remained basically unchanged throughout the remainder of the 1980s. Although President Ronald Reagan and his advisers took office at the beginning of the decade committed to strengthening the U.S. military posture across the board after what they called a "decade of neglect" and secured substantial real increases in defense spending in each of the next four years, these efforts had surprisingly little impact on the alliance's conventional capabilities in Central Europe. Compared with its investment in U.S. strategic and naval forces, the Reagan administration paid relatively little attention to NATO's conventional forces.[118] And it asserted much less leadership in this area than had the Carter administration.[119]

The two principal efforts to modify the status quo during this period originated elsewhere. In the early 1980s, several prominent U.S. figures called on the alliance to adopt a doctrine of no first use of nuclear weapons, which would have greatly enhanced the role of conventional forces in NATO strategy. The decade also saw renewed congressional interest in reducing the number of U.S. forces stationed in Europe if the allies did not increase their defense contributions. Both of these challenges were beaten back, although the congressional pressure did stimulate a new comprehensive initiative to improve the alliance's conventional forces. This collective NATO effort, however, was even more modest than its 1970s predecessors, resulted in few concrete achievements, and soon dropped from sight.

DEMISE OF THE LTDP

One of the first casualties of the relatively little attention the United States devoted to NATO's conventional force posture in the 1980s was the LTDP. Although the program had encountered resistance from European allies and NATO's international staff, it had been kept alive by the strong support of Carter and his advisers. In contrast, the incoming Reagan administration, though generally quite bullish on defense, displayed no special enthusiasm for conventional force improvements in Europe. In particular, it had no attachment to the LTDP, which it had inherited from a discredited predecessor.

Consequently, after the installation of the new government in Washington, the LTDP began to fade noticeably. Its functions were gradually assumed by established NATO bodies, and the force improvement measures it had established were increasingly incorporated into the force goals. In 1982, the special emphasis on these measures was dropped, and when the last progress report was issued the following year, the program had ceased to exist.[120]

In the end, the record of the LTDP continued to be mixed. Some headway had been made in measures involving little or no cost. Elsewhere, however, progress toward correcting NATO's deficiencies had been unsatisfactory, especially with regard to measures requiring significant financial contributions.[121] Perhaps the most that could be said of the LTDP was that the allies had devoted more resources to defense than if the program had not existed.[122]

A more lasting legacy of the Carter years was the allied commitment to seek annual real increases in defense spending in the region of 3 percent, which was regularly reaffirmed during the 1980s. This goal, however, proved to be a constant source of intra-alliance friction and recrimination. Beginning in 1982 and continuing through the rest of the decade, most of the European allies consistently failed to achieve increases of the desired magnitude, even though the American defense budget grew at a record peacetime clip through 1986. Moreover, U.S. and NATO studies suggested that sustained growth in spending of 5 percent per year over a period of five years would be needed to achieve a confident conventional force posture.[123] Yet the Reagan administration was reluctant to press the West Europeans to devote considerably more resources to defense than they had already committed themselves to do, in part because it was already placing heavy demands on them to ensure the success of NATO's nuclear modernization plans.[124] Nevertheless, as was true of the LTDP, the very existence of the 3 percent goal probably resulted in higher levels of allied spending than would have been expected otherwise.[125]

The Reagan administration also sustained its predecessor's efforts to enhance U.S. rapid reinforcement capabilities. It continued to seek congressional appropriations for the prepositioning of equipment in Europe, and in April 1982 it successfully concluded negotiations with West Germany to establish a force of 93,000 German reservists that would provide logistics support for U.S. combat units in Europe.[126] Finally, the new administration accelerated the conventional force modernization programs that it had inherited, increasing the rate of weapons procurement by approximately 25 percent. Although not specifically directed at NATO, this acquisition program would help to strengthen Western Europe's defenses.[127]

THE DEBATE OVER NO FIRST USE

Just as the LTDP was winding down, the strategy of flexible response underwent its greatest challenge prior to the dramatic revolutions of 1989–90. In early 1982, four prominent former U.S. officials called upon NATO to adopt a doctrine of no first use (NFU) of nuclear weapons.[128] This proposal touched off a fierce debate over the wisdom of NATO strategy that raged for the next several years in the press and the pages of academic journals. Nevertheless, NATO authorities never seriously considered NFU as an alternative to flexible response.

The proponents of NFU argued that NATO's long-standing reliance on the threat of nuclear retaliation to deter conventional attacks had lost its former utility and had even become counterproductive.[129] First, such threats were no longer credible. The proliferation of nuclear weapons of all types on both sides had made it more difficult than ever to construct rational plans for their use. Second, whatever benefits might still be gained in deterrence were no longer worth the risk of escalation. Because there was no guarantee that even the limited use of nuclear weapons would remain so, the only meaningful firebreak lay between conventional hostilities and any use of nuclear weapons whatsoever. Finally, NATO's reliance on nuclear weapons threatened the very unity of the alliance.

Behind the call for NFU was the familiar belief that an adequate conventional deterrent was within NATO's grasp. Indeed, one of the reasons cited for adopting such a policy was that the alliance's reliance on nuclear weapons actually reduced its conventional capabilities and thus increased the likelihood that they would have to be used. Optimism about the feasibility of an effective conventional defense sprang from a combination of recent developments. New military technologies, such as precision-guided munitions (PGM), seemed to hold the promise of enabling NATO to neutralize the long-standing Soviet advantage in armor at relatively little cost.[130] In combination with advanced systems for target acquisition and long-range delivery platforms, these weapons made possible new, high-leverage tactics, such as attacking Warsaw Pact follow-on forces before they could reach the battlefield.[131] Indeed, the prospective use of sophisticated conventional weapons for deep strikes against columns of vehicles, choke points, and airfields meant that they might someday assume roles that had been traditionally reserved for nuclear weapons.

The opponents of NFU were no less outspoken and ultimately prevailed, if only because it was easier to maintain the status quo than to change it. Like the advocates, they could be found on both sides of the Atlantic, although the strongest reactions were voiced in Europe.[132] As in the 1960s, many Europeans, but especially West Germans, felt that placing greater reliance on conventional forces was undesirable, even if a conventional balance could have been achieved. For them, deterrence of any type of aggression remained the overriding objective. It made little difference if NATO were able to defeat a Warsaw Pact invasion without nuclear weapons in the event that deterrence failed because sustained conventional combat was thought to be just as likely as a nuclear exchange to destroy Europe.

Indeed, the critics argued, excessive reliance on conventional forces could make war even more likely. The threat of unacceptable punishment, which only nuclear weapons could pose, rather than the prospect of being denied any conceivable gains, constituted the most effective deterrent to aggression. Consequently, the Soviet Union had to be constantly confronted with the possibility that nuclear weapons might be used in response to even a conventional attack. If NATO were to raise the nuclear threshold too high or, even worse, to adopt a policy of NFU, however, Soviet decision makers would no longer have to fear that nuclear weapons might be brought into play. They would be better able to calculate the risk of aggression and might conclude that the gains would outweigh the costs. Thus there was no substitute for the deterrent effect of the nuclear threat, however diminished its credibility might be.[133]

Even the most bitter critics of NFU, however, acknowledged the importance of reducing the alliance's dependence on the first use, especially the early first use, of nuclear weapons. Many shared the concern of NFU proponents about the adequacy of NATO's conventional capabilities and their faith in the promise of new military technologies, and most advocated strengthening the alliance's conventional forces. Thus the debate over NFU helped to set the stage for a further force improvement effort.

CONGRESSIONAL CHALLENGES

The NFU challenge to NATO strategy was paralleled by a revival of congressional sentiment in favor of reducing the U.S. military contribution to NATO, which had been dormant since the early to mid-1970s.[134] Many members of Congress once again began to express the view that the United States was bearing a disproportionately large share of the burden of Western security. After all, the United States was spending approximately twice as much of its gross national product on defense as were its allies—6 percent versus an average of 3 percent—and the U.S. share was increasing as a result of the Carter-Reagan buildup. If the Europeans were unwilling to do more in their own defense, why should the United States continue to station so many troops on the continent?

The resentment fostered by the perception that the defense burden was not being shared equitably was compounded by sharp intra-alliance differences over East-West relations in the early 1980s. On one issue after another—Afghanistan, Poland, and the Siberian gas pipeline—the United States and its major European allies pursued conflicting policies toward the Soviet Union. In general, the United States emphasized taking a hard line on what it viewed as unacceptable Soviet behavior while the allies were anxious to preserve as many of the benefits of detente as possible.[135]

Congress began to reassert itself on the issue in 1982, when the Senate Appropriations Committee voted to reduce the number of U.S. military personnel in Europe by 19,000 the following year. Although that particular measure went no further, Congress did establish a ceiling of 316,000 on the number of American troops that could be stationed on the continent, and it continued to refuse to appropriate the money necessary to complete the prepositioning of additional equipment for U.S. reinforcements that had begun under Carter.[136]

The most serious congressional challenge occurred two years later, when Senator Sam Nunn introduced an amendment mandating the withdrawal of 90,000 U.S. troops from Europe over three years unless the allies achieved 3 percent real annual growth in defense spending. The withdrawal could be avoided, or at least limited, if the allies increased their stockpiles of ammunition and the number of airfields and hardened aircraft shelters in Europe sufficiently and otherwise took significant measures to improve their conventional capabilities.[137]

The significance of the proposal was underscored by its authorship. Nunn had been a staunch supporter of NATO throughout his years in the Senate and had played a leading role in the defeat of the 1973 Mansfield amendment. In fact, Nunn viewed his purpose as entirely constructive. In contrast to the Mansfield amendments, which had simply sought to reduce the number of U.S. troops deployed overseas, his proposal was intended "to prod the Allies into doing more, and thereby establish conditions in which American troops would contribute to a sustainable conventional defence."[138] Nevertheless, the amendment provoked considerable consternation in Europe, where it was seen as an unacceptably heavy-handed attempt to dictate the terms of intra-alliance relations.

The 1984 Nunn amendment was eventually defeated, although by a narrow margin. Instead, the Senate passed by a nearly unanimous vote a version lacking the troop withdrawal clause but placing a permanent cap of 326,000 on the number of American servicemen stationed in Europe. And later that year, the Senate Appropriations Committee established a separate ceiling of 90,000 for air force personnel.[139] Nevertheless, the threat of legislated U.S. troop reductions had once again subsided. Although congressional interest in the adequacy of NATO's conventional forces and the burden-sharing issue continued, Congress chose to play a more low-key role through the end of the decade.[140]

THE CONVENTIONAL DEFENSE IMPROVEMENT INITIATIVE

Although congressional disgruntlement did not result in any U.S. troop reductions, it did help to stimulate NATO's most comprehensive effort to strengthen its conventional force posture in the 1980s, the Conventional Defense Improvement (CDI) initiative. Congressional pressure was not the only factor, however. Also important was the renewed NATO consensus on the need to bolster the alliance's conventional capabilities to strengthen deterrence and to reduce NATO's dependence on the early use of nuclear weapons. Despite the LTDP and other efforts, the conventional balance in Europe had continued to deteriorate because of steady improvements in Warsaw Pact forces.[141] As a result, a growing number of officials on both sides of the Atlantic reached the conclusion that in the event of a major conventional attack, the alliance would quickly be faced with a decision to escalate to the nuclear level.[142]

The CDI grew out of a 1984 Pentagon report, "Improving NATO's Conventional Capabilities," that was prepared in response to a congressional request the previous year and briefed to the allies that summer. The report concluded that flexible response continued to provide a sound basis for meeting the alliance's security requirements. Because of the worsening military balance in Europe, however, NATO would have to strengthen its conventional forces substantially if the strategy were to remain effective.[143] The report went on to outline suggestions for achieving the necessary improvements.

Although American officials were finally willing to take the lead, they were anxious to avoid the impression that the United States was once again imposing its own agenda on the alliance. Thus they tried to obscure the U.S. role and sought to make the origins of the initiative appear as multilateral as possible. A willing partner was quickly found in West Germany, which agreed to help spearhead the effort.[144] This low-profile approach seemed to pay off in December 1984, when the NATO defense ministers asked the secretary general and the NATO ambassadors to develop proposals for a comprehensive effort to improve the alliance's conventional defenses.[145]

During the following five months, the NATO International Staff and the national delegations to NATO hastily devised an action plan, which was adopted by the defense ministers in May 1985. The resulting CDI consisted of two related approaches for strengthening the alliance's conventional forces.[146] Procedurally, it called for improving NATO defense planning, both through better coordination among the various planning bodies and by extending more planning activities farther into the future. Substantively, it proposed to eliminate critical deficiencies in the alliance's conventional capabilities in nine key areas that had been identified during the preparation of the initiative.[147] The NATO commanders were invited to "highlight" the force goals that they considered most relevant to the agreed areas of deficiency, while each country committed itself to make a "special effort" to implement the highlighted force goals.

Effectively, the CDI sought to address many of the same problems that had been the concern of the LTDP.[148] Even by the standards of the LTDP, however, the objectives of the CDI were extremely modest. It did not establish any new force goals, let alone call for any additional forces. Moreover, it contained no provision for increasing the amount of resources devoted to defense. Instead, it was exclusively concerned with making better use of the resources that were already available through greater prioritization and coordination. Even then, some of the most promising low-cost measures for improving NATO's conventional strength, such as correcting maldeployment, making better use of reserves, terrain enhancement, and national role specialization, were excluded. European allies once again claimed that they could not afford to equip additional reserve brigades. And the German government vetoed proposals for building defensive positions and otherwise preparing the terrain along the inter-German border in peacetime to strengthen the defense, for fear that such efforts would stir a public outcry or symbolize the permanent division of the country.[149] Finally, and in sharp contrast to the LTDP, the CDI was intended to take place entirely within the alliance's existing force planning process.[150]

Even when measured against its own modest objectives, the initiative enjoyed only limited success. Between 1986 and 1988, only the United States, Britain, West Germany, and the Netherlands implemented more than half of their highlighted force goals, and these countries were already fulfilling most of their NATO requirements. The remaining countries performed poorly. Thus the CDI appeared to have little positive impact,[151] and by the summer of 1987 the Pentagon was already looking for ways to revitalize the effort.

The shortcomings of the CDI were strikingly similar to those that had hobbled the LTDP. First, the agreed areas of critical deficiency were too broad. Consequently, almost any force goal could be construed as relevant. Second, too many force goals were highlighted—as many as one-third of the total in some cases—while no effort was made to calculate the cost of implementation. As a result, some countries, particularly the smaller ones, were asked to make efforts that were far beyond their means.

The experience of the CDI must nevertheless be placed in the broader context of the times. Overall, NATO's conventional force posture received relatively little attention during the 1980s, when it was constantly overshadowed by other defense issues. In the early years of the decade, the alliance was preoccupied with the deployment of intermediate-range nuclear forces (INF) in Europe, the question of how to respond to new threats in the Persian Gulf, and its own internal conflicts. In the mid-1980s, these issues were supplanted by new concerns, including strategic defenses, U.S.-Soviet summitry, and the negotiations leading to the INF treaty. And as the decade moved to a close, the need for significant conventional force improvements was called into question when the Soviet Union and its Warsaw Pact allies suddenly displayed a willingness to scale back their forces substantially through a combination of unilateral and negotiated reductions.

Analysis

How are we to explain the high degree of stability that characterized NATO's conventional force posture during the two decades following the adoption of flexible response? Why was there no further revision of the strategic concept? Why did so little change occur in the alliance's conventional force levels in the Central Region?

From the balance-of-power perspective, this stability is somewhat problematic. From the late 1960s until the early 1980s, substantial increases in Soviet nuclear and conventional capabilities created pressure for a further significant modification of NATO's conventional force posture. The Soviet Union's achievement of strategic nuclear parity and its deployment, beginning in the late 1970s, of a much more capable generation of theater nuclear forces appeared to set even more restrictive limits on the degree to which NATO could rely on the threat of nuclear retaliation to deter non-nuclear forms of aggression. Consequently, many observers concluded that NATO should strengthen its conventional forces to compensate for the diminished utility of nuclear weapons, and some even advocated revising the alliance's military strategy to reflect the new situation. Conventional force improvements were made even more imperative by the striking growth of Soviet conventional capabilities in Eastern Europe during the same period.

Concerns about a deterioration of the East-West military balance were reinforced by increasing Soviet activism in the Third World, which suggested a greater Soviet propensity to take risks. This trend culminated with the invasion of Afghanistan, which represented the most brazen use of force by the Soviet Union outside its own borders since its intervention in Czechoslovakia more than a decade before.

Of course, until the late 1970s, the doubts that these events inspired about Soviet intentions were largely counterbalanced by the lingering atmosphere of detente produced by the striking improvements in East-West relations of the late 1960s and early 1970s. The series of negotiations and agreements leading up to the 1975 Helsinki Accords and the widespread feeling that these resulted in a true relaxation of tension in Europe weakened the case for NATO conventional force improvements.[152]

Efforts to strengthen the alliance's conventional capabilities were further hampered by recurring global economic crises, which spared few NATO members during this period and reduced the alliance's military potential. The first oil shock in the mid-1970s put pressure on countries to reduce their defense budgets, and the economic downturn of the early 1980s made it all but impossible for most to achieve even the modest objective of annual 3 percent real increases in defense spending that accompanied the LTDP.

Nevertheless, given the magnitude of the increases in Soviet capabilities that took place in the late 1960s and the 1970s, it would not have been unreasonable to expect a more substantial NATO response than actually occurred. Indeed, the alliance would seem to have reacted less forcefully to adverse shifts in the military balance than it had at any previous time. Thus while balance-of-power considerations may explain the recurrence of attempts to strengthen NATO's conventional forces during this period, they cannot account for the persistent failure of these efforts.

The intra-alliance perspective is more helpful in this regard. Force improvement efforts were hamstrung because the NATO countries remained deeply divided on several fundamental issues.[153] As a result, they found it virtually impossible to reach a consensus in support of any course of action that departed significantly from the status quo. For example, the allies still differed in their assessments of Warsaw Pact capabilities and thus the feasibility of conventional defense. Although the United States once again argued that a rough balance of forces already existed on the continent, the Germans in particular believed that Western Europe could not be defended against a large-scale attack.

More important, the allies continued to disagree on the desirability of placing greater reliance on conventional forces, either formally or implicitly by strengthening NATO's non-nuclear capabilities, although the debate was much more muted than it had been in the 1960s. In combination with flexible response's high degree of ambiguity, such differences ruled out any revision of the strategic concept. As before, the United States stressed the importance of being able to fight conventionally for a prolonged period, while the Europeans tended to emphasize the threat of deliberate escalation to the nuclear level, a move they felt would become necessary soon after the outbreak of hostilities in any event. Although these conflicting strategic preferences can be understood largely on the basis of geographical differences, the fact that the burden of conventional force improvements would have fallen most heavily on the Europeans almost certainly played a role in shaping them as well.

Beginning in the late 1970s, NATO force improvement efforts were further complicated by increasingly divergent allied views of Soviet intentions and how to deal with the Soviet Union more generally.[154] The United States, which took a more global view, believed that Soviet activities in the Middle East, Africa, and Afghanistan demanded a firm Western response across the board. The Europeans, in contrast, sought to differentiate between Soviet behavior in other regions of the world and in Europe. From their more regional perspective, the Soviet invasion of Afghanistan, though deplorable, did not necessarily imply a greater risk of war on the continent. Consequently, they resisted responses to Soviet actions elsewhere that threatened to increase the level of tension in Europe.

This difference in approach reflected an asymmetry in relations with the Soviet Union and Eastern Europe that had developed as a result of detente and, especially, West Germany's Ostpolitik. By virtue of their geographical proximity and historical ties, West Germany and other European NATO countries were fashioning an increasingly dense set of economic and human links with their Warsaw Pact counterparts by the late 1970s. Consequently, they stood to lose much more than the United States did from a renewed cooling of East-West relations.[155]

Institutional factors also contributed to the stability of NATO's conventional force posture during this period. In the early 1970s, the obligation to maintain national force contributions at existing levels was reinforced by the initiation of the MBFR talks. Henceforth, any unilateral reduction could be portrayed as weakening the West's bargaining position and thus reducing the chances of achieving a successful outcome. This argument was used by the Nixon administration especially effectively against the successive Mansfield amendments. In addition, independent NATO authorities, notably SACEUR Goodpaster and Secretary General Brosio, played key roles in heading off the force reductions that had been mooted in 1969 and 1970. Years later, the NATO bureaucracy demonstrated its ability to emasculate attempts to strengthen the alliance's conventional forces as well when it resisted the procedural and organizational innovations of the LTDP.

The tendency of the NATO countries not to alter their force contributions was reinforced in some cases by the domestic and cognitive roots that the norm of preserving the status quo had acquired. Although the U.S. Congress imposed an additional and more explicit cap on the size of the American military presence in Europe in the early 1980s, it proved unwilling on several occasions seriously to challenge the administration's prerogative to maintain these forces at their existing levels. Continuity was also fostered by the rigid allocation of budgetary shares in certain countries that reflected a careful balancing of competing interests and priorities. The difficulty of quickly altering levels of defense spending was especially acute in West Germany, where firm budget ceilings were established years in advance.[156]

Finally, attempts to alter NATO's conventional force structure were frequently stymied and constrained by deeply held beliefs. The strongly negative attitudes of administration officials and a substantial bipartisan group of supporters toward the Mansfield amendments were as predictable as their grimly deterministic views of the likely consequences of unilateral cuts.[157] During this period, moreover, U.S. efforts to strengthen the alliance's conventional capabilities were increasingly hindered by the assumption that ambitious initiatives would be doomed to failure. Thus U.S. proposals became ever more modest. That there would be no increase in the basic size of each country's contribution was taken for granted; qualitative force improvements were all that could be expected. The LTDP at least attempted to establish new institutional mechanisms to facilitate such improvements, but the CDI took place entirely within the existing force planning system. Although this assumption may have had a strong basis in reality, U.S. officials would never know for certain how much NATO's conventional forces could be strengthened if it was never challenged.

CHAPTER 7

Explaining the Evolution
of NATO's Conventional
Force Posture, 1949–1989

A central theme of the preceding chapters is that through the late 1980s, NATO's conventional force posture had become increasingly stable over time. Until the upheavals that ended the Cold War in Europe, NATO's formal strategic concept had not changed in more than two decades. The differences between the strategy of flexible response and its predecessor, MC 14/2, which had been adopted ten years before, moreover, were not as great as has often been suggested. In contrast, NATO strategy and the role it defined for conventional forces were volatile during the early years of the alliance, undergoing significant modification on several occasions during the 1950s.

A similar pattern characterized the alliance's conventional capabilities. During the 1950s, NATO conventional force levels in the Central Region fluctuated dramatically. During the following three decades, however, they were remarkably stable, especially following the completion of West German rearmament in the early 1960s. From the mid-1960s, NATO fielded between 26 and 28 nominal divisions and between 71 and 78 active brigades in Germany and the low countries in all but a few years. During the same period, the number of active NATO military personnel in the region remained between 1.0 million and 1.1 million. Thus these key indices of overall conventional strength varied for the most part within a range of less than 10 percent. The individual contributions of each of the countries with forces in the region were almost as stable[1] (see Table 7.1).

Given these similarities, one might expect that the two aspects of NATO's conventional force posture emphasized in this book, strategy and force structure, would lend themselves to similar explana-

TABLE 7.1
NATO Force Levels in the Central Region, 1955–1990

Units	United States	United Kingdom	France	Germany	Belgium	Nether- lands	Canada	Total
				1955				
Divs	5⅔	4	3	0	3	1	⅓	16⅓
Bdes	18	12	8	0	9	3	1	51
MP	261	107	n/a	20	n/a	n/a	n/a	n/a
				1960				
Divs	5⅔	3	2	7	2	2	⅓	21⅓
Bdes	18	7	6	19	6	5	1	62
MP	237	64	72	270	120	135	9	907
				1965				
Divs	5⅔	3	2	12	2	2	⅓	26⅓
Bdes	18	6	6	33	6	6	1	76
MP	262	60	72	441	107	135	12	1,089
				1970				
Divs	4⅔	3	2	12	2	2	⅓	25⅓
Bdes	15	5	6	33	4	6	1	70
MP	213	60	62	466	95	121	5	1,022
				1975				
Divs	4⅓	3	2	12	2	2	⅓	25⅔
Bdes	16	6	6	36	4	6	1	75
MP	218	64	58	495	87	112	5	1,039
				1980				
Divs	4⅔	4	3	12	2	2	⅓	27⅔
Bdes	17	8	6	36	4	6	1	78
MP	244	66	47	495	88	115	5	1,060
				1985				
Divs	4⅔	3	3	12	2	2	⅓	26⅔
Bdes	17	8	6	36	4	6	1	78
MP	246	66	48	478	92	106	5	1,041
				1990				
Divs	4⅓	3	3	12	2	2	⅓	26⅔
Bdes	16	8	6	36	4	6	1	77
MP	244	64	53	461	92	103	7	1,024

SOURCES: IISS, *Military Balance* (various years); Golden, *Dynamics of Change*, 126; Harrison, *Reluctant Ally*, 153; *Congressional Record*, 90th Cong., 1st sess., vol. 12, pt. 1, p. 1003; Yost, *France and Conventional Defense*, 22; Ruiz Palmer, "National Contributions," 27ff.; Honig, *Defense Policy in the North Atlantic Alliance*, 41; and DC(56)15, "Costs of British Forces in Germany," June 11, 1956, PRO, CAB 131/17.

NOTE: MP = total military personnel (×1000), n/a = not available. Figures do not include reserve or territorial forces. U.S. figures are for forces in Germany only. The number of U.S. troops in the low countries, however, has never exceeded a few thousand.

After 1980, the brigade level of organization was eliminated in French divisions. Because the overall size of the French contingent in Germany remained roughly constant, however, it is treated as consisting of six brigades during this period.

tions. Consideration of the relationship between them would point to the same conclusion. As noted in Chapter 1, alliance strategy and actual conventional capabilities are logically linked. The force requirements derived from the strategy may be expected to exert at least some influence over the decisions made by alliance members concerning the number of forces to provide. They may engender feelings of obligation, or, by shaping expectations, they may result in allied pressure to honor them. Conversely, member states will actively seek to shape strategy in ways that help to justify the forces they are willing or able to provide, while the demonstrated failure of states to generate the required forces will create pressure for strategic adjustment.

There are also good reasons for expecting somewhat different explanations, however. Despite their interrelationship, alliance strategy and force levels are determined by distinct political processes and are subject to divergent pressures and constraints. NATO's formal military strategy has had to be approved by all members. The alliance's overall force structure, in contrast, has been simply the amalgamation of individual countries' contributions, which have ultimately been determined on a national basis. In addition, the provision of conventional forces has entailed material sacrifices and opportunity costs that the articulation of new strategies has not.

Indeed, this study finds that to explain these two aspects of NATO's conventional force posture, one must draw upon the three perspectives in different ways. NATO strategy and the changing role of conventional forces in it can be largely understood in terms of the balance-of-power and intra-alliance perspectives. A satisfactory account of the evolution of NATO conventional force levels in the Central Region, however, must also make use of the institutional perspective.

Explaining the Evolution of NATO Military Strategy

During NATO's first four decades of existence, the alliance's formal military strategy passed through at least four distinct phases. The principal objectives of NATO military preparations remained virtually constant: to preserve the political integrity and territorial status quo of Western Europe by preventing Soviet political intimidation, deterring aggression, and defending the region by force of arms if necessary. But each of the other main elements of NATO strategy—its portrayal of the military threat, the indicated military responses, and force requirements—varied over the years, often substantially.

Such changes, however, occurred with declining frequency. If the abrupt revision of threat assessments that accompanied the outbreak of the Korean War is included, NATO strategy was revised three times during the alliance's first decade. In contrast, between the late 1950s and the beginning of the post–Cold War era, the strategic concept was modified only once, when flexible response was adopted in 1967. At the same time, NATO strategy became increasingly ambiguous. A significant degree of ambiguity was first injected into the strategic concept in 1957, when MC 14/2 was approved, and flexible response further broadened the range of interpretations that could be placed on it.

The evolution of NATO military strategy and the role it has assigned to conventional forces can be largely understood in terms of the variables emphasized by the balance-of-power and intra-alliance perspectives. In this case, there is little need to invoke the institutional perspective to provide a satisfactory explanation.[2]

CONTRIBUTIONS OF THE BALANCE-OF-POWER PERSPECTIVE

Consideration of the balance-of-power perspective is essential to account for the rapid changes in NATO military strategy that occurred during the early years of the alliance. This perspective also provides a useful starting point for understanding the subsequent lack of variation in the strategic concept, although it alone is not sufficient to provide a complete explanation.

During NATO's first year of existence, the threat posed by the Soviet Union was viewed primarily as a political one. Although the Soviets appeared to enjoy a preponderance of military power on the continent, armed aggression did not seem imminent. Instead, the Soviet Union was regarded as largely content to pursue its objective of greater political influence in Western Europe through a combination of external pressure and internal subversion designed to exploit the political and economic weaknesses of the countries of the region.

Accordingly, the purpose of the alliance initially was defined in political rather than military terms. The immediate aim was to create an atmosphere of security in Western Europe so that the people of the region could approach the task of postwar reconstruction with confidence.[3] In the short term, this objective would largely be served simply by associating the substantial military power and war-making potential of the United States with Western Europe through the commitment embodied in the North Atlantic Treaty. Because the threat of retaliation inherent in the U.S. atomic arsenal was widely regarded as sufficient to deter any conceivable Soviet act of aggression, the process of preparing detailed military plans for the defense of the region was undertaken with little sense of urgency. Consequently, the role of conventional forces in NATO strategy was defined only in general terms, and the alliance's force requirements remained to be determined.

The initial inattention to the details of NATO's military posture ended abruptly after the Soviet demonstration of a nuclear capability in the second half of 1949 and the outbreak of the Korean War the following summer. The unexpected Soviet atomic test signaled the end of the U.S. nuclear monopoly and suggested that as the Soviet arsenal grew the United States would be increasingly less able to deter aggression in Europe with the threat of nuclear retaliation. More important, these two events were widely seen as heralding a sharp change in Soviet intentions. Even before the Korean War began, some U.S. officials had concluded that the acquisition of atomic weapons would embolden the Soviet Union to use force to achieve its objectives. The outbreak of the war dispelled most remaining doubts about the immediacy of the Soviet military threat to Western Europe. The North Korean invasion was widely regarded in the West as Soviet-inspired, and many unsettling parallels could be drawn between Korea and Germany.

These changes in Western estimates of Soviet capabilities and intentions caused NATO authorities to take the possibility of aggression in Western Europe much more seriously. Visible preparations for the defense of the region were now regarded as indispensable to deter a Soviet attack while maintaining the confidence of the European allies and their support for the alliance. Although strategic nuclear strikes might help retard a Soviet advance, the defense of Western Europe would depend primarily on conventional forces. Because any future Soviet attack was expected to be all-out, moreover, substantial conventional capabilities would be necessary. NATO planners subsequently calculated that in the Central Region alone, the alliance would need more than 30 fully ready divisions and a total of some 65 divisions within a month of the outbreak of hostilities.

The next shift in NATO strategy can also be explained largely through the balance-of-power perspective. This time, however, the change was not primarily the result of revised estimates of Soviet capabilities and intentions. That the Korean War was not followed by a Soviet move made the eruption of hostilities in Europe seem increasingly unlikely, and Stalin's death in 1953 raised hopes that an East-West accommodation might be reached. Yet Soviet aggression remained plausible enough to demand a high degree of military preparation, in contrast to the first year of the alliance. Any future war, moreover, was still expected to involve all-out conventional and nuclear attacks from the outset.

More important were the changes that had occurred in Western leaders' assessments of the military capabilities likely to be available to NATO and thus the range of military responses that would be feasible. By 1953, the European allies had demonstrated their inability to provide the conventional forces estimated to be necessary to conduct a successful defense of Western Europe. The rearmament efforts that had been undertaken following the outbreak of the Korean War had imposed tremendous strains on the economies of many European countries, perpetuating economic weakness and, in some cases, even causing living standards to fall, as some critics had feared. The resulting gap between conventional force requirements and actual capabilities generated considerable pressure for the adoption of a new strategy that would be more compatible with the alliance's military potential.

Notwithstanding this pressure, the revision of NATO strategy in 1954 might not have taken place but for a further development emphasized by the balance-of-power perspective: the availability of tactical nuclear weapons in increasing numbers for use on the battlefield. By that time, it was widely believed in both the United States and Europe that these new weapons could be substituted to a substantial degree for conventional forces. As a result, NATO would no longer have to rely primarily on conventional forces to defend Western Europe, and their role in NATO strategy could be reduced accordingly.[4]

This change in how NATO would respond to Soviet aggression, which was embodied in MC 48, in turn precipitated a downward revision of the alliance's conventional force requirements. Because tactical nuclear weapons would surely be used, any future war was expected to be shorter and more intense than before. Consequently, NATO planners estimated that the number of reserve units needed to implement the strategy could be reduced substantially. They also concluded, however, that there could be no relaxation of the requirement for ready divisions. Substantial in-place conventional ground forces would continue to be necessary, if only to make the use of tactical nuclear weapons more effective by compelling the Soviet Union to concentrate its forces before an attack. In particular, the long-awaited German contribution of twelve divisions, which had been delayed by the impasse over the European Defense Community, was still deemed vital.

The last two revisions of NATO strategy can be largely attributed to the growth of Soviet nuclear capabilities. Although the possibility of a general nuclear war could not be excluded, by the late 1950s such an eventuality was regarded as increasingly improbable. Because of the tremendous destructive power of thermonuclear weapons, which were becoming available in large numbers, Soviet leaders were expected to be especially anxious to avoid total war. At the same time, however, they would be likely to assume that the United States would itself be increasingly reluctant to use nuclear weapons in response to minor provocations because the Soviet Union had developed long-range delivery vehicles capable of striking American territory. In short, Soviet leaders might perceive both the need and the opportunity to pursue their objectives through lesser military means. Consequently, NATO leaders expected Soviet political intimidation intended to exploit Western fears of nuclear war and limited forms of Soviet-sponsored aggression to be more likely than an all-out attack.

As a result of these altered strategic conditions, it became necessary, well before the adoption of MC 14/3, to modify NATO's formal military strategy to allow for a broader range of possible military responses than was provided in MC 48. Although the ability to make an all-out response with both conventional and nuclear forces was still deemed vital, prompt strategic retaliation would no longer be appropriate in all cases. Such a reaction could needlessly turn what might otherwise be a minor conflict into a full-scale war. Thus the new strategic concept, MC 14/2, indicated that the alliance should also be prepared to make limited responses confined to the local forces in Western Europe. But because even the use of tactical nuclear weapons might not always be advisable, a place was created for purely conventional responses in certain circumstances. In short, the need for a wider variety of military options translated into greater reliance on conventional forces and a larger role for them in NATO strategy. During the following years, this need increased as Soviet nuclear capabilities continued to grow, setting the stage for the development of flexible response.

Although both MC 14/2 and MC 14/3 sought to address a variety of possible forms of aggression, in contrast to earlier versions of NATO strategy, they differed in ways that can also be attributed to balance-of-power considerations. Most important, Western perceptions of Soviet intentions had shifted substantially from the late 1950s to the late 1960s. The adoption of MC 14/2 followed the Soviet intervention in Hungary, which, like the outbreak of the Korean War, was interpreted as suggesting greater Soviet willingness to use force. In contrast, the development of flexible response occurred after the beginning of East-West detente, by which time the USSR seemed increasingly content simply to maintain the political and territorial status quo in Europe.

Consequently, MC 14/3 took a different view of the military threat. The possibility of deliberate aggression seemed even more remote to NATO leaders than it had a decade before. They now thought conflict was more likely to arise from some unexpected event or miscalculation during a crisis. Thus an all-out surprise attack, whether conventional or nuclear, was regarded as extremely improbable. Instead, flexible response placed more emphasis on the deterrence of limited and especially conventional forms of aggression than had its predecessor, which had continued to stress the danger of general nuclear war.

Consonant with its revised view of the threat, MC 14/3 gave somewhat greater prominence to limited, and especially conventional, responses designed to deny the Soviet Union any military advantage while deemphasizing all-out retaliatory responses largely intended to punish aggression. It also provided much less guidance on force requirements. Following the adoption of MC 14/2, NATO military planners had developed a detailed statement of requirements that were closely attuned to Soviet capabilities. Flexible response, in contrast, was predicated on the assumption that any future hostilities were likely to be preceded by a prolonged period of political warning, during which time the alliance could mobilize sufficient forces to deter or, if necessary, defeat an attack. As a result, less importance could be attached to ensuring that the required forces were ready and in place at all times.

Finally, the longevity of flexible response can be ascribed in part to balance-of-power influences. In short, it could be argued that NATO strategy did not change for more than two decades because there was little pressure for adjustment. By the late 1960s, the Soviet military posture had largely matured. The Soviet Union had achieved a strong second-strike nuclear capability. Similarly, Western perceptions of Soviet intentions, at least in Europe, had settled into a steady state characterized by varying degrees of mistrust but not alarm. Even when the Soviet Union pursued adventuresome policies elsewhere in the world, it seemed content to maintain the status quo on the continent, as indicated by Soviet interest in a European security conference and the resulting 1975 Helsinki Accords. By this time, moreover, changes in military technology of the magnitude that had characterized the first decade of the postwar era seemed unlikely. The technologies associated with nuclear weapons and their means of delivery in particular were nearing the top of the learning curve.

By itself, however, the balance-of-power perspective does not fully account for the evolution of NATO strategy. For example, an exclusive focus on the causal factors emphasized by this perspective might have led one to expect some subsequent variation in NATO strategy, especially in response to the dramatic increase in the strategic capabilities of the Soviet Union that took place in the 1970s and 1980s. The growth in the number of Soviet strategic delivery vehicles and the proliferation of Soviet warheads appeared to neutralize the U.S. nuclear advantage that had still seemed to exist at the time flexible response was adopted and upon which the U.S. threat to escalate to the strategic level if necessary was predicated. Second, the balance-of-power perspective is unable to account for the growing ambiguity in NATO strategy. For explanations of these puzzles, we must turn to the intra-alliance perspective.

CONTRIBUTIONS OF THE INTRA-ALLIANCE PERSPECTIVE

The intra-alliance perspective also helps to account for the increasing stability of NATO strategy over time. In particular, this focus suggests how diverging strategic preferences made agreement on a new strategic concept ever more difficult to achieve. As a result, there was little choice but to retain the old strategy, even though it came to seem increasingly out of step with external strategic circumstances.

The intra-alliance perspective also explains the growing ambiguity of NATO strategy. Beginning with MC 14/2, the strategic concept failed to provide clear answers to several questions with important implications for conventional force planning. Under what circumstances would NATO make only a limited response to aggression? When would limited responses be confined to the use of conventional forces? In the event that a limited conventional response was unsuccessful, at what point would a decision be made to escalate to the nuclear level? In brief, an element of ambiguity became necessary to bridge a widening gap in the strategic preferences held by the allies. This ambiguity in turn contributed to the lack of change in NATO strategy. As the strategic concept became more indeterminate, its modification became increasingly unnecessary because it could accommodate an expanding range of interpretations.

Until the mid-1950s, the strategic preferences of the NATO countries were largely identical. Each of the allies sought to place as much emphasis as possible in NATO strategy on nuclear weapons, first strategic and then tactical, once the latter became available. In addition to sharing the view that nuclear weapons offered the most effective deterrent, the allies were universally eager to reduce the cost of NATO military preparations by limiting the role of conventional forces, which were regarded as relatively expensive, to the greatest extent possible. Although for lack of alternatives, NATO's initial strategy called for substantial conventional forces to defend Western Europe in the event of an attack, it was widely hoped by 1953 that the availability of tactical nuclear weapons would reduce NATO's conventional force requirements to a minimum.

The allies' strategic preferences began to diverge markedly only in the mid- to late 1950s. These differences were rooted in geography, which ensured that the allies would be unevenly exposed to the threat. For the Europeans, and especially the West Germans, who were on the front line, any war would assume near total proportions. Consequently, these allies were anxious to minimize the possibility that hostilities of any kind would occur. For the United States, however, which was separated from Europe by a wide oceanic expanse, not all wars would have the same consequences. In particular, an all-out nuclear war would be much more devastating than hostilities confined to the continent. Hence it was only natural for American leaders to give roughly equal weight to the potentially conflicting goals of deterring war and keeping any military action limited should deterrence fail.

During the first decade of NATO's existence, these geographically rooted differences in perspective remained largely dormant and did not have to be addressed. As long as the Soviet Union lacked the ability to devastate the United States with nuclear weapons, any war would remain confined to Europe, and U.S. decision makers would not have to confront the dilemma posed above. The costs and potential risks to the United States of a high degree of nuclear reliance seemed small compared with the benefits. The United States could threaten nuclear retaliation with relative impunity.

Developments in military technology soon exposed this underlying cleavage within the alliance. When it became clear in the late 1950s that the Soviet Union was acquiring a large strategic nuclear retaliatory capability, the old cost-benefit calculations were revised. The fundamental differences in perspective began to translate into conflicting strategic preferences as the United States and its allies sought to shift the potential costs and risks of NATO strategy onto each other.

As the United States became vulnerable to Soviet nuclear attack, U.S. officials increasingly emphasized the importance of being able to make limited and even non-nuclear responses to aggression in Europe. They regarded such threats as highly credible, and to the degree that NATO could deny an aggressor any possible gains, war would become an unattractive option. This approach had the added advantage of making it possible to keep any conflict limited, should aggression nevertheless occur. In contrast, the threat of immediate U.S. nuclear retaliation in response to virtually any form of aggression, as mandated by MC 48, was viewed in Washington as increasingly incredible because carrying out the threat was likely to result in the devastation of the United States. Thus continued reliance on such threats would make conflict at all levels—not just general nuclear war—more likely.

Because their overriding concern was to avoid war of any kind, the West Europeans continued to emphasize the threat of an all-out response to aggression. From their perspective, effective deterrence required making the consequences of war appear as terrifying as possible to the Soviet Union to ensure that the costs of aggression would outweigh any possible gains. In contrast, the suggestion that NATO might limit its response to conventional forces would reduce the potential costs and risks of aggression as calculated by the Soviet Union and thus was bound to make limited conflicts more likely.

This view was not held by all the European allies with equal intensity at all times. For a short period after it formally joined the alliance in 1955, West Germany was also a proponent of limited options. The Germans had not been involved in the drafting of MC 48, and they feared that the strategy's emphasis on tactical nuclear weapons would be used to justify U.S. troop withdrawals while guaranteeing the destruction of their country in the event of even a minor military engagement. Thus they initially sought to build greater flexibility into NATO strategy, providing considerable impetus for the adoption of MC 14/2.

Later, in the early 1960s, the West Europeans parted company over the form that NATO's initial nuclear response should take. The West Germans favored the early use of tactical nuclear weapons, while the French insisted on retaliating with strategic nuclear forces from the outset. Nevertheless, from the late 1950s, the principal schism over the proper military strategy for NATO was between Western Europe and the United States.[5] While American leaders hoped to be able to postpone a decision to use nuclear weapons as long as possible in the event of hostilities, the European allies generally preferred to emphasize the threat to make a nuclear response very early—possibly immediately—in a conflict.

These fundamental differences in strategic preferences were reinforced by several other intra-alliance factors. One was the uneven distribution of nuclear capabilities within the alliance. The United States maintained a substantial lead over its allies in the technology of nuclear weapons and delivery systems, and none of them could ever hope to match the size of the U.S. nuclear arsenal. Nevertheless, the major European allies sought to use NATO strategy as a vehicle for guaranteeing themselves important nuclear roles. For Britain and subsequently France, a nuclear role in NATO strategy would provide an important additional justification for the national nuclear forces they sought to build. For West Germany, the situation was more complicated. Although the country had forsworn the manufacture of nuclear weapons in 1954, it insisted on being treated as an equal partner by its allies. Symbolically and for practical reasons, equality required access to nuclear weapons, even if they were provided by the United States and retained under tight American control in peacetime. These concerns contributed to the allies' insistence on a greater role for nuclear weapons in NATO strategy than the United States desired.

The respective positions of the allies were also influenced by differences in their overall military potential and their out-of-area interests. Because of their limited resources, Britain and France, which maintained substantial forces outside of Europe into the 1960s, were particularly anxious to minimize the costs of their NATO commitments. Yet the burden of providing the greater conventional capabilities that the more flexible strategy favored by the United States seemed to imply would have fallen disproportionately on the Europeans because of their proximity to the threat and their relatively modest nuclear contributions.

Of course, in the determination of alliance military strategy, differences in strategic preferences do not tell the whole story. As suggested in Chapter 1, variations in bargaining power among allies may also shape the outcome. If the distribution of influence is highly asymmetrical, disparities in preferences will have little bearing on the result of intra-alliance negotiations. Given the size of American military capabilities relative to those of the other NATO countries and thus the unparalleled ability of the United States to contribute to the security of its allies, it would seem to have enjoyed a disproportionate amount of leverage in the policy-making process.

In a bipolar system, such as that which characterized the East-West confrontation in Europe during the Cold War, however, allies may have relatively little influence over one another because the threat to withdraw one's support lacks credibility. Thus the strong

U.S. interest in preserving Western Europe's independence undercut its bargaining power.[6] The potential leverage provided by U.S. military capabilities, moreover, was largely counterbalanced by other considerations that enhanced the influence of its allies. First, the critical location of a number of European countries made their willing participation essential if the entire region was to be made defensible. Most obviously, West Germany provided the territory on which the bulk of the allied forces were stationed and where a high percentage of NATO military operations would have been conducted in the event of war. Britain, France, and the low countries, though more removed from the potential battlefield, nevertheless served as important staging areas and hosted lines of communication.

The influence of the United States also remained limited because it was never willing to provide more than a fraction of the conventional forces required to defend Western Europe. As a result, substantial allied contributions were essential, ensuring that Britain, France, and above all West Germany enjoyed greater potential bargaining power.[7] Finally, a good deal of influence flowed from the fact that the security of certain European countries depended more critically on the details of NATO strategy because of their proximity to the threat. These countries, again led by West Germany, could be justified in taking a firmer stand in intra-alliance negotiations because they would be affected more directly by the outcome.

How, then, were conflicting strategic preferences to be reconciled? The only recourse in such circumstances was ambiguity. When a consensus was unattainable, NATO strategy had to accommodate divergent and even contradictory allied positions. It had to be presented in such a way that each country could plausibly interpret the document as satisfying its minimal strategic requirements, which typically necessitated couching the strategy in vague language. The wider the divergence of strategic preferences, the greater the ambiguity that was required. Hence the increasing indeterminacy of NATO's strategic concept as the growth in Soviet strategic capabilities caused allied preferences to diverge.

Explaining the Evolution of NATO Conventional Force Levels

A satisfactory account of the evolution of NATO's conventional force structure in the Central Region, as in the case of alliance strategy, must begin with several of the factors emphasized by the balance-of-power and intra-alliance perspectives. These variables are particularly useful for explaining the fluctuations in both national and overall force levels over the years, especially in the 1950s. The first two perspectives, however, cannot account for the generally stable record of NATO's conventional force structure during the following three decades. Rather, this record is best understood in terms of the institutional perspective, which postulates that the provision of conventional forces in the Central Region was strongly influenced by an international regime. This approach is especially helpful for explaining the failure of force levels to decline more than they actually did at various times.

CONTRIBUTIONS OF THE BALANCE-OF-POWER AND
INTRA-ALLIANCE PERSPECTIVES

The balance-of-power perspective accounts for many of the changes that have occurred in the alliance's conventional forces in the Central Region. The initial rapid buildup of NATO forces beginning in 1950 owed primarily to the sharp revision of Western perceptions of Soviet intentions that followed the outbreak of the Korean War. Conversely, the subsequent decline in fears of Soviet aggression in combination with resource constraints caused most NATO countries to relax their rearmament efforts, creating the seemingly unbridgable gap that emerged between the alliance's force requirements and the number of divisions that NATO military commanders could reasonably expect to have at their disposal in the following years. Although the increasing availability of tactical nuclear weapons in the mid-1950s had no direct impact on existing NATO conventional force levels, it did reduce the pressure to attain further increases.

Such factors continued to have an impact on NATO force levels into the 1960s. The heightened risk of war that accompanied the 1961–62 Berlin crisis triggered modest yet abrupt increases in the number of U.S. and German military personnel in the region. Conversely, the declining sense of threat that followed improvements in East-West relations and the downward revision of estimates of the Warsaw Pact's conventional capabilities that took place in the early to mid-1960s set the stage for the force reductions that resulted from the trilateral negotiations.

While the balance-of-power perspective explains increases and decreases in overall NATO force levels, an examination of the determinants suggested by the intra-alliance perspective best helps to account for variations in the size of national contributions. The distribution of military power among the allies provides a useful starting point. The most important contingents in the Central Region were provided by the countries—the United States, Britain, France, and West Germany—with the greatest capacity to generate forces. Nevertheless, an exclusive focus on relative military power suggests several paradoxes. For instance, although the United States remained by far the largest member of the alliance, its conventional forces in Europe were surpassed in size by those of West Germany in the early 1960s. Similarly, the conventional force contributions of Britain and France in the Central Region were comparable in size to those of Belgium and the Netherlands, even though the first two countries greatly exceeded the last two in overall military power.

To account for the relative sizes of national contributions with greater precision, we must consider several other factors emphasized by the intra-alliance perspective, beginning with differences in geographical location. A focus on geography explains why West Germany, the NATO member that was most exposed to a Soviet invasion, provided the greatest number of conventional forces while the contribution of the United States, which was located far from the Central Region, was small in comparison with that country's relative military potential. Although Britain was much closer than the United States to the potential battlefield, its physical separation from the continent exerted a similar influence. At first glance, the relatively small size of France's contribution after the mid-1950s seems puzzling in view of that country's greater proximity to the threat. The presence, however, of substantial allied forces in West Germany—between France and the Red Army—diminished France's incentive to maintain a large conventional presence of its own.

Closely related to the geographical factor were differences among NATO members in their extraregional interests and commitments. Lacking such diversions, West Germany was able to dedicate all of its forces to the defense of the Central Region. In sharp contrast, the United States maintained substantial military commitments all over the world throughout the postwar era, which prevented it from devoting more than a fraction of its conventional forces to NATO.

Although British and French out-of-area interests declined significantly over the years, they nevertheless exerted a powerful influence at times on those countries' contributions to NATO. France's involvement in Indochina in the early 1950s contributed substantially to its failure to achieve its initial force goals. France consistently accorded the war effort priority over the fulfillment of its alliance obligations,[8] and its subsequent engagement in Algeria resulted in the withdrawal of much of its NATO force contribution. Similarly, Britain's continuing global commitments as well as the demands of its nuclear program played key roles in the decision to reduce the size of the British Army on the Rhine in the late 1950s.

Finally, differences in recent national historical experiences also shaped NATO's conventional force structure. First, they played an important role in conditioning how each country viewed the threat. Although the primary purpose of the alliance was to address the dangers posed by the Soviet Union, an important secondary objective, at least initially, was the management of the German problem. During the late 1940s and early 1950s, France in particular was concerned about possible German revanchism and consequently sought to place sharp limits on German military power. These efforts contributed greatly to the long delay in the start of West German rearmament.

For its part, West Germany was anxious to be treated as an equal partner by the Western countries following the war and occupation. The Germans were highly sensitive to any suggestion that their country would occupy a subordinate role in the alliance. Consequently, they reacted with particular vehemence to the leak of the Radford plan, which implied that West Germany would be limited to contributing conventional forces while the United States and possibly other countries provided the alliance's nuclear backing. Subsequently, West German leaders placed great emphasis on obtaining a nuclear role and deemphasized conventional forces, further delaying the completion of the German buildup.

LIMITATIONS OF THE BALANCE-OF-POWER AND INTRA-ALLIANCE PERSPECTIVES

Although the balance-of-power and intra-alliance perspectives account for many of the fluctuations in NATO's overall conventional capabilities as well as the relative size of national force contributions, they do not adequately explain the stability of NATO force levels from the late 1950s until the late 1980s. To the contrary, changes in the determinants emphasized by these perspectives would lead us to expect somewhat greater and more frequent variation in NATO's conventional force structure than actually occurred.

NATO conventional force levels failed to respond to changes in Soviet military capabilities and to developments that suggested altered Soviet intentions on several occasions. For example, there was no immediate increase in the wake of the Soviet invasion of Hungary, which revived Western fears of aggression, or of the demonstration of a Soviet intercontinental nuclear capability in 1957, which clearly called into question the willingness of the United States to respond to lesser forms of aggression with strategic nuclear retaliation. To the contrary, the pace of the West German military buildup actually slowed and the British, and the United States to a lesser extent, even reduced their forces in the region during this period. Although arrangements for making tactical nuclear weapons available to the European allies were approved in the late 1950s and the number of such weapons on the continent more than doubled in the 1960s, these responses were largely offset by the growth of the Soviet theater nuclear arsenal.[9]

Similarly, no significant upward movement of NATO force levels occurred in response to the substantial increases in Warsaw Pact conventional capabilities in the late 1960s and 1970s, the Soviet Union's achievement of parity first in strategic nuclear delivery vehicles in the early 1970s and subsequently in nuclear warheads, or the invasions of Czechoslovakia and Afghanistan, which, like Korea and Berlin before, could have been viewed as signaling more aggressive Soviet intentions or at least a heightened risk of war. Although NATO's 1979 dual-track decision may have been expected to neutralize the Soviet deployment of SS-20s, it did nothing to rectify what was widely perceived as a growing conventional imbalance. The modest increase in the number of American, British, and French combat units that did occur during this period, moreover, was largely the result of force restructuring and had little direct impact on personnel strengths. At the same time, apart from the reductions agreed to in the trilateral talks, which can largely be attributed to demands the Vietnam War placed on U.S. military resources, the downward revision of American estimates of Warsaw Pact conventional capabilities in the early to mid-1960s and the more benign NATO assessments of Soviet intentions in the late 1960s and early 1970s produced no significant decline in alliance force levels.[10]

A focus on variations in the distribution of military power among the NATO allies, as suggested by the intra-alliance perspective, confronts similar problems. In the period from 1965, when the West German buildup was completed, to 1990, little positive relationship can be found between changes in the relative ability to contribute forces, as measured by gross domestic product (GDP), of the countries with forces in the Central Region and changes in their relative force contributions, as measured in military personnel and numbers of active brigades (see Table 7.2). Only in the case of the United States, and possibly West Germany, is there a clear positive association. For Britain, France, and Canada, the association is negative. And for Belgium and the Netherlands, changes in relative force con-

TABLE 7.2
*Changes in National Shares of Gross Domestic Product,
Military Personnel, and Major Combat Units, 1965–1990*

	Change in GDP as pct. of total	Change in military personnel (× 1,000)		Change in units (brigades)	
		Absolute	As pct. of total	Absolute	As pct. of total
United States	−1.29	−18	−0.23	−2	−2.90
United Kingdom	−1.40	+4	+0.74	+2	+2.50
France	+1.13	−9	−1.43	0	−0.10
Germany	+0.62	+20	+4.52	+3	+3.33
Belgium	−0.09	−15	−0.85	−2	−2.70
Netherlands	+0.05	−32	−2.34	0	−0.10
Canada	+0.87	−5	−0.42	0	−0.02

SOURCES: GDP figures are from Summers and Heston, "New Set of International Comparisons," and IISS, *Military Balance, 1990–91*. Figures for military personnel and brigades are from Table 7.1.

tributions were accompanied by virtually no change in relative GDP, suggesting no relationship whatever.[11]

How, then, are we to account for the failure of NATO conventional force levels in the Central Region to increase or to decrease in response to balance-of-power and intra-alliance pressures for change? The availability of nuclear weapons and the growing ambiguity of NATO strategy provide a partial explanation. These conditions allowed allies to argue that stronger conventional forces were unnecessary and even undesirable.

Domestic politics must also be considered.[12] Unlike decisions about military strategy, the provision of conventional forces entails government outlays and, when forces are stationed abroad, expenditures of foreign exchange. Because national resources are finite, increases in a country's contribution to the alliance will entail substantial opportunity costs and are likely to be highly contested. Domestic interest groups that stand to lose from greater military spending will oppose additional outlays for defense. Even interests that are not adversely affected in a direct way may very well resist such changes because of their uncertain impact on carefully crafted national budget agreements. Or national budgets may be so inflexible that increases in force structure require years to implement.[13] As a result of such domestic constraints, states will tend to postpone increasing their alliance force contributions for as long as possible, often until the pressure to do so has subsided.

Many of the same domestic factors that militate against increases in force levels, however, should cause force levels to decline whenever international conditions, the presence of nuclear weapons, or strategic ambiguity permit. In short, force levels should be sticky upward but not downward. Thus attention to these factors can provide only a partial explanation of the stability of NATO force levels.

CONTRIBUTIONS OF THE INSTITUTIONAL PERSPECTIVE

To develop a more comprehensive explanation, we must turn to the institutional perspective. As discussed in Chapter 1, international institutions such as regimes tend to promote regularity in state behavior and international outcomes. The existence of a regime may alter the costs and benefits associated with different courses of action in ways that increase the likelihood of compliance with its injunctions. In addition, a presumption in favor of compliance may become embedded in domestic institutions and even in the belief systems of national decision makers. Although it may not be necessary to invoke regime theory to explain the absence of increases in NATO force levels, this approach is useful for explaining why alliance members often did not decrease their NATO contributions when international conditions allowed or domestic pressures militated for force reductions.

This section elaborates the argument that a regime strongly influenced NATO conventional force levels from the late 1950s until the end of the Cold War in Europe. It first summarizes the institutional rules that governed the provision of conventional forces in the Central Region during this period and then discusses why NATO countries complied with these rules. It indicates the general causes of compliance, including the incentives for compliance created by the very existence of the regime and reinforcing domestic and cognitive sources.[14]

Norms and Rules of the Regime. A combination of general norms and more specific rules evolved during the 1950s that subsequently served as guidelines for the provision of conventional forces in the Central Region. The highest norm was that, in the absence of an agreement to the contrary, states should seek to provide previously agreed force levels or at least to preserve the status quo. If a state concluded that it could no longer maintain its force contribution at the existing level, it was expected to refrain from altering its contribution unilaterally and to consult with its allies before taking any significant actions. The norm of consultation reinforced the regime by allowing allies to express any objections to proposed force reductions that they may have had, thereby underscoring the costs of proceeding to implement them.[15]

TABLE 7.3
NATO Force Goals and Requirements
(Divisions)

	1952	1953	1957	1961	(Actual) 1962
United States	6⅔	6⅔	6	6	6
United Kingdom	4⅔	4⅔	4	3	3
France	7⅓	7⅓	5	4	2⅓
Germany	12	12	12	12⅓	8
Belgium	n/a	3	2	2	2
Netherlands	n/a	1	2	2	2
Canada	n/a	⅓	⅓	⅓	⅓
Total	n/a	35⅓	31⅓	29⅔	23⅔

SOURCES: Poole, *History of the JCS*; Memorandum for the JCS, April 16, 1953, RG 59, 740.5/4-1553; House of Representatives, Foreign Affairs, *Selected Executive Session Hearings, 1957–60*, 19: 28; and DOD, "Remarks by Secretary McNamara."
NOTE: n/a = not available.

NATO also offered specific rules concerning the forces that member countries should provide. These rules primarily took the form of national force goals and requirements for major ground force units and aircraft, which were based on NATO's formal military strategy, estimates of Soviet and Warsaw Pact military capabilities, and assessments of alliance resources. Formal force goals or requirements were adopted on a more or less annual basis between 1950 and 1953 and again in 1958 and 1961 (see Table 7.3).[16]

NATO force goals and requirements were never binding, however. Their fulfillment, moreover, often lagged by several years and, in many cases, was never achieved. This was especially true of the requirements for mobilizable reserve units. Nevertheless, most of the goals for ready forces in the Central Region were eventually realized. By the early 1960s, NATO requirements closely corresponded with the forces then being provided.[17]

NATO last formulated divisional force requirements in 1961. In principle, the task of formally determining such requirements, although never again attempted, was greatly complicated in 1967, when the alliance adopted the concept of political warning in conjunction with the strategy of flexible response. As explained in Chapter 5, this concept largely decoupled the requirement for ready forces from the capabilities of the Warsaw Pact, allowing NATO countries to rationalize a much wider range of possible contributions of standing forces in peacetime. Yet force requirements continued to exist in the form of the status quo. In institutional terminology, the maintenance of existing force levels became a social convention. And in the words of Thomas Schelling, the status quo became a salient "focal point" on which expectations of future behavior naturally converged. In the absence of jointly agreed force goals, the status quo was the one point at which a line could be drawn that would not seem entirely arbitrary, of which one could ask, "If not here, where?"[18]

After the late 1950s, several changes occurred in the status quo, but these were negotiated or received multilateral endorsement. The British withdrawals in 1957 and 1958 were discussed and approved by both the Western European Union and NATO, and the modest U.S. and British reductions of the late 1960s took place only after lengthy negotiations with the West Germans and then only after they had received the concurrence of the alliance as a whole. Thus in each case, the status quo was effectively renegotiated, and a new but only slightly different consensual norm regarding force levels replaced the old.

In addition to explicit and implicit force goals, the provision of conventional forces in the Central Region was shaped by several other conventions. For example, upper limits on national force contributions were developed in 1951 for the ill-fated European Defense Community. After the EDC was scuttled in 1954, these restrictions were adopted by the parties to the Brussels Treaty as one of the interlocking agreements that paved the way for West German rearmament. In practice, however, only the German military contribution, which was capped at twelve divisions, was affected because the forces of the other WEU countries were never expected to exceed the ceilings set for them.

At the same time that West Germany was admitted to NATO, Britain stated its intention to maintain the equivalent of four divisions and a tactical air force on the continent to calm French anxieties about German rearmament. The British agreed not to reduce these forces without the approval of a majority of the Brussels Treaty countries. As discussed in Chapter 4, this declaration became an important constraint on subsequent British behavior.

Rational Incentives for Compliance. The provision of conventional forces in the Central Region was highly consistent with these norms and rules. Until the end of the Cold War, NATO force levels had fluctuated very little during more than three decades, although there were occasional strong pressures for change. This continuity can be understood largely as resulting from both the incentives for compliance created by the very existence of the regime and internal factors that reinforced the status quo.

Several aspects of the regime helped to foster compliance. Most obviously, the existence of concrete force goals provided a clear yardstick for assessing national contributions. In addition, the NATO force planning process required member countries to furnish each other with detailed information about their existing and planned force structures. As a result, intra-alliance relations were highly transparent.

More important, by maintaining their forces at the indicated levels, NATO countries theoretically improved the prospects for deterrence and, in the event that deterrence failed, a successful defense of the region. Observing established force goals thus contributed directly to the security of individual countries. It also resulted in subsidiary benefits such as enhanced domestic and international prestige and the appreciation of grateful allies, which might yield greater influence with them on other issues. Finally, states regarded steadfast adherence to the status quo as necessary to foster a continued military contribution by other countries. They feared that a sustained failure to provide the forces expected of them could result in similar behavior by their allies, leading to a weakening of NATO and the consequent loss of some if not all of the benefits of alliance membership.

It is worth examining these incentives in greater detail from the perspectives of both sides of the Atlantic. To the European members of NATO, the presence of American troops in Europe served as a potent symbol of the U.S. commitment to their security, and the continuation of this presence at approximately the same level was necessary to sustain the credibility of the American nuclear guarantee. As a result, the European allies tended to maintain their forces at established levels largely to ensure that no U.S. forces were withdrawn from the continent. For the most part, European leaders calculated that the benefits to be derived from unilateral reductions would not outweigh the resulting costs in diminished security if U.S. force levels were also reduced.[19]

On several occasions, the United States as well refrained from deviating from alliance norms by reducing its forces in the Central Region out of concern for the possible negative repercussions that such a move might have on the behavior of its allies. American leaders feared that U.S. troop withdrawals would precipitate corresponding allied force reductions, rather than compensatory allied increases, further weakening deterrence, making Western Europe even more vulnerable to Soviet political intimidation or aggression, and forcing the United States to face the decision to use nuclear weapons earlier in the event of hostilities. In the worst case, allies would have lost confidence in the U.S. security guarantee and would have sought an accommodation with the Soviet Union.

Internalization of the Regime. Although it might be tempting to view compliance with regime rules simply as a rational response to the external opportunities and constraints created by the regime, the outcomes of cost-benefit calculations are ultimately profoundly influenced by the values that decision makers hold and their beliefs about the likely consequences of different actions.[20] In this sense, cognitive factors also played an important role in shaping decisions concerning the provision of conventional forces in the Central Region in ways that reinforced the regime. Support for maintaining the status quo followed logically from a set of mutually reinforcing attitudes and assumptions.[21] First, the preservation of the political independence of Western Europe was a vital interest not only of the countries in the region but of the United States as well. Second, the Soviet Union sought to limit the political independence of Western Europe and, through the maintenance of substantial, offensively postured military capabilities, possessed the means to influence the politics of—if not to intimidate—the countries of the region. Third, to neutralize the potential political influence afforded by these military capabilities, the NATO countries had to maintain substantial conventional and nuclear forces in Western Europe, including a large U.S. component.[22] Fourth, up to a certain point, the details of NATO's force structure were less important than the preservation of alliance cohesion, especially if disagreement threatened the U.S. commitment to defend the political independence of Western Europe as represented by the American military presence on the continent. Finally, by undermining alliance cohesion, which presumed a high degree of cooperation and self-restraint, unilateral force reductions would vitiate the effectiveness of the alliance in countering Soviet political pressure and should therefore be strongly resisted.

Such general beliefs may nevertheless place only loose constraints on state behavior. Careful adherence to regime norms and rules, however, was also promoted by more specific ideas and assumptions that were broadly shared.[23] At many times, for instance, existing NATO force levels were widely regarded as the minimum necessary to preserve deterrence, given Warsaw Pact capabilities and the state of military technology, or, in the case of the U.S. contribution, to reassure nervous allies. Also important was the common view that if one country reduced its forces, its allies would follow suit rather than increase their contributions.[24]

Another important aspect of the internalization of the regime was the redefinition of organizational interests around its rules. Most national military organizations developed strong interests in maintaining NATO force levels at their existing levels. In many cases, the defense of Western Europe became a central if not the primary organizational mission, and those national forces stationed in the region came to constitute a large share if not all of the assets of many military organizations. Consequently, attempts to reduce a country's NATO force contribution automatically became attacks on the "essence" of the threatened military organizations, prompting fierce resistance.[25] This regime-reinforcing behavior stands in marked contrast to the initial reluctance of some military organizations, notably in the United States, to station additional forces in the Central Region.

Other organizations with a less direct stake in NATO force levels also became staunch advocates of compliance with regime rules. Foreign ministry budgets were little affected by changes in force levels, but diplomats in regular contact with their allied counterparts had an interest in maintaining good working relations. Consequently, their positions often took into account allied views, which were likely to be opposed to any reductions.[26]

Finally, the regime was reinforced by some features of national policy-making processes. Congressional-executive relations in the United States are an important example. The process by which U.S. policy regarding NATO's conventional force posture was made usually was marked by a high degree of congressional restraint. Typically, the legislative branch deferred to the president and his advisers on the details of U.S. force levels in Europe, even though the issue had substantial budgetary implications. Although Congress occasionally placed upper limits on U.S. force levels and even threatened to impose reductions from time to time, it ultimately proved loath to do so in the face of administration opposition. Another factor was the distribution of power within successive U.S. administrations in which the views of the State Department, the Defense Department, and the military, which usually favored maintaining the status quo, carried more weight than those of the agencies with overall responsibility for U.S. fiscal policy, which were more inclined to support reductions.

The Role of the NATO Organizational Structure. Finally, NATO's organizational structure helped to shape the alliance's conventional force levels. Most obviously, the major NATO military commanders had primary responsibility for determining the alliance's force requirements. Beyond that, NATO officials regularly used their considerable moral authority and powers of persuasion to promote compliance with alliance norms concerning force levels, and they were able to prevent or at least to minimize imminent national force reductions on several occasions. SACEUR Lauris Norstad played a leading role in convincing British leaders to limit and to stretch out the cuts they had hoped to make in the late 1950s. Similarly, SACEUR Andrew Goodpaster and Secretary General Manlio Brosio led the effort to head off a further decline in NATO force levels in the early 1970s. On the other side of the ledger, the NATO bureaucracy tended to resist U.S. proposals to strengthen the alliance's conventional forces, as occurred with the initiatives of the Kennedy and Carter administrations. In both cases, however, the existence of the organizational structure generally served to reinforce the status quo.

Summary

The preceding analysis suggests that all three of the theoretical perspectives introduced in Chapter 1 are essential for understanding the evolution of NATO's conventional force posture. Taken together, they can be used to construct comprehensive explanations of the role of conventional forces in NATO strategy and the alliance's actual conventional force levels that require little augmentation by other possible perspectives, such as those emphasizing domestic-level factors alone. Each of the three perspectives, however, makes a somewhat different contribution.

The balance-of-power perspective plays a major role in both explanations. The variables it stresses—the adversary's capabilities, perceptions of the adversary's intentions, the state of military technology, and the alliance's military potential—determined the broad outlines of NATO's conventional force posture. Variations in those causal factors frequently created pressures for change in both NATO strategy and the alliance's force levels, and these pressures often resulted in significant adjustments, especially during the first decade of the alliance.

The intra-alliance perspective also contributes to both explanations. Consideration of the variables it emphasizes—differences in military power, geography, out-of-area interests, and national historical experience—is necessary to account for differences in the strategic preferences of the allies as well as in the sizes of their respective conventional force contributions. These determinants also help to explain the allies' differing reactions to variations in the factors stressed by the balance-of-power perspective. Finally, the intra-alliance perspective helps in understanding the increasing stability of NATO strategy over time, notwithstanding pressures for change, as manifested in the growing difficulty that the allies encountered in reaching agreement as their strategic preferences diverged and in the consequent ambiguity of NATO strategy, which made successive strategic concepts ever less sensitive to external conditions.

In contrast, there is little need to draw upon the institutional perspective to account for the evolution of NATO strategy. This perspective, however, is indispensable for explaining the stability that characterized NATO's conventional force structure during much of the postwar era. Institutional factors were especially critical for limiting reductions in force levels in response to downward revisions of estimated Warsaw Pact military capabilities, increasingly benign assessments of Soviet intentions, and other demands on alliance resources.

Epilogue: The Transformation of NATO's Conventional Force Posture

NATO's conventional force posture underwent a profound transformation during the three years following the opening of the Berlin Wall in 1989. The alliance's strategic concept was revised for the first time in nearly two and a half decades, and flexible response was replaced by a new strategy that placed even greater emphasis on conventional forces. At the same time, the alliance's conventional capabilities in the Central Region were considerably reduced, and the remaining forces were substantially reorganized.

These abrupt changes in NATO's conventional force posture were a consequence of a dramatic shift in the policies and capabilities of the alliance's principal adversary, the Soviet Union, that resulted in a drastic decline in the political and military threat to Western Europe, putting an end to the Cold War. Indeed, the impact of these momentous developments was not limited to the alliance's military arrangements. They precipitated a fundamental reorientation of NATO's overall grand strategy, only the second such occurrence in the history of the alliance.

These events have important theoretical implications. First, they point to the limits of institutional explanations, underscoring the fact that a given set of institutional arrangements cannot exist indefinitely in the face of profound variations in their environment. The main contours of alliance policy are ultimately determined by external and intra-alliance factors. At the same time, however, the end of the Cold War in Europe reveals the considerable extent to which the international environment must change before institutional structures are altered. NATO's survival as a highly integrated military alliance, moreover, demonstrates the persistence of basic institutional elements even as their details are modified.

The Decline of the Traditional Threat to Western Europe

The Atlantic alliance was formed primarily in response to the threat posed by the ideology and policies of the Soviet Union, while its military posture was greatly influenced by Soviet and Warsaw Pact military capabilities. Thus it is not surprising that a drastic decline in the traditional threat to Western Europe, which radically changed the alliance's political and strategic environment, precipitated a fundamental modification of NATO's conventional force posture.

The main features of the changes in the threat are easy to describe. As a result of Soviet troop reductions and withdrawals, the demise of the Warsaw Pact, and the dissolution of the Soviet Union itself, the number of potentially hostile forces facing NATO would fall by more than one million by the mid-1990s. The residual forces of the alliance's most formidable potential adversary, Russia, would now have to cross unfriendly territory in Eastern Europe and even the western regions of the former Soviet Union before they could attack NATO. Perhaps most important, it would be more difficult to imagine circumstances in which Russia would even consider using force against Western Europe, given the sweeping political reforms occurring in the country.

CHANGES IN SOVIET INTENTIONS: GORBACHEV'S NEW THINKING

The collapse of the threat stemmed from the new thinking in Soviet foreign and security policy that emerged in the mid- to late 1980s under the leadership of Mikhail Gorbachev, who became head of the Soviet Communist Party in March 1985.[1] The new thinking had interrelated domestic and international roots. Externally, it was a response to the failure of the foreign policy of Leonid Brezhnev, which had occasioned a sharp downturn in U.S.-Soviet relations in the late 1970s and a potentially destabilizing acceleration of the strategic arms race in the early 1980s. Internally, the new thinking dovetailed with Gorbachev's efforts to deal with the deep-seated economic and social crisis that he had inherited from his predecessors. If his far-reaching political and economic reforms were to succeed, Gorbachev would have to reduce the heavy burden imposed by Soviet military policy on the economy and to foster a more stable international environment, which in turn required establishing better relations with the West.

The new thinking was based on two fundamental tenets. First, it placed the maintenance of peace and the avoidance of war, especially nuclear war, decidedly ahead of the spread of socialism in the hierarchy of Soviet foreign policy objectives. Second, it recognized that in the nuclear age, real security could not be achieved unilaterally through the acquisition of ever greater military power. Exclusive reliance on deterrence and defense only stimulated the arms race and fostered instability. Instead, security would have to be mutual and pursued increasingly through political rather than military means. The principal consequences of the new thinking for Soviet military posture were twofold: Soviet strategy would henceforth be based on "defensive" defense, not offensive capabilities and operations, and Soviet force requirements would be shaped by the principle of "reasonable sufficiency."

The new thinking was quickly put into practice in Europe. Under Gorbachev, the Soviet Union exhibited a renewed interest in European arms control. Beginning in late 1985, Soviet leaders offered previously unimaginable concessions in the negotiations on intermediate-range nuclear forces, making possible the signature of a treaty just two years later. Also in 1987, the Soviet Union and its allies began to call for joint reductions in NATO and Warsaw Pact forces to the point where neither side could launch a surprise attack or engage in offensive operations.[2]

Nevertheless, these seemingly positive developments in Soviet policy had little impact on NATO's conventional force posture through the late 1980s. For a variety of reasons, many Western observers remained skeptical of Soviet intentions.[3] For example, Soviet military leaders argued that a defensive doctrine did not rule out counteroffensive operations, which allowed them to continue to develop offensive concepts and capabilities. Similarly, through 1988, there was little evidence that the new emphasis on reasonable sufficiency would result in any significant reduction in the Soviet conventional forces facing Western Europe.

REDUCTIONS IN SOVIET AND WARSAW PACT
CONVENTIONAL CAPABILITIES

These voices of skepticism were largely silenced, however, by a series of startling Soviet initiatives beginning at the end of 1988. In an address to the United Nations that December, Gorbachev announced that the Soviet Union would undertake deep unilateral force reductions over the next two years. Then at the opening of the new negotiations on Conventional Armed Forces in Europe (CFE) the following March, the Warsaw Pact tabled a proposal calling for deep and largely asymmetrical cuts in its own forces, setting the stage for an agreement the following year.

Although there had been previous hints of possible Soviet reductions,[4] Gorbachev's UN speech took the West largely by surprise. In it, he announced that the Soviet Union would reduce the size of its armed forces by 500,000 men by the end of 1990. West of the Ural Mountains, the cuts would amount to 240,000 troops as well as 10,000 tanks, 8,500 artillery pieces, and 800 combat aircraft. Of even greater interest to NATO planners was his promise to withdraw 50,000 troops, 6 tank divisions, and 5,000 tanks—equivalent to the holdings of 15 to 16 divisions—from the territory of its Warsaw Pact allies and to restructure the divisions remaining in Eastern Europe along defensive lines. The Soviet Union's allies quickly followed suit. During the next several months, five other Warsaw Pact countries announced unilateral reductions totaling over 80,000 troops, 2,750 tanks, 1,500 artillery pieces, and 200 combat aircraft.[5]

The unilateral reductions signaled in Gorbachev's speech marked the first significant practical step toward a more defensive Soviet military posture in Europe. More important, when completed, they would all but eliminate the possibility of a short-warning attack in the Central Region. Henceforth, an attack would require the dispatch of substantial reinforcements from the USSR if it were to have any chance of success.[6]

Even this dramatic gesture did not dispel all doubts about Soviet intentions.[7] The Soviet leadership took another long step toward mollifying Western skeptics in March 1989, however, when it presented the initial Warsaw Pact position in the CFE negotiations. The mandate for the talks, which was agreed on the previous January, set forth the objective of strengthening stability and security in Europe through the establishment of a balance of forces at lower levels, the removal of destabilizing disparities, and the elimination of the capability for launching a massive surprise attack and for initiating large-scale offensive action.[8] Consistent with the mandate, the Warsaw Pact paper called for deep cuts in the most destabilizing categories of armaments down to equal ceilings for the two alliances. In practice, the proposal meant highly asymmetrical reductions by the Pact in tanks, artillery, and armored vehicles.[9]

Although the Warsaw Pact proposal differed from NATO's opening bid in several particulars,[10] the initial positions of the alliances were surprisingly close, and the differences were further narrowed when the Pact provided more detailed figures during the first round of negotiations the following May. Indeed, the Soviet Union and its allies had accepted virtually all of the basic principles espoused by the West, namely, the elimination of asymmetries, reductions to equal levels, the establishment of zonal sublimits, the exclusion of naval forces, and the need for stringent verification measures.[11] This broad consensus made it possible for the two sides to overcome their remaining differences and to reach an agreement in November 1990.[12]

The CFE treaty established specific limits on five categories of armaments—battle tanks, artillery, armored combat vehicles, combat aircraft, and attack helicopters—and mandated the destruction or the removal of tens of thousands of pieces of military equipment from Europe. The vast majority of the reductions fell upon the members of the Warsaw Pact. Most important, the treaty required a 26 percent cut in the Soviet Union's declared holdings of military equipment west of the Urals, including 36 percent of its tanks. When combined with the reductions and withdrawals already undertaken by the Soviets on a unilateral basis, the treaty represented a 67 percent decline in treaty-limited Soviet armaments in the region since 1988, including two-thirds of the Soviet tank holdings, or a total of more than 100,000 major items of equipment.[13]

By themselves, then, the unilateral reductions undertaken by the Warsaw Pact countries after 1988 and the new limits imposed by the CFE treaty had important implications for NATO's conventional force posture. The resulting substantial decline in the Warsaw Pact's military capabilities would enable the alliance to reduce its dependence on nuclear weapons and to place greater reliance on conventional forces in its military planning, although such strategic adjustments could probably be accommodated within the broad parameters of flexible response. More concretely, the decrease in the threat would allow NATO countries to reduce the size of the conventional forces they maintained in the Central Region or at least the readiness of the existing forces. Despite their potential importance, however, these remarkable developments were largely overshadowed by an even more breathtaking set of events.[14]

THE END OF THE COLD WAR IN EUROPE

Of even greater significance for NATO's conventional force posture were the political revolutions that shook the Soviet Union and Eastern Europe after 1988. These changes, any one of which would have been inconceivable just a few years if not months before, put a definitive end to the postwar East-West confrontation in Europe. Consequently, they largely eliminated the traditional rationale for the very existence of the Atlantic alliance.

The first step in this historical transformation was the demise of communist regimes throughout Eastern Europe in 1989. Beginning in Hungary and then Poland, the process gradually gained momentum and did not subside until the governments in East Germany, Czechoslovakia, Bulgaria, and Romania had all been either toppled or fundamentally reformed by the end of the year. Throughout these tumultuous events, Gorbachev's Soviet Union stood passively by, making little effort to halt the process and, on occasion, even seeming to encourage it.[15]

Once in place, the new governments promptly moved to adopt new defense policies that turned the once feared Warsaw Pact into an empty shell.[16] Several countries announced further substantial unilateral reductions and the defensive restructuring of their forces.[17] Even more important, Czechoslovakia and Hungary demanded the removal of all Soviet forces from their territory. Although a small number of units had already departed in accordance with the reductions announced by Gorbachev in 1988, some 74,000 Soviet troops remained in Czechoslovakia and some 50,000 in Hungary. Despite initial misgivings, the Soviet Union quickly agreed to withdraw all of its forces from the two countries by mid-1991.[18]

Even this concession did not satisfy the new East European governments, which began to press for the transformation of the Warsaw Pact into a truly democratic alliance, if not its abolition.[19] Under pressure from Hungary and Czechoslovakia, the alliance's military structure was dissolved on April 1, 1991. The following July, the remaining Warsaw Pact members signed a protocol calling for the complete dismantlement of the organization by the end of the year.[20]

The next significant change in the European strategic landscape following the revolutions of 1989 was German unification. Although the West German government had initially proposed a gradual approach toward unity involving the creation of confederal structures during a transition period of indeterminate length,[21] the crumbling of authority in East Germany in combination with the freedom of movement between the two countries made their rapid fusion virtually inevitable. The internal and external terms of the union were largely worked out by July 1990, when West Germany and the Soviet Union reached a historic accord on the matter, enabling unification to take place formally in October of that year.[22]

Unification was tantamount to the wholesale absorption of East Germany by its larger Western counterpart. As a result, the East German military, previously 170,000 strong and regarded as "the most effective East European armed force," disappeared in one fell swoop.[23] Under the terms of the German-Soviet agreement, moreover, the Soviet Union was to withdraw all of the approximately 400,000 troops remaining on German territory by the end of 1994.[24]

The final stage in this remarkable series of events was the breakup of the Soviet Union itself, which took place with equally remarkable speed during the second half of 1991. Although fissiparous tendencies had already become apparent, the process of dissolution was greatly accelerated by the failure of the August putsch. By simultaneously undermining the authority of the central government and discrediting the remaining conservative hard-liners, the coup emboldened separatist movements in the various republics and enabled progressive elements in Russia to move ahead with further-reaching political and economic reforms, thereby ensuring that a return to a Stalinist system would be extremely unlikely.

The logical consequence of these developments was a further decline in the capabilities of the armed forces of the Soviet Union and, with the exception of the strategic arsenal, their rapid dissolution into separate and smaller republican forces. By April 1992, even Russia, which had been the most fervent advocate of maintaining a unified commonwealth force, announced that it would establish an independent military of only 1.5 million troops, a far cry from the formidable Red Army of the past.[25]

Thus in the short span of three years, the traditional military threat to Western Europe, which had played a central role in shaping NATO's conventional force posture, had all but disappeared. Not only was a massive surprise attack no longer possible, but the Soviet capacity for large-scale conventional aggression in the Central Region had been virtually eliminated. Henceforth any deliberate attack would have to be preceded by months, if not years, of preparation, and any force sufficiently large to pose a threat to the West would first have to cross hundreds of miles of hostile territory in Central and Eastern Europe. Although Russia continued to maintain a strategic nuclear arsenal second in size only to that of the United States, the possibility that it would use these forces in an attempt to coerce or to intimidate NATO countries appeared extremely remote in view of the sweeping political reforms that were occurring in the country. The intentions of leaders in the Kremlin would not soon again be perceived in the same hostile light. Indeed, most if not all of NATO's former adversaries had been transformed into potential partners in cooperative efforts to achieve security.

The Reorientation of NATO Grand Strategy

The impact of the collapse of the traditional threat to Western Europe was not limited to NATO's military posture. The change in the external security environment precipitated a fundamental re-orientation of the alliance's overall grand strategy for ensuring the security of its members.[26]

NATO GRAND STRATEGY BEFORE 1989

Previously, NATO grand strategy had passed through two phases.[27] During the first phase, which lasted from the signing of the North Atlantic Treaty until the 1960s, it emphasized the military confrontation with the Soviet Union. East-West relations were conceived primarily as adversarial, and NATO countries duly assigned top priority to creating an effective collective deterrent and defense while maintaining alliance cohesion. During the second phase, NATO grand strategy increasingly included an element of cooperation with the Soviet Union as the alliance sought to ameliorate if not eliminate the underlying causes of East-West conflict, although deterrence was not neglected. Thus greater emphasis was placed on political means.

This second NATO grand strategy was formalized in a report, "The Future Tasks of the Alliance," adopted by the North Atlantic Council in 1967, at the same time that flexible response was approved.[28] Better known as the Harmel Report, after its principal author, Belgian foreign minister Pierre Harmel, the paper outlined two main functions for the alliance. The first was to maintain adequate military strength and political solidarity to deter aggression and other forms of pressure and to defend the territory of members if attacked. The second and more novel function was to pursue better relations with the Soviet Union and Eastern Europe in the hope of resolving the outstanding political issues of the region. In particular, the report called upon the allies to explore the possibility of achieving agreements on disarmament and arms control, especially balanced force reductions in Europe. In short, this grand strategy viewed military security and detente as complementary.

The Harmel Report contained two important caveats, however. First, the pursuit of detente was not to be allowed to split the alliance or to undermine its cohesion. An effective collective defense was regarded as a necessary precondition for cooperative policies directed toward the relaxation of tensions. Second, detente was not to be pursued for its own sake. Rather, its purpose was to foster a just, lasting, and peaceful European settlement.

In some important respects, the military and political components of NATO grand strategy during this second phase were highly interdependent. For example, the 1979 LRTNF decision explicitly linked the deployment of a new generation of nuclear weapons to the initiation of arms control negotiations with the Soviet Union that, it was hoped, might eventually make the deployment unnecessary. Nevertheless, the new emphasis on cooperation had little or no impact on NATO's conventional force posture. It necessitated no modification of the alliance's military strategy, and through the late 1980s, the resulting East-West negotiations on conventional forces produced no change in NATO force levels in the Central Region. To the contrary, the existence of the talks on Mutual and Balanced Force Reductions was used as an argument to dissuade member countries from reducing their contributions unilaterally.

NATO'S NEW GRAND STRATEGY

In response to the upheavals that convulsed the Soviet Union and Eastern Europe beginning in 1989, NATO quickly adopted a new grand strategy.[29] The essential purpose of the alliance remained the same: to safeguard the security, freedom, and territorial integrity of its members; to prevent war; and to establish a just and lasting peaceful order in Europe. NATO's more concrete aims and the hierarchy among them, however, were considerably revised. The long-standing objectives of preventing intimidation and coercion, deterring aggression, and, if necessary, defending allied territory and restoring the peace were restated, as was the related goal of maintaining alliance solidarity. The alliance would also be prepared to manage any crises that might arise.

Nevertheless, the previous element of confrontation in NATO grand strategy was almost entirely absent. Virtually no attention was given to the possibility of deliberate aggression. Instead, the alliance's principal security concerns were defined in terms of the unforeseeable consequences of instability in the former Soviet bloc and elsewhere. Consequently, primary emphasis was placed on the objective of preventing crises and instability. The achievement of this goal would require ensuring the success of the political and economic reforms being undertaken in the former Soviet republics and the countries of Central and Eastern Europe while addressing the external security concerns of the new democracies to the greatest extent possible. Thus the conciliation and reassurance of former adversaries had emerged as a central tenet of NATO policy.

This reorientation of the ends of alliance grand strategy occasioned an equally fundamental change in the definition of the appropriate means. NATO would still need to preserve an effective collective defense capability to be ready to meet any potential military threats. Yet such preparations would serve primarily as a hedge against the failure of reform in the former Soviet bloc and the consequent emergence of instability. Instead, primary emphasis would be placed on the use of political instruments intended to ensure that such contingencies would not arise in the first place.

To this end, the alliance quickly devised new mechanisms for greatly expanding dialogue, cooperation, and consultation with its former adversaries on security matters and related issues. These measures included formal diplomatic liaison, meetings of high-level officials, extensive military contacts and exchanges, and, most notably, the creation of the North Atlantic Cooperation Council (NACC) in November 1991 and the Partnership for Peace in January 1994. The NACC provided for annual gatherings of the foreign ministers of the NATO and former Warsaw Pact countries and established permanent bodies in which lower-level national representatives could discuss a wide range of security issues on a regular basis. The Partnership for Peace offered Central and East European countries the opportunity to participate in a variety of joint NATO military activities and operations.[30]

The alliance also sought to bring ongoing arms control talks to a successful conclusion, to secure the ratification and implementation of the resulting CFE treaty, and to initiate follow-on negotiations to achieve even greater force reductions. And to facilitate the difficult reforms being undertaken in Central and Eastern Europe, the NATO countries offered a variety of forms of economic, financial, and technical assistance, although many of these were provided through other fora such as the European Community and the European Bank for Reconstruction and Development.

NATO stopped short of extending formal security guarantees to its former adversaries.[31] The alliance did, however, seek to demonstrate in the clearest possible terms its interest in sheltering the fledgling democracies from outside pressures. In June 1991, the North Atlantic Council stated that "our own security is inseparably linked to that of all other states in Europe" and that "their freedom from any form of coercion or intimidation [is] therefore of direct and material concern" to the members of NATO.[32] During the attempted coup in

Moscow two months later, the NATO foreign ministers pointedly declared that they expected the Soviet Union "to respect the integrity and security of all states in Europe."[33] And the Partnership for Peace offered former Soviet bloc countries formal consultation with the alliance in the event of an external threat as well as increasingly strong de facto ties through practical military cooperation.

Finally, NATO strongly supported the further institutionalization of the Conference on Security and Cooperation in Europe (CSCE). Not only was it hoped that the CSCE might eventually be developed into an effective mechanism for conflict prevention, crisis management, and the peaceful resolution of disputes, but, like the NACC, it could be used to integrate the new democracies into a common European home and otherwise to help address their security concerns.

The Transformation of NATO's Conventional Force Posture

In sum, NATO's new grand strategy contained a much reduced military component. With the end of the traditional threat to Western Europe, military means were no longer appropriate for addressing the alliance's principal security concerns. Military forces would still play an important role, but unlike the Harmel Report, the shift to a post–Cold War grand strategy and the profound alteration of NATO's security environment that this shift represented dictated significant changes in both the alliance's military strategy and its force structure.

The critical steps in the transformation of NATO's military posture were completed in well under two years. Initially, the process was driven largely by the need to make German unification and continued German membership in NATO acceptable to the Soviet Union.[34] In early February 1990, as the headlong rush toward German unification accelerated, German chancellor Helmut Kohl called for a thorough review of the alliance's military strategy and force structure.[35] Then in July, when the negotiations with the Soviet Union reached a critical stage, the alliance announced that it would prepare a new military strategy and elaborate new force plans.[36]

The role of such tactical considerations should not be exaggerated, however. The negotiations over German unification may have influenced the timing of the process of modifying NATO's military posture, but the need to do so was dictated by the larger political and military changes taking place in Europe. Thus the process continued unabated even after Soviet agreement to unification largely on

Western terms had been secured. By June 1991, the NATO military authorities had developed detailed recommendations for the future structure of the alliance's forces.[37] And at their Rome summit in November of that year, the NATO heads of government approved a new strategic concept, which formally replaced flexible response.

THE ALLIANCE'S NEW STRATEGIC CONCEPT

The revised strategic concept clearly reflected the objectives established in NATO's new grand strategy.[38] The altered security environment posed a serious problem, however, when it came to identifying the specific military threats that the strategy was intended to address. The end of the East-West confrontation had virtually eliminated the danger of major conflict in Europe. In particular, the threat of a full-scale attack had effectively been removed, while the danger of a surprise attack on any scale had declined substantially, especially in the Central Region. Instead, the possibility of deliberate aggression instigated by the Soviet Union, or even an inadvertent conflict growing out of interbloc tension, had given way to an undefined set of diverse and multidimensional "risks."

Although unprecedented attention was paid to possible problems emanating from the southern Mediterranean and the Middle East, the principal risks were viewed as most likely to result from the instability that might arise out of the serious political, economic, and social difficulties faced by many countries in Central and Eastern Europe. These difficulties could spawn tensions that could in turn lead to crises and even to armed conflict. In this connection, the residual military capability and buildup potential of the Soviet Union or its successor states was singled out as the most significant factor that NATO would have to take into account in preserving an overall military balance on the continent. Just when and how a direct threat to the alliance might arise, however, could not be specified.

The responses the strategy directed the alliance to prepare were equally indeterminate. The most that could be said was that the allies "must maintain the forces necessary to provide a wide range of conventional response options." A significant shift did take place in the relative emphasis placed on nuclear and conventional forces, however. Nuclear weapons would continue to play several roles in NATO strategy. First, they would neutralize the coercive potential of the Russian nuclear arsenal, thereby shielding non-nuclear alliance members from intimidation. Second, they would deter nuclear attacks by nuclear-armed adversaries in a crisis or during conventional hostilities. Third, because of the possibility of escalation inherent in their presence, they would help to deter non-nuclear aggression by making the risks of conventional operations incalculable and unacceptable. Indeed, the alliance specifically ruled out a policy of no first use, declaring that "there are no circumstances in which nuclear retaliation in response to military action might be discounted."[39]

Nevertheless, NATO's reliance on nuclear weapons, especially substrategic systems of the shortest range, would be reduced as the alliance sought to make "nuclear forces truly weapons of last resort."[40] Accordingly, greater relative emphasis would be placed on conventional forces, which would assume primary responsibility for deterring and defending against non-nuclear attacks. In addition, they would play the leading role in managing crises and containing regional conflicts. Subsequently, the alliance approved in principle the use of its forces in support of peacekeeping operations outside the NATO area under the auspices of the Conference on Security and Cooperation in Europe and of the United Nations.[41]

Not surprisingly, the new strategic concept offered little detailed guidance for setting conventional force requirements. Indeed, the alliance's new force structure was largely worked out before the strategic concept was formally adopted. Nevertheless, the concept did establish several general principles concerning the form that NATO's conventional forces would take. First, and most important, they would be reduced in size and readiness. The alliance would maintain smaller active forces, moving away from "forward defense" toward a reduced forward presence in the Central Region, while placing greater reliance on mobilizable reserves, reinforcements, and prepositioned equipment. The remaining forces, however, would require greater mobility, flexibility, and versatility. At the same time, considerable emphasis was placed on maintaining the alliance's integrated military structure and even furthering the degree of integration through increased use of multinational units.[42] Finally, the strategic concept called for the continued presence of significant numbers of U.S. conventional forces in Europe.[43]

CHANGES IN NATO'S CONVENTIONAL CAPABILITIES

Consistent with the altered strategic environment and NATO's revised military strategy, the alliance's conventional forces also underwent considerable change. The contributions of each of the countries with forces in the Central Region declined substantially, and those units that remained were integrated into the new NATO force structure.

Ironically, the CFE treaty had relatively little impact on NATO force levels. The treaty itself contained no limits on manpower, and the restrictions it placed on major items of military equipment required only that the alliance reduce its tank holdings by approximately 10 percent.[44] Of much greater significance were the reductions announced by Germany during the negotiations over the terms of unification. Germany agreed to limit the number of military personnel to 370,000, including no more than 345,000 in the ground and air forces, within three to four years. Subsequently, the German army developed a new force structure consisting of 260,000 men in eight divisions and 28 brigades. Only seven brigades would be fully active while most of the rest would be maintained at only 50 percent strength in peacetime.[45] In the meantime, the Luftwaffe would be reduced to 500 combat aircraft, well below Germany's CFE ceiling of 900.[46] And in 1993, German leaders began to indicate that further cuts in the armed forces might be necessary after 1995 for financial reasons.[47]

The number of U.S. forces in the Central Region declined even more in relative terms. The Bush administration decided to reduce the size of the American military presence from two army corps and five full divisions to one corps and two divisions and from eight to three tactical fighter wings by the mid-1990s. The overall number of troops would be cut from over 300,000 to just 150,000, while the strength of the U.S. Army in Europe would drop from 200,000 to about 92,000 personnel.[48] Congressional sentiment, moreover, favored further reductions. Several prominent members of Congress indicated that U.S. troop strength on the continent would or should be brought down to between 75,000 and 100,000,[49] and in mid-1992, the House of Representatives approved a ceiling of 100,000 military personnel by 1995.[50] In early 1993, this figure was adopted by the new Clinton administration.[51] Thus even barring further cuts, the size of the American presence would be trimmed by at least two-thirds.

Each of the other countries with forces in the Central Region took similar actions. In July 1990, Britain announced plans to reduce its military presence on the continent by some 50 percent. Specifically, it would withdraw two of the three divisions and some 30,000 of the 55,000 men in the British Army on the Rhine, although the remaining armored division would have a full complement of three brigades. The Royal Air Force would redeploy nearly half of the fifteen squadrons in Germany to the British Isles.[52]

Almost simultaneously, France indicated that it would withdraw its force of three armored divisions and 50,000 military personnel from Germany. The first phase of the redeployment, involving 20,000 men, would be completed in 1992. The 4,000 troops in the Franco-German brigade would not be affected by the withdrawal, however, and following the October 1991 decision to form a joint Franco-German corps, France agreed to retain one complete division in Germany as well.[53]

Following the British and French examples, Belgium announced a major restructuring of its armed forces in December 1990. All but one brigade (3,500 men) of the 25,000-man Belgian corps based in Germany would be withdrawn, one active brigade would be disbanded, and the number of military personnel would be trimmed by some 20 percent. Then in July 1992, the government announced even deeper cuts that would result in a halving of the armed forces, to a total of only 40,000 men, and would reduce from seven to four the number of active squadrons.[54]

The Netherlands also moved to slash its military establishment by more than 50 percent during the 1990s. Initial plans called for reducing the armed forces by one-third. The army would consist of two smaller divisions of two brigades each, although an independent brigade would be maintained in Germany and a new airmobile brigade would be formed. A further review of Dutch defense priorities, completed in early 1993, however, set the stage for even deeper cuts, to as few as four brigades and 40,000 military personnel.[55] Finally, Canada announced that its entire contingent of one brigade and three squadrons would leave Germany by 1994.[56]

The remaining national contributions would be incorporated into the new force structure devised by the NATO commanders in 1991, which had three principal components: reaction forces, main defense forces, and augmentation forces. The centerpiece of the reaction forces would be a newly organized Rapid Reaction Corps under a British commander and consisting of four division-size units and 50,000 to 70,000 troops. The British would contribute one division based in the United Kingdom as well as the division remaining on the continent, and there would be two multinational divisions. Eventually, the corps would be deployable to potential trouble spots anywhere in NATO Europe on short notice.[57]

The main defense forces in Central Europe would consist of up to seven corps of 50,000 to 70,000 troops each. All but one of the corps would be multinational, each containing divisions or brigades from two or more countries. The remaining, all-German division would be based in eastern Germany.[58] Finally, the augmentation forces would come primarily from the United States, which was establishing a contingency corps on its own territory.

Theoretical Implications

This book has sought to demonstrate, among other things, the utility of the institutional perspective for explaining the high degree of stability that characterized NATO conventional force levels in the Central Region of Europe during the 1960s, 1970s, and 1980s. It has argued that the balance-of-power and intra-alliance perspectives alone cannot provide a satisfactory account of this history. Rather, these traditional approaches for understanding alliance behavior must be supplemented by an explanation grounded in institutional considerations that emphasizes the persistence of established patterns of behavior.

The sharp decline in alliance conventional force levels after 1989, however, demonstrates the limits of the institutional perspective. For the first time in many years, NATO's conventional force posture varied directly in response to changes in the alliance's external environment, as the balance-of-power perspective would predict. The regime that had governed the provision of conventional forces in the Central Region for more than three decades unraveled when the disappearance of the traditional threat to Western Europe caused the alliance's existing force levels, so often bemoaned as inadequate, suddenly to seem excessively high. In this case, the institutional pressures for continuity were overwhelmed.

Nevertheless, the experience of the end of the Cold War does not completely invalidate the conclusions of Chapter 7 about the important influence of institutional factors. First, the modification of NATO's conventional force posture only followed a dramatic decline in the threat. A more modest shift in Soviet capabilities and perceived intentions might not necessarily have resulted in regime change, as was often the case in the past.

This observation corresponds with previous findings, which suggest that international institutions enjoy considerable autonomy and are difficult to modify.[59] They do not respond in a rapid and fluid way to changes in their environment. To the contrary, institutional arrangements are likely to persist even though the underlying distribution of power and interests that created them change. Adjustment will occur only when substantial disjunctures develop between prevailing institutional structures and their environments. Thus institutional change is episodic and dramatic rather than continuous and incremental.[60]

Moreover, NATO's conventional force posture was not completely dismantled, as a simplistic application of the balance-of-power perspective might predict. As discussed in Chapter 1, states typically form alliances in response to commonly perceived threats.[61] When a threat dissipates, it is reasonable to doubt whether the countering alliance will be maintained. From this perspective, the virtual disappearance of the Soviet threat could have heralded NATO's demise.

Not only did the alliance survive, but it retained many of its traditional military trappings. Both a common strategy and the integrated military structure were maintained, and the United States and Britain continued to station significant forces on the continent. Thus this episode also demonstrates the persistence of institutional arrangements, even in the face of considerable external change.

To some extent, this persistence can be explained in functional terms. Even at the height of the Cold War, deterrence of Soviet aggression was not NATO's only purpose. The integrated military structure also served to manage the German problem.[62] Thus its preservation was seen as helping to reassure the countries of the region with respect to a more powerful and less constrained Germany.

A complete explanation, however, must also include an institutional component. Past institutional choices tend to limit future institutional alternatives. As John Ikenberry has observed, "Institutional reform is carried out within an existing array of organizations and structures that shape and constrain any efforts at change."[63] In addition, as argued above, international institutions may put down domestic and cognitive roots that are not easily extirpated.

These observations suggest the usefulness of conceptualizing the relationship between the balance-of-power and intra-alliance perspectives, on the one hand, and the institutional perspective, on the other hand, in the following manner.[64] The factors emphasized by the first two perspectives set general limits on alliance behavior as well as the possible forms that alliance institutions may take.[65] Thus a variety of specific institutional arrangements may be consistent with a given international distribution of power, for example. Rarely if ever, however, will institutions promote behavior that is sharply at odds with the dictates of the external and intra-alliance environments. Ultimately, institutional content is conditioned by international structure and national interests.

Yet the international and intra-alliance settings do not dictate the details of the institutions that arise within them. And once institutions form, they further direct and constrain behavior within the broad parameters established by these settings. As long as external and intra-alliance conditions do not change substantially, institutional rules will have a more direct bearing on state action. Thus, returning to the subject of this book, the balance-of-power and intra-alliance perspectives account for the range of NATO conventional forces levels that were possible, while the institutional perspective is necessary to explain why a particular set of force levels within that range was adhered to.

Reference Matter

Abbreviations

ACDA	Arms Control and Disarmament Agency
AIR	Air Ministry
AWF	Ann Whitman File
BP	Bulky Package
CAB	Cabinet Office
CBO	Congressional Budget Office
CCS	Combined Chiefs of Staff
CJCS	Chairman of the Joint Chiefs of Staff
CNP	Committee on Nuclear Proliferation
COS	Chiefs of Staff
DC	Defence Committee
DDE	Dwight D. Eisenhower
DDEL	Dwight D. Eisenhower Library
DDEP	Dwight D. Eisenhower Papers
DDRS	*Declassified Documents Reference System*
DEFE	Ministry of Defence
DMP	Draft Memorandum for the President
DOD	Department of Defense
DSB	*Department of State Bulletin*
EDC	European Defense Community
FO	Foreign Office
FRUS	*Foreign Relations of the United States*
FY	Fiscal Year
GAO	General Accounting Office
HSTL	Harry S. Truman Library
HSTP	Harry S. Truman Papers

IMTF	International Meetings and Travel File
IISS	International Institute for Strategic Studies
JCS	Joint Chiefs of Staff
JFDP	John Foster Dulles Papers
JFKL	John F. Kennedy Library
JIC	Joint Intelligence Committee
LBJL	Lyndon Baines Johnson Library
LTDP	Long Term Defense Program
MC	Military Committee
MSA	Mutual Security Agency
MTDP	Medium Term Defense Plan
NA	National Archives
NAC	North Atlantic Council
NAT	North Atlantic Treaty
NATO	North Atlantic Treaty Organization
n.d.	no date
NFC	*NATO Final Communiques*
NIE	National Intelligence Estimate
NIS	NATO Information Service
NSC	National Security Council
NSF	National Security File
OSANSA	Office of the Special Assistant for National Security Affairs
OSD	Office of the Secretary of Defense
OSS	Office of the Staff Secretary
OTA	Office of Technology Assessment
POF	President's Office Files
PPP	*Public Papers of the Presidents of the United States*
PREM	Prime Minister's Office
PRO	Public Record Office
RG	Record Group
SACEUR	Supreme Allied Commander Europe
SDAH	State Department Administrative History
SHAPE	Supreme Headquarters Allied Powers Europe
TCC	Temporary Council Committee
VPSF	Vice Presidential Security File
WHCF	White House Central Files
WHO	White House Office

Notes

Introduction

1. Some of the best surveys are Barnet, *Alliance*; Calleo, *Beyond Hegemony*; Cleveland, *NATO*; Grosser, *Western Alliance*; and Joffe, *Limited Partnership*.

2. Important studies include Buteux, *Strategy, Doctrine, and the Politics of Alliance*; Daalder, *Nature and Practice of Flexible Response*; Kelleher, *Germany*; and Schwartz, *NATO's Nuclear Dilemmas*.

3. Of particular value are Dockrill, *Britain's Policy*; Ireland, *Creating the Entangling Alliance*; Kaplan, *Community of Interests*; Osgood, *NATO*; Riste, *Western Security*; and Stromseth, *Origins of Flexible Response*.

4. Useful monographs include Raj, *American Military*; Treverton, "*Dollar Drain*"; and Williams, *Senate and U.S. Troops*.

5. See, for example, Pierre, ed., *Conventional Defense of Europe*; Mako, *U.S. Ground Forces*; Steering Group, *Strengthening Conventional Deterrence*; Flanagan, *NATO's Conventional Defense*; and Epstein, *Conventional Force Reductions*. Numerous journal articles have also addressed these questions.

6. For comprehensive surveys of the literature, see Holsti, Hopmann, and Sullivan, *Unity and Disintegration*; and Ward, "Research Gaps."

7. See, for example, Walt, *Origins of Alliances*.

8. Most of the literature on alliance behavior concerns the single issue of burden sharing among alliance members. An important exception is the work of Glenn H. Snyder, especially "Security Dilemma in Alliance Politics" and "Alliance Theory."

9. Other examples include the Entente Cordial and British and French military cooperation from early 1939 until the German attack on France in May 1940. On the Entente Cordial, see Williamson, *Politics of Grand Strat-*

egy. On the later Anglo-French staff talks, see Young, *In Command of France,* chap. 9; and Mearsheimer, *Conventional Deterrence,* chap. 3.

10. This purpose corresponds to what Sidney Verba, Harry Eckstein, and Alexander George have termed the disciplined-configurative case study, which employs available theories to explain the outcome of a particular case. For further discussion, see Eckstein, "Case Study and Theory," and George, "Case Studies and Theory Development."

It should be emphasized that the primary theoretical purpose of this study is not theory testing per se. Rather, the goal is to explain the evolution of NATO's conventional force posture. The empirical focus was chosen because of its intrinsic importance, not because it would serve as a crucial case study that would invalidate or confirm one theory or another.

11. For an insightful application of these theories to a related set of issues, see Kupchan, "NATO and the Persian Gulf."

12. See, esp., Krasner, ed., *International Regimes;* Keohane, *After Hegemony;* Aggarwal, *Liberal Protectionism;* and Moravcsik, "Disciplining Trade Finance."

Chapter 1

1. Before German unification, the NATO Central Region was defined as including Belgium, the Netherlands, Luxembourg, and the Federal Republic of Germany, less its territory north of the Elbe River. Although West Germany traditionally stationed a small fraction of its forces outside of the Central Region in Schleswig-Holstein, the figures provided in this book are for all West German forces. The small Berlin-based forces of the United States, Britain, and France, however, are excluded.

Little attention is paid to other aspects of the alliance's overall military posture, such as its nuclear forces and the doctrine governing their use, which have already been the subject of considerable analysis, naval forces, and those conventional forces that have been stationed on NATO's northern and southern flanks. Similarly, it does not explicitly consider alliance diplomacy or arms control policy except when these have had a direct bearing on NATO's conventional force posture.

2. For similar definitions, see Posen, *Sources of Military Doctrine,* 13–14; and Osgood, *NATO,* 5.

3. For a similar decomposition, see Osgood, *NATO,* 7.

4. Ibid., 9.

5. For a discussion of the distinction between denial and punishment, see Snyder, *Deterrence and Defense,* 14–16. Thomas Schelling analyzes the third type of response, which he terms the "manipulation of risk," in *Arms and Influence.*

6. Such a strategy is proposed in Huntington, "Conventional Deterrence and Conventional Retaliation."

7. For instructive examples of efforts to compare the explanatory power of different theories and models, see Posen, *Sources of Military Doctrine;* Al-

lison, *Essence of Decision;* and Walt, *Origins of Alliances.* Posen, for example, deduces specific propositions from balance-of-power theory and organization theory and applies them to British, French, and German interwar military doctrine "to see which theory better explains/predicts what happened" (p. 8).

8. Indeed, where the dependent variable can be disaggregated, as with NATO's conventional force posture, pitting the perspectives against one another in an attempt to establish a monocausal explanation is likely to be both logically and methodologically unsound. See Moravcsik, "Disciplining Trade Finance," 174–75.

9. Analogously, Jack Snyder has sought to explain military strategy in terms of the interaction of three general determinants—motivational biases, doctrinal simplifications, and rational calculations—which are in turn shaped by a variety of organizational, cognitive, and strategic variables. See his *Ideology of the Offensive.*

10. Given the complexity of the subject, one could argue that they are required to provide a satisfactory explanation. As a leading student of alliances has noted, "The specific commitments that allies accept will reflect a host of idiosyncratic features that are unlikely to be easily generalized" (Walt, *Origins of Alliances,* 13).

11. On this distinction, see Gilpin, *War and Change,* chap. 2.

12. For a clear articulation of balance-of-power theory, see Waltz, *Theory of International Politics,* chap. 6.

13. See, for example, Posen, *Sources of Military Doctrine,* 61; and Waltz, *Theory of International Politics,* 131. On some of the difficulties of measuring military power, see Daalder, *Nature and Practice of Flexible Response,* 23–25.

14. A more general formulation of balance-of-power theory is offered by Walt, *Origins of Alliances,* who emphasizes the role of threats, rather than power alone, in determining international behavior. Walt defines threats as a function of offensive capabilities, geographical proximity, and perceived intentions as well as aggregate power.

15. Perceptions of an adversary's intentions may be highly subjective. Without access to the thoughts and deliberations of foreign leaders, their intentions may be extremely hard to discern. The difficulties of fathoming Soviet intentions toward Western Europe—and even Soviet military capabilities—were especially formidable because of the closed nature of Soviet society during most of the postwar era. Because more reliable forms of evidence are frequently lacking, perceptions of intentions are often simply inferred from behavior. For further discussion of the problems of threat perception, see Knorr, "Threat Perception," 78–119.

16. For example, Posen has argued that the influence of technology on strategy is seldom direct (*Sources of Military Doctrine,* 236).

17. The following discussion is based on Knorr, *Power of Nations,* esp. chap. 3; and Aron, *Peace and War,* chap. 2.

18. For a perceptive discussion of how successive U.S. administrations

have assessed the trade-off between costs and risks in American grand strategy, see Gaddis, "Containment and the Logic of Strategy."

19. In alliances characterized by a single dominant member, this distinction corresponds roughly to that between benevolent and coercive hegemonic leadership. See Snidal, "Limits of Hegemonic Stability Theory," 585–90.

20. The Soviet-dominated Warsaw Pact approximated this type of alliance.

21. Snyder, "Alliance Theory," 113.

22. The seminal article is Olson and Zeckhauser, "Economic Theory of Alliances." For a valuable recent analysis, see Oneal, "Testing the Theory of Collective Action."

23. For further discussion, see Duffield, "International Regimes and Alliance Behavior," 827–28.

24. See Waltz, *Theory of International Politics*, 169; and Morgenthau, *Politics Among Nations*, 178–79.

25. Thus in the case of a consensual alliance such as NATO, a focus on the distribution of military power yields somewhat contradictory expectations. The more powerful countries should enjoy greater influence over the details of alliance strategy, but they will find that they must provide the bulk of the forces required to implement the strategy.

26. See, for example, Daalder, *Nature and Practice of Flexible Response*, 3, 22.

27. One should be careful not to rely too heavily on geography to infer strategic preferences, however. Such predictions are necessarily somewhat indeterminate. One might expect to find that the states that are physically closest to the potential adversary and thus most subject to invasion and occupation will demand a relatively high degree of alliance military preparation. But if the adversary is relatively well behaved in that region and less so elsewhere, states with a purely regional perspective may perceive less of a threat than those with a broader outlook and may thus be satisfied with a lower level of military effort.

28. See Morgenthau, "Alliances in Theory and Practice," 190.

29. Osgood, *NATO*, 20.

30. For useful discussions of both the impact of institutions and the sources of their persistence, see Krasner, "Approaches to the State"; Krasner, "Sovereignty"; and Ikenberry, "Conclusion."

31. Young, *International Cooperation*, 5, 6, 13.

32. Ibid., 15–16. Similarly, Robert Keohane has defined international institutions, including regimes, as "persistent and connected sets of rules, formal and informal, that prescribe behavioural roles, constrain activity, and shape expectations" ("Multilateralism," 732). And an early consensual definition of regimes described them as "sets of implicit or explicit principles, norms, rules, and decision-making procedures around which actors' expectations converge in a given area of international relations" (Krasner, "Structural Causes and Regime Consequences," 2).

Norms and rules are also central in the sense that in their absence, other regime elements, such as those creating incentives to comply, would serve no meaningful purpose. In what follows, the terms "rules," "norms," "conventions," "standards," "injunctions," and "guides to behavior" will be used interchangeably.

33. Surprisingly little of the regime literature addresses the question of whether and how regimes actually influence state conduct. See Haggard and Simmons, "Theories of International Regimes." The best general treatments of this question are Young, *International Cooperation*, 70–80; and Keohane, *After Hegemony*, 98–106, 237–40.

34. The first approach is most fully developed in Keohane, *After Hegemony*. The second is suggested by Rosenau, "Before Cooperation." For similar distinctions, see Young, *International Cooperation*, 210–12; and Bull, *Anarchical Society*, 139. Below, I seek to identify more explicitly the domestic and cognitive sources of state "habits."

35. For further discussion of the role played by concern for reputation, see Axelrod, "Evolutionary Approach to Norms," 1107–8.

36. Keohane, *After Hegemony*, 103–6.

37. Young, *International Cooperation*, 20, 77–82; and Nye, "Nuclear Learning." On the internalization of norms, see Alexrod, "Evolutionary Approach to Norms," 1104. See also Keohane's discussion of the implications of bounded rationality in *After Hegemony*, 111–16.

38. For a useful general discussion of organizational behavior in the national security arena, see Halperin, *Bureaucratic Politics*.

39. For an analysis of how regimes can empower new groups, see Haas, "Do Regimes Matter?" The role of firms in promoting compliance with the principle of free trade is discussed in Milner, *Resisting Protectionism*.

40. Chayes, "Inquiry into the Workings of Arms Control Agreements."

41. In addition to Halperin, *Bureaucratic Politics*, see Allison, *Essence of Decision*, and Chayes, "Inquiry into the Workings of Arms Control Agreements," 935–46.

42. Useful overviews of the impact of cognitive factors on international relations are provided by Jervis, *Perception and Misperception*, esp. chap. 4, and Steinbruner, *Cybernetic Theory of Decision*, chap. 4. The role of the development of shared knowledge and cause-and-effect beliefs in reinforcing regime compliance is discussed in Haas, "Do Regimes Matter?"

43. Young, *International Cooperation*, 71.

44. See Haggard and Simmons, "Theories of International Regimes," 513–14.

45. On the relationship between regimes and organizations, see Young, *International Cooperation*, chap. 2.

46. This step corresponds to the "congruence procedure" for assessing the impact of beliefs on decisional choices, which is discussed in George, "Causal Nexus," 105–13.

47. This step corresponds to the "process-tracing procedure," which is also discussed in George, "Causal Nexus," 113–19.

Chapter 2

1. The strategic concept, which was formally designated Defense Committee document 6/1, or DC 6/1, is reprinted in *FRUS, 1949*, 4: 352–56. Its development is described in Rearden, *History of the OSD*, 481–82; K. Condit, *History of the JCS*, 399–400; and *FRUS, 1950*, 3: 1–3. On the development of the MTDP, see K. Condit, *History of the JCS*, 400–407; and D. Condit, *History of the OSD*, 311–14.

2. U.S. war plans are described in K. Condit, *History of the JCS*, chap. 9.

3. D. Condit, *History of the OSD*, 312. The section of the MTDP concerned with Western Europe was probably based on the plans previously prepared by the Chiefs of Staff Committee of the Western Union, which was established in 1948 and became the NATO planning group for the region in 1949 (K. Condit, *History of the JCS*, 398). These plans are discussed in Kugler, *Laying the Foundations*, 38–40.

4. Wampler, "NATO Strategic Planning," 8; and Trachtenberg, *History and Strategy*, 159.

5. K. Condit, *History of the JCS*, 405–6; and Poole, *History of the JCS*, 184–85.

6. *FRUS, 1949*, 4: 355. The United States would be primarily responsible for strategic bombing and defense of sea lines of communication.

7. "USSR and Satellite Ground Forces Strength and Disposition Report," Apr. 21, 1950, RG 218, CCS 319.1 USSR (7-12-49), sec. 1. Another two Soviet divisions were stationed in Romania at the time—for a total of 30 in all of Eastern Europe—but these were unlikely to be used in an attack in the Central Region. Despite the apparent magnitude of these forces, their effectiveness could be questioned on several grounds. For further analysis, see Duffield, "Soviet Military Threat."

8. The senior Western Union military official, British Field Marshal Bernard Montgomery, felt that only the two U.S. divisions were capable of staging effective resistance (*FRUS, 1950*, 3: 143).

9. Williams, *Senate and U.S. Troops*, 23; and Ireland, *Creating the Entangling Alliance*, 184.

10. On Truman's fiscal conservatism, see Jervis, "Impact of the Korean War," 568.

11. Mako, *U.S. Ground Forces*, 4, 7. According to some analysts, the United States would have been unable to dispatch more than a single division anywhere without resorting to partial mobilization. See Osgood, *NATO*, 29; and Huntington, *Common Defense*, 40.

12. Wells, "First Cold War Buildup," 182; and JCS 1868/191, June 22, 1950, RG 218, CCS 092 Western Europe (3-12-48), sec. 49.

13. Jervis, "Impact of the Korean War," 571; and Osgood, *NATO*, 38.

14. Osgood, *NATO*, 37.

15. Kaplan, *Community of Interests*, 71. See also Osgood, *NATO*, 37, n. 16. U.S. military planners estimated that even if continued through 1954 at its original level, the program would have provided for only a small increase in

the number of major European combat units, both ready and mobilizable, although the combat effectiveness of the existing forces would have been improved considerably. See Memorandum, "Approximation of Friendly Forces," June 7, 1950, RG 218, CCS 092 Western Europe (3-12-48), sec. 49.

16. Memorandum for the Standing Group of NATO, Mar. 22, 1950, RG 218, CCS 092 Western Europe (3-12-48), sec. 44.

17. Martin, "American Decision," 648–49; Memorandum for the Secretary of Defense, "Protest to the Soviet Government Concerning East German Militarized Police," May 2, 1950, RG 218, CCS 092 Germany (5-4-49), sec. 1; and Poole, *History of the JCS,* 192.

18. *FRUS, 1950,* 4: 680; Martin, "American Decision," 646–50; Ireland, *Creating the Entangling Alliance,* 186; Trachtenberg, " 'Wasting Asset,' " 48; and *FRUS, 1950,* 3: 1015, 1021, 1064. On Acheson's views, see Leffler, *Preponderance of Power,* 345, 349–51. The definitive State Department position was set forth in NSC 71/1, "Department of State Views on German Rearmament," July 3, 1950, in *FRUS, 1950,* 4: 691–95, which was drafted before the Korean War.

19. Dockrill, "Evolution of Britain's Policy," 38–40. For the French perspective, see McGeehan, *German Rearmament Question,* 14–15.

20. For a more detailed analysis, see Leffler, *Preponderance of Power,* 325–33.

21. *FRUS, 1950,* 1: 141–42.

22. NSC 68 is reprinted in *FRUS, 1950,* 1: 234–92. For conversations and commentary associated with the drafting of NSC 68, see ibid., 142–226. Two important but somewhat conflicting interpretations of the document are Gaddis, *Strategies of Containment,* chap. 4, and Leffler, *Preponderance of Power,* 355–60.

23. NSC 68 estimated that at that time, the Soviet Union would be able to drop 100 atomic bombs on the United States, a number sufficient to destroy the latter's superior war-making potential (*FRUS, 1950,* 1: 251; and Trachtenberg, *History and Strategy,* 107–8).

24. *FRUS, 1950,* 3: 110–11, 1007, 1019, 1034–35. These objectives were clearly spelled out in two State Department position papers prepared for the NATO meeting: "Building Up the Defensive Strength of the West," May 3, 1950, ibid., 85–90, and "U.S. Objectives and Courses of Action in the May Meetings," Apr. 28, 1950, ibid., 1001–6. They were also embodied in two resolutions that Acheson placed before the council. See ibid., 87–88.

25. Warner, "British Labour Government," 257; and *FRUS, 1950,* 3: 94–95, 109–11, 119–20.

26. *FRUS, 1950,* 1: 234–35, 293–96.

27. Ibid., 293–98; Poole, *History of the JCS,* 16; and Rearden, *History of the OSD,* 534–35. Although the JCS plan would have required doubling the size of the defense budget, the army force structure would have increased by only two divisions, or approximately 20 percent.

28. See, for example, the discussions in *FRUS, 1950,* 1: 298–313.

29. On the belief that the Soviet Union was behind the North Korean

action, see ibid., 332, and Leffler, *Preponderance of Power*, 366–67. For the reactions of top U.S. officials, including Truman, Acheson, Secretary of Defense Louis Johnson, and Chairman of the Joint Chiefs of Staff Omar Bradley, see Truman's special message to Congress, in *PPP: Harry S. Truman, 1950*, 527–37; Osgood, *NATO*, 69; D. Condit, *History of the OSD*, 10–11; Martin, "American Decision," 651; and *FRUS, 1950*, 3: 232–37.

30. *FRUS, 1950*, 3: 1667. See also NIE-3, "Soviet Capabilities and Intentions," Nov. 15, 1950, *FRUS, 1950*, 1: 415.

31. On Soviet cautiousness in the short term, see *FRUS, 1950*, 1: 333, 367; and Leffler, *Preponderance of Power*, 367, 383–84. On the parallels seen between Korea and Germany, see Schwartz, "'Skeleton Key,'" 373. On the threat posed by the East German security forces, see Senate, Foreign Relations, *Reviews of the World Situation, 1949–1950*, 320; *FRUS, 1950*, 4: 717; and CP(50)210, Sept. 18, 1950, PRO, CAB 129/42.

32. See, for example, *FRUS, 1950*, 1: 342–46. For a summary of French views, see *FRUS, 1950*, 3: 1383–87.

33. The British cabinet, for example, decided to spend an additional £100 million on defense over the next two years. See CP(50)181, July 31, 1950, PRO, CAB 129/41. France increased its planned defense budget for 1951 from $1.2 billion (of which approximately $400 million was being spent in Indochina) to $1.43 billion and declared that it would accelerate the pace of the program to rebuild its air force. See *FRUS, 1950*, 3: 1386, 1444.

34. *FRUS, 1950*, 3: 133, 136.

35. Memorandum, Lemnitzer to Burns, July 17, 1950, RG 330, CD 092.3 NATO (General) 1950; and Memorandum, Lemnitzer to Bradley, "Estimate of Additional Funds for Mutual Defense Assistance," July 19, 1950, RG 218, Bradley Files, CJCS 092.2 NAT (Apr. 7–July 1950).

36. *FRUS, 1950*, 3: 130, 133, 136–37; and Kaplan, *Community of Interests*, 104. On July 31, Truman formally requested that Congress appropriate an additional $4 billion in military assistance, $3.5 billion of which would be earmarked for NATO. See *FRUS, 1950*, 3: 168; and *DSB* 23 (Aug. 14, 1950): 247.

37. *FRUS, 1950*, 3: 138–41.

38. See Acheson's remarks of July 24 in Senate, Foreign Relations, *Reviews of the World Situation, 1949–1950*, 320, 329.

39. CP(50)181; *FRUS, 1950*, 3: 1669–73; Dean and Gary, "Military and Economic Strength," 121–22; and Wall, *United States*, 195–96.

40. *FRUS, 1950*, 3: 184, 209.

41. Ibid., 228–30, 241–44. The British, for example, were counting on £550 million in aid, which meant that they intended to increase their annual defense spending by just 20 percent over the pre-Korea levels, from £780 million to £950 million. See CP(50)181.

42. *FRUS, 1950*, 3: 162.

43. Ibid., 184–89, 232–36.

44. Ibid., 131, 148, 166, 170–71, 206.

45. Ibid., 182–83. On August 10, the president indicated that he was favorably inclined toward the eventual creation of a unified NATO command and

further U.S. troop deployments. See ibid., 205. On the evolution of the secretary of state's views, see Acheson, *Present at the Creation*, 435–37.

46. *FRUS, 1950*, 3: 167–68, 157, 180–82, 190–95, 206.

47. Ibid., 211–19. The State Department's decision to propose the establishment of a European defense force may have been influenced by the call of the Council of Europe's Consultative Assembly for the immediate creation of a unified European army under a European minister of defense on August 11. See Acheson, *Present at the Creation*, 438; and *FRUS, 1950*, 3: 261.

48. *FRUS, 1950*, 3: 211, 226–27.

49. JCS 2124/18, Sept. 1, 1950, RG 218, CCS 092 Germany (5-4-49), sec. 3. Another JCS study objected that the discussion of military arrangements did not fall within the competence of the State Department. See JCS 2124/16, Aug. 26, 1950, RG 218, CCS 092 Germany (5-4-49), sec. 1.

50. *FRUS, 1950*, 3: 250–51.

51. JCS 2116/20, "Position on Recommendations to Be Submitted to the President Regarding a European Defense Force and Related Matters," Aug. 29, 1950, RG 218, CCS 337 (4-19-50), sec. 3.

52. *FRUS, 1950*, 3: 273–78. See also JCS 2073/61, "U.S. Participation in the Defense of Western Europe," Sept. 7, 1950, *Records of the JCS*, reel 5, 618–31. On the details of the negotiations, see McLellan, *Dean Acheson*, 329; and Martin, "American Decision," 657. Truman approved the State-Defense paper on September 11, just a day before a preliminary meeting of the U.S., British, and French foreign ministers. See *FRUS, 1950*, 3: 273.

53. Acheson subsequently wrote that the State Department's concessions to the JCS position were necessary to obtain Pentagon acceptance of a unified command (*Present at the Creation*, 440).

54. *FRUS, 1950*, 3: 261–62, 267–68.

55. Ibid., 285–88, 1191–1201; and CP(50)223, Oct. 6, 1950, PRO, CAB 129/42.

56. *FRUS, 1950*, 3: 1208, 293–94, 306, 310.

57. CP(50)223; and *FRUS, 1950*, 3: 302, 1221.

58. *FRUS, 1950*, 3: 150; *FRUS, 1950*, 4: 716–21; DO(50)17th Meeting, Minute 4, "UK Contribution to the Defence of Western Europe," Sept. 1, 1950, PRO, CAB 131/8; and Dockrill, *Britain's Policy*, 24–28.

59. CP(50)223; *FRUS, 1950*, 3: 1120; and Dockrill, *Britain's Policy*, 34–35.

60. *FRUS, 1950*, 3: 287–88, 296–99, 312, 339, 1120; and CP(50)223. The French government informed its allies that it would not be ready to consider the problem for nine months. See *FRUS, 1950*, 3: 342.

61. CP(50)223; *FRUS, 1950*, 3: 344; and Ismay, *NATO*, 185–86. For the text of the council's resolution, see *FRUS, 1950*, 3: 350–52.

62. McGeehan, *German Rearmament Question*, 63–64; and Dockrill, *Britain's Policy*, 41–42. The text of the Pleven Plan is reprinted in Carlyle, ed., *Documents*, 339–44.

63. *FRUS, 1950*, 3: 1442; Dockrill, "Evolution of Britain's Policy," 41; Harrison, *Reluctant Ally*, 28; and McGeehan, *German Rearmament Question*, 62–67.

64. *FRUS, 1950*, 3: 382–85, 405, 411–12, 1690; Dockrill, *Britain's Policy*, 42–43; and McGeehan, *German Rearmament Question*, 67–74. For a detailed statement of the U.S. position, see *FRUS, 1950*, 3: 428–31.

65. For records of the meetings, see *FRUS, 1950*, 3: 415–28. See also DO(50)21st Meeting, Minute 4, Nov. 8, 1950, PRO, CAB 131/8, for the British perspective. Marshall had replaced Johnson on September 21.

66. Acheson, *Present at the Creation*, 444, 459; and D. Condit, *History of the OSD*, 323–25.

67. *FRUS, 1950*, 3: 382, 497; and Senate, Foreign Relations, *Reviews of the World Situation, 1949–1950*, 372. A British official reported that the United States was convinced that war was inevitable and that it was almost certain to occur within the next eighteen months. See Warner, "British Labour Government," 259.

68. *FRUS, 1950*, 3: 394–95.

69. For a discussion of German attitudes, see McGeehan, *German Rearmament Question*, 67–74.

70. *FRUS, 1950*, 3: 394–95.

71. Ibid., 1442, 497, 505–6, 520; Senate, Foreign Relations, *Reviews of the World Situation, 1949–1950*, 377–78, 417; and Dockrill, *Britain's Policy*, 45–46.

72. *FRUS, 1950*, 3: 457–60, 471–72, 501–5, 515–17, 531–38.

73. Ibid., 185–88; and JCS 2073/81, Oct. 17, 1950, RG 218, CCS 092 Western Europe (3-12-48), sec. 60. The 23-division figure assumed a British contribution of at least 2⅓ divisions.

74. DO(50)19th Meeting, Oct. 16, 1950, PRO, CAB 131/8.

75. Ibid.; and DO(50)84, Oct. 13, 1950, PRO, CAB 131/9.

76. According to one estimate, the success of the revised program would depend on increasing total exports by £100 million, notwithstanding the diversion of industrial capacity and resources to the defense effort, and would require a reduction in the standard of living. See Warner, "British Labour Government," 259–60.

77. *FRUS, 1950*, 3: 1396–1401.

78. Ibid., 1398, 1418–19, 1444, 1448; and Wall, *United States*, 200–203.

79. For the minutes of the meeting, see *FRUS, 1950*, 3: 584–604.

80. Poole, *History of the JCS*, 221.

81. The document was a composite of two reports that treated the political and military aspects separately. For the texts of the two original reports, see *FRUS, 1950*, 3: 531–47. See also Poole, *History of the JCS*, 218–19.

82. Bell, *Negotiation from Strength*, 53; Poole, *History of the JCS*, 216–18; and *FRUS, 1950*, 3: 1748.

83. McGeehan, *German Rearmament Question*, 85.

84. *FRUS, 1951*, 3: 578–82; and Senate, Foreign Relations, *Reviews of the World Situation, 1949–1950*, 446.

85. For firsthand accounts, see Goodpaster, "Development of SHAPE"; and Knowlton, "Early Stages."

86. JCS 2073/201, Sept. 7, 1951, RG 218, CCS 092 Western Europe (3-12-48), sec. 93; and Poole, *History of the JCS*, 274, 276.

87. Much of the Great Debate is reprinted in Senate, Foreign Relations and Armed Services, *Assignment of Ground Forces.* The most thorough analyses of the Great Debate are Raj, *American Military*, 26–47; and Williams, *Senate and U.S. Troops*, chap. 3.

88. See Williams, *Senate and U.S. Troops*, 48–49.

89. Ibid., 48, 54–55; Ireland, *Creating the Entangling Alliance*, 209–10; and Senate, Foreign Relations and Armed Services, *Assignment of Ground Forces*, 683–85.

90. Williams, *Senate and U.S. Troops*, 59–62.

91. A related argument was that the president lacked the power to assign U.S. forces to an integrated international army. Because participation in such a force involved international agreement, the consent of the Congress was once again required (ibid., 62, 64–65, 81–82).

92. Ibid., 64; Ireland, *Creating the Entangling Alliance*, 210; and Senate, Foreign Relations and Armed Services, *Assignment of Ground Forces*, 89–93.

93. Senate, Foreign Relations and Armed Services, *Assignment of Ground Forces*, 64, 81.

94. See esp. Acheson's testimony, ibid., 79–80.

95. Poole, *History of the JCS*, 222–23.

96. See esp. the testimony of Eisenhower and Marshall in Senate, Foreign Relations and Armed Services, *Assignment of Ground Forces*, 10, 19, 40–41, 49–50; and Senate, Foreign Relations, *Reviews of the World Situation, 1949–1950*, 431–32.

97. Ireland, *Creating the Entangling Alliance*, 209; and Williams, *Senate and U.S. Troops*, 48.

98. Ireland, *Creating the Entangling Alliance*, 211–12.

99. Raj, *American Military*, 49–50.

100. Poole, *History of the JCS*, 243, 302.

101. In May, U.S. officials continued to estimate gaps of 4 ready, 11 D+30, and 12 D+90 divisions in 1954, not including prospective German forces, against requirements of 49⅓, 79⅓, and 95⅓ divisions, respectively. In the Central Region, the differences were expected to be 1⅓ ready, 9⅔ D+30, and 2⅔ D+90 divisions, against requirements of 32, 54, and 58 divisions, respectively. See Memorandum for the Secretary of Defense, "Forces for the Medium Term Defense Plan," May 28, 1951, RG 218, CCS 092 Western Europe (3-12-48), sec. 82.

102. *FRUS, 1950*, 1: 469–71; and *FRUS, 1951*, 3: 30, 63.

103. *FRUS, 1951*, 3: 59.

104. The increases in the D-Day and D+30 gaps nevertheless owed to different causes. The former was caused primarily by a downward revision in estimated national contributions, by ten divisions in the Central Region and fifteen overall. The latter was largely the result of an upward revision of NATO's D+30 force requirements, by eleven divisions in the Central Region and eighteen overall. SHAPE's estimate of national force contributions at D+30 had also increased, although not nearly as much as the requirements. See JCS 2073/201 and Tables 2.2 and 2.3.

105. Specifically, Western Europe was expected to spend $45 billion on defense, assuming $2.0 billion per year in U.S. economic support. Of this total, however, some $5.0 billion would be used outside of the NATO area. Thus to prevent a financial gap, the United States would have to provide a total of $25.4 billion. See *FRUS, 1950*, 1: 434.

106. *FRUS, 1951*, 3: 3–5, and 29.

107. Ibid., 193–96, 248–52; and *FRUS, 1951*, 1: 360–74.

108. Poole, *History of the JCS*, 241.

109. *FRUS, 1951*, 3: 5. France, for example, experienced five changes of governments between the outbreak of the Korean War and the end of 1952.

110. For example, the British program was expected to absorb as much as 14 percent of national income.

111. French officials expressed concern about the "growing economic disequilibrium" caused by the rearmament effort as early as February 1951. See *FRUS, 1951*, 3: 66–67.

112. Osgood, *NATO*, 81.

113. By May 1951, import prices had risen an average of 42 percent over 1950 while export prices had increased by only 18 percent. As a result, Britain experienced a severe balance-of-payments crisis in the second half of 1951, when its gold and dollar reserves fell by £600 million, while a shortage of labor caused overall industrial production to fall in 1951–52. See *FRUS, 1951*, 3: 67; Bartlett, *Long Retreat*, 62–63; Warner, "British Labour Government," 261; and Gordon, "Economic Aspects," 533.

114. Poole, *History of the JCS*, 241.

115. *FRUS, 1951*, 3: 5; Kaplan, *Community of Interests*, 156; and Melandri, "France," 272. The quotation is from Osgood, *NATO*, 77.

116. Warner, "British Labour Government," 250; and Healey, "Britain and NATO," 211.

117. Poole, *History of the JCS*, 241, n. 42; Harrison, *Reluctant Ally*, 34; and ALO 110, Mar. 17, 1951, RG 218, CCS 092 Western Europe (3-12-48), sec. 74. For further discussion of U.S. involvement in the conflict, see Wall, *United States*, chap. 8.

118. *FRUS, 1951*, 3: 161–62; *FRUS, 1951*, 1: 360–61, 365; and House of Representatives, Foreign Affairs, *Selected Executive Session Hearings, 1951–56*, 15: 280. For details of the military assistance program, see Kaplan, *Community of Interests*.

119. *FRUS, 1951*, 3: 210, 216, 262.

120. For the text of the NAC resolution establishing the TCC, see *FRUS, 1951*, 3: 677–78. On the origins and responsibilities of the committee, see Acheson, *Present at the Creation*, 569–71; and Gordon, "Economic Aspects."

121. Gordon, "Economic Aspects," 538.

122. For summaries of the report, see *FRUS, 1951*, 3: 373–74, 389–92.

123. Ibid., 389–90.

124. Ibid., 372, 390–91; and Gordon, "Economic Aspects," 538–39. See Poole, *History of the JCS*, 277, for details of the defense spending increases recommended by the TCC.

125. The 1952 force goal for M+30 was reduced somewhat because of France's inability to provide its share of the recommended forces. See *FRUS, 1952–54,* 5: 147; and House of Representatives, Foreign Affairs, *Selected Executive Session Hearings, 1951–56,* 15: 279.

126. *FRUS, 1952–54,* 6: 295; and Kaplan, *Community of Interests,* 166.

127. *FRUS, 1952–54,* 5: 214–15. Canada and Belgium in particular were unwilling to make the additional financial contributions asked of them. See C(52)49, Feb. 19, 1952, PRO, CAB 129.

128. *FRUS, 1952–54,* 5: 213–14, 220.

129. C(51)1, Oct. 31, 1951, PRO, CAB 129. Overall, Britain's dollar reserves fell by 40 percent and its current account registered a deficit of $1.5 billion in 1951. See *FRUS, 1952–54,* 6: 788.

130. C(52)10, Jan. 19, 1952, PRO, CAB 129.

131. C(51)48, Dec. 17, 1951, PRO, CAB 129; and *FRUS, 1952–54,* 6: 868.

132. C(52)4, Jan. 11, 1952, PRO, CAB 129; C(52)49, Feb. 19, 1952, PRO, CAB 129; and CC(52)5th Conclusions, Jan. 22, 1952, PRO, CAB 128. Churchill informed Parliament of these plans in early February, shortly before the Lisbon meeting.

133. C(52)173, May 23, 1952, PRO, CAB 129; C(52)253, July 22, 1952, PRO, CAB 129; C(52)316, Oct. 3, 1952, PRO, CAB 128; and CC(52)94th Conclusions, Nov. 7, 1952, PRO, CAB 128.

134. *FRUS, 1952–54,* 6: 146, 273, 1171.

135. The United States would also spend approximately $1 billion on troops and bases in France, thereby helping to bolster France's dollar reserves. See ibid., 1154; *FRUS, 1952–54,* 5: 142; House of Representatives, Foreign Affairs, *Selected Executive Session Hearings, 1951–56,* 15: 249; and Wall, *United States,* 224–26. U.S. officials were unable to make a commitment for the following fiscal year, although they hoped to provide a comparable level of assistance. See *FRUS, 1952–54,* 6: 1172–74.

136. *FRUS, 1952–54,* 6: 1157. At least one U.S. official, however, thought that the French economy could not support more than ten divisions on a long-term basis. See ibid., 1149.

137. *FRUS, 1952–54,* 5: 142–43. In 1952, the annual cost of the Indochina war reached approximately $1.3 billion. See Osgood, *NATO,* 90.

138. *FRUS, 1952–54,* 5: 144, 147.

139. Ibid., 143; and *FRUS, 1952–54,* 6: 1179. See also Osgood, *NATO,* 89.

140. The U.S. ambassador in Paris viewed the crisis as "the most critical and delicate in French-American relations since the beginning of NATO" (Wall, *United States,* 228). For a brief catalog of France's economic troubles, see Calvocoressi, *Survey,* 37.

141. *FRUS, 1952–54,* 6: 1204, 1217.

142. Ibid., 1221, 1223.

143. Ibid., 1236, 1251.

144. Ibid., 459, 483, 1239, 1243, 1249. The bulk of the end-item reduction occurred in the equipment intended for French ground forces, which was slashed from $390 million to $174 million.

145. Osgood, *NATO*, 90; Harrison, *Reluctant Ally*, 34; and Wall, *United States*, 227–32.

146. For more detailed analyses, see McGeehan, *German Rearmament Question*; Fursdon, *European Defence Community*; and Dockrill, *Britain's Policy*.

147. The United States had proposed a figure of nearly 200,000 men but was unable to obtain British and French agreement on the higher number. See *FRUS, 1951*, 3: 992–93.

148. Ibid., 993–94.

149. See McGeehan, *German Rearmament Question*, 117.

150. For a summary of the points of agreement and disagreement, see *FRUS, 1951*, 3: 1044–47.

151. Ibid., 801–5. Beginning in March, the JCS pressed for the creation of the equivalent of eight German divisions by the end of the year. See Poole, *History of the JCS*, 258–59.

152. *FRUS, 1951*, 3: 805–12, 838–42; and Poole, *History of the JCS*, 259–61.

153. Memorandum from the Secretary of State and the Acting Secretary of Defense to the President, July 30, 1951, *FRUS, 1951*, 3: 849–52. This memorandum was approved by Truman on August 2 and circulated as NSC 115. On the shift in U.S. policy, see also Dockrill, *Britain's Policy*, 68–71; Leffler, *Preponderance of Power*, 413–15; and Schwartz, *America's Germany*, chap. 8.

154. *FRUS, 1951*, 3: 790–91.

155. Ibid., 843–46, 862–65; Poole, *History of the JCS*, 261; and Stebbins, *United States in World Affairs, 1951*, 346–47.

156. Poole, *History of the JCS*, 204, 208; and *FRUS, 1951*, 3: 695–96.

157. The Netherlands became a full participant in October.

158. *FRUS, 1951*, 3: 933–46; and Poole, *History of the JCS*, 288.

159. *FRUS, 1951*, 3: 723–75.

160. Ibid., 905–32; and Dockrill, *Britain's Policy*, 90.

161. *FRUS, 1951*, 3: 985–89.

162. *FRUS, 1952–54*, 5: 597–605.

163. Poole, *History of the JCS*, 289–90; Acheson, *Present at the Creation*, 608–10; and McGeehan, *German Rearmament Question*, 180–81.

164. The resolution also recommended the approval of reciprocal security guarantees between NATO and the EDC, which were intended to satisfy the German desire to be included in NATO without offending French sensibilities about German membership in the alliance. See Acheson, *Present at the Creation*, 615–20, 623–26; and Poole, *History of the JCS*, 290–92. The text of the NAC resolution is in *FRUS, 1952–54*, 5: 252–55.

165. *FRUS, 1952–54*, 5: 627–34.

166. Ibid., 684–88, which contains the text of the declaration; Fursdon, *European Defence Community*, 143–47; and Dockrill, *Britain's Policy*, 98–99.

167. *FRUS, 1952–54*, 5: 610–11; *FRUS, 1952–54*, 6: 1266; Memorandum, Parsons to Bruce, Jan. 27, 1953, RG 59, 740.5; Melandri, "France," 275–76;

Harrison, *Reluctant Ally*, 29–30; and Fursdon, *European Defence Community*, 199–202. See also the analysis in Aron and Lerner, eds., *France Defeats EDC.*

168. Poole, *History of the JCS*, 327; and Fursdon, *European Defence Community*, 205–6.

169. Another five or so allied divisions existed but would require more than 30 days of mobilization. See Poole, *History of the JCS*, 306, n. 13; Watson, *History of the JCS*, 283, 319; Memorandum for Mr. N. E. Halaby, May 4, 1953, RG 330, Assistant Secretary of Defense (Comptroller) Subject File, 1947–55, Entry 4, Box 34, "NATO 1953"; AOC(52)109, Dec. 11, 1952, PRO, CAB 134/764; and M. B. Ridgway, Memorandum for the Secretary of State, "Allied Command Europe," Apr. 26, 1953, RG 218, CCS 092 Western Europe (3-12-48), sec. 216. Ridgway provides slightly different figures for the end of 1952.

170. Osgood, *NATO*, 88; Hilsman, "NATO," 22–23; and Wilmot, "If NATO Had to Fight," 200–204.

171. According to one estimate, no more than 64 divisions could be expected by the end of 1953, and many of these would fall short of NATO's readiness standards. See AOC(52)109.

172. Poole, *History of the JCS*, 306. For the text of the NAC communique, see *DSB* 28 (Jan. 5, 1953): 3–4.

173. *FRUS, 1950*, 3: 1386.

174. Osgood, *NATO*, 77.

Chapter 3

1. Some NATO leaders feared that if nothing were done to close the gap, the resulting sense of frustration might lead to yet a further relaxation of effort by the European allies. See, for example, "Certain European Issues Affecting the United States," May 15, 1953, DDEL, DDEP, AWF, Administration Series, Box 13, "William H. Draper, Jr. (2)," 38.

As long as the alliance lacked the forces needed to implement a truly forward strategy, moreover, its cohesion remained vulnerable to external threats and intimidation. As late as 1954, SACEUR planned to fall back on the Rhine-Ijssel line in the event of hostilities, leaving most of Germany, Denmark, and significant parts of the Netherlands undefended. (COS[54]186, "The Strategic Position of the Netherlands," June 9, 1954, PRO, DEFE 5/53). Were the details of this plan to become publicly known, some officials worried, these countries might begin to question their commitment to NATO and to seek an accommodation with the Soviet Union.

2. Wampler, "From Lisbon to M.C. 48," 12–14. See also Gowing, *Independence and Deterrence*, 439–41; Baylis and Macmillan, "British Global Strategy Paper"; and C(52)253, "The Defence Programme," July 22, 1952, PRO, CAB 129.

3. Wampler, "From Lisbon to M.C. 48," 17–19; Poole, *History of the JCS*, 309–10; and AOC (52)106, "Draft Brief for December Meeting of the Council

on Determination of the N.A.T.O. Defence Effort," Dec. 5, 1952, PRO, CAB 134/764. Concerned that the British thesis might gain currency among the other allies, chairman of the JCS General Omar Bradley made a determined effort to dispel any doubts about the need to persevere with the planned conventional buildup during a subsequent trip to Europe. See Wampler, "From Lisbon to M.C. 48," 19–20; Poole, *History of the JCS*, 310; and Statement to Be Included in Press Conference by General Bradley at SHAPE Headquarters, Attachment to Letter, Blanchard to Short, Sept. 12, 1952, HSTL, HSTP, President's Official File.

4. Memorandum, Merchant to Murphy, "The Problem of NATO Force Requirements and the Annual Review," Nov. 2, 1953, RG 59, 740.5; Wampler, "From Lisbon to M.C. 48," 20–21; and *FRUS, 1952–54*, 6: 652.

5. COS(53)108, "UK Force Contributions to NATO, 1954," Feb. 19, 1953, PRO, DEFE 5/45; and TF Exec-R/7, Jan. 21, 1953, RG 330, Entry 44, "TF-Exec." See also Letter, Draper to the President, Mar. 8, 1953, RG 59, 740.5.

6. U.S. officials were aware that Ridgway's study would probably not result in any reduction of the force requirements. See Memorandum, Wolf to Ohly, "Planning for the 1953 Annual Review," Apr. 4, 1953, RG 59, 740.5.

7. AOC(52)104, "NAC Meeting—Brief on MC 14/1," Dec. 5, 1952, PRO, CAB 134/764.

8. COS(53)108; D(53)5th Meeting, Minute 4, Mar. 26, 1953, PRO, CAB 131/13; and Memorandum, Brownjohn to the Prime Minister, "NATO Annual Review for 1953," Mar. 25, 1953, PRO, PREM 11/369. See also CC(53) 28th Meeting, Minute 6, Apr. 21, 1953, PRO, CAB 128/23; and Polto 2085, Apr. 19, 1953, RG 59, 740.5.

9. In his opinion, the economy had suffered from the high level of government spending necessitated by the Korean War buildup, which had resulted in both higher taxes and substantial government borrowing. For Eisenhower's views on this subject, see his letter to his former chief of staff at SHAPE, General Alfred Gruenther, May 4, 1953, DDEL, DDEP, AWF, Eisenhower Diary Series, Box 3, "Eisenhower Diary, 12/52–7/53 (2)," and his remarks to key congressional leaders on April 30, 1953, in *FRUS, 1952–54*, 2: 317. For a good general discussion of the president's attitudes and policies on national security, see Kinnard, *President Eisenhower*.

10. See Watson, *History of the JCS*, 3–4.

11. *FRUS, 1952–54*, 2: 244–45; Memorandum for the Joint Chiefs of Staff, "Proposed Expenditure Limitations, Mutual Security Program," Mar. 10, 1953, *DDRS*, 1986, no. 3502; and National Security Council—Meeting with Consultants, Outline for R. Cutler, Mar. 31, 1953, *DDRS*, 1986, no. 1570.

12. The JCS concluded that such cuts "would so increase the risk to the United States as to pose a grave threat to the survival of our allies and the security of the nation" (Watson, *History of the JCS*, 6–7). The NSC discussion of the proposed reductions is in *FRUS, 1952–54*, 2: 258–64. See also the interview-based account of Snyder, " 'New Look,' " 394–95.

13. *FRUS, 1952–54*, 2: 278–79.

14. See ibid., 281–87, 305–16. The report recognized that a balanced fed-

eral budget could be approached only gradually. Hence it envisioned a more incremental decline than had previously been suggested in the size of the military budget, which would reach a level that could be sustained indefinitely in FY 1957. Similarly, the Mutual Security Program would be reduced by only $1.5 billion in each of the next two years. For the council's discussions of the earlier versions of NSC 149/2, see ibid., 287–305.

15. For the text of the president's remarks, see Snyder, " 'New Look,' " 400.

16. Despite Stalin's death, the official State Department position was that the threat of Soviet aggression had in no way diminished. This view followed from the assumption that Soviet policies were determined by the totalitarian nature of the Soviet state structure and communist doctrine, which had not changed, rather than by individual leaders. See Topol 1195, Apr. 13, 1953, RG 59, 740.5.

17. See Secretary of State John Foster Dulles's remarks at the special meeting of the NSC on March 31, 1953, in *FRUS, 1952–54,* 2: 266.

18. Ibid.; Memorandum, Merchant to Moore, Apr. 2, 1953, RG 59, 740.5; and Polto Circular A-26, Apr. 17, 1953, RG 59, 740.5. These conclusions were subsequently reflected in NSC 149/2, which called for placing less stress on the expansion of NATO forces and for cutting future force levels while accelerating the process of bringing existing divisions to a high level of combat effectiveness. Similarly, NSC 149 had urged decreased emphasis on the expansion of NATO forces-in-being and cutting NATO force goals. See *FRUS, 1952–54,* 2: 283, 308, 312.

19. Memorandum, Merchant to Moore, Apr. 2, 1953, RG 59, 740.5.

20. For records of the meetings, see *FRUS, 1952–54,* 5: 369–70, 373–78. See also the statements of Dulles and Secretary of Defense Wilson before the North Atlantic Council on April 23 and 24, 1953, respectively, in RG 59, 740.5.

21. The final communique adopted by the ministers stated that "the development of sound national economies and the increase of military forces should be pursued concurrently." See *DSB* 28 (May 11, 1953): 673–74.

22. *FRUS, 1952–54,* 5: 406. See also Dulles's statement after the NATO meeting in *DSB* 28 (May 25, 1953): 738. On the U.S. desire to place greater emphasis on qualitative improvements, see *FRUS, 1952–54,* 5: 367, 371.

23. Dulles estimated that NATO combat effectiveness would increase by as much as 30 percent in 1953 alone (*DSB* 28 [May 11, 1953]: 671). See also Wilson's statement to the NATO Council, Apr. 24, 1953, RG 59, 740.5. Interestingly, the council, at British urging, did not adopt force goals for 1955. The British government still hoped that the Ridgway study might result in lower force requirements by that time, allowing the force goals to be reduced commensurately.

24. See, for example, *FRUS, 1952–54,* 5: 385.

25. See, for example, ibid., 408–9, 412–14. In fact, new force requirements developed by the NATO commanders indicated a sharp increase in the number of aircraft needed in the Central Region. See COS(53)158, "1954 Force Requirements," Mar. 27, 1953, PRO, DEFE 5/45.

26. *FRUS, 1952–54*, 5: 407, 427–32, 435.

27. On the positive side, the study concluded that the alliance could successfully defend Western Europe along the inner German border, provided that NATO commanders were given sufficient tactical nuclear weapons and the authority to use them immediately upon the outbreak of hostilities. Specifically, it assumed the availability of 1,000 nuclear weapons, some 700 of which would be used in the first two days of combat as against 200 to 300 Soviet atomic strikes. See COS(53)490, "SACEUR's Estimate of the Situation and 1956 Force Requirements," Oct. 2, 1953, PRO, DEFE 5/49.

28. In the spring, Eisenhower designated Ridgway as the next U.S. Army chief of staff.

29. COS(53)158; COS(53)585, "Briefs for the Military Committee Meeting," Dec. 4, 1953, PRO, DEFE 5/50; COS(53)609, "SACEUR's Estimate of the Situation and 1956 Force Requirements," Dec. 23, 1953, PRO, DEFE 5/50; Memorandum, Merchant to Murphy, Nov. 2, 1953, RG 59, 740.5; and Richardson, "NATO's Nuclear Strategy," 38–40. See also USM-2-53, "1956 NATO Force Requirements," Oct. 13, 1953, RG 218, CCS 092 Western Europe (3-12-48), sec. 244.

30. C(53)234, "A Revised Political Directive for NATO," Aug. 17, 1953, PRO, CAB 129/62; AOC(53)29, "New Directive for NATO," July 25, 1953, PRO, CAB 134/766; and Merchant to Nash, Ohly, Overby, and Macy, Nov. 17, 1953, RG 59, 740.5. See also AOC(53)50, "Bermuda Conference: New Policy Directive for NATO," Nov. 26, 1953, PRO, CAB 134/766; and COS(53)536, "Meeting with U.S. Chairman of the JCS," Oct. 30, 1953, PRO, DEFE 5/49. On the development of the British proposal, see AOC(53)8th Meeting, July 6, 1953, and AOC(53)9th Meeting, July 29, 1953, both in PRO, CAB 134/765.

31. The British Chiefs of Staff proposed, moreover, that SHAPE examine what could be done with both the conventional forces and the new weapons likely to be available in 1956. See COS(53)490 and COS(53)110th Meeting, Minute 4, Sept. 30, 1953, PRO, DEFE 4/65.

32. Memorandum, Merchant to Murphy, Nov. 2, 1953, and Memorandum, Merchant to Nash, et al., Nov. 17, 1953, both in RG 59, 740.5; and COS(53)609. See also JCS 2073/683, "Proposed Supplementary Planning Project by SACEUR," Nov. 30, 1953, RG 218, CCS 092 Western Europe (3-12-48), sec. 247.

33. AOC(53)50.

34. *FRUS, 1952–54*, 5: 440–46; and Memorandum, Merchant to Murphy, Nov. 2, 1953, RG 59, 740.5. Some U.S. officials seemed to endorse the British approach without reservation. See, for example, Polto 312, Sept. 1, 1953, RG 59, 740.5.

35. Memorandum, Merchant to the Undersecretary of State, et al., Aug. 31, 1953; Topol 278, Sept. 26, 1953; Memorandum, Merchant to Murphy, Nov. 2, 1953; Memorandum of Conversation, Oct. 21, 1953; Memorandum, Merchant to Nash, et al., Nov. 17, 1953, all in RG 59, 740.5; Memorandum for the Secretary of Defense, "U.S. Guidance for the 1953 Annual Review," Oct. 22, 1953, RG 218, CCS 092 Western Europe (3-12-48), sec. 244; and

COS(53)121st Meeting, Minute 1, Oct. 27, 1953, PRO, DEFE 4/66. U.S. officials also sought to take greater account of the effect of U.S. strategic bombing on a European campaign and to reappraise Soviet capabilities in the wake of the mid-June disturbances in East Germany. See *FRUS, 1952–54,* 5: 431, 443.

36. The JCS had generally discounted the use of atomic weapons when determining force levels and requirements. See Memorandum for Admiral Radford, "Review of Events Leading up to JCS Action of 10 December 1953 Concerning *Military Strategy and Posture,*" Nov. 7, 1955, RG 218, Radford Files, CJCS 381 (Military Strategy & Posture).

37. Report to the National Security Council by the Executive Secretary, NSC 162/2, Oct. 30, 1953, *FRUS, 1952–54,* 2: 577–97 at 593. For earlier drafts of NSC 162/2 and National Security Council discussions thereof, see ibid., 489–576. See also Snyder, "'New Look.'"

38. *FRUS, 1952–54,* 2: 593.

39. For summaries of the discussions at the Bermuda conference and North Atlantic Council meetings, see *FRUS, 1952–54,* 5: 466–67, 476–81, 1786–91, 1846–48.

40. "Resolution on the 1954 Annual Review and Related Problems," cited in Watson, *History of the JCS,* 301. See also COS(53)144th Meeting, Minute 1, Dec. 22, 1953, PRO, DEFE 4/67.

41. At the same time, the NATO Council adopted the U.S. recommendation that the force goal for ready divisions in 1954 (less Greece and Turkey) be reduced from 31⅔ to 29⅔, and it set the force goals for 1955 and 1956 at approximately the same level. Because the number of combat-ready divisions had increased from 25⅓ to 29⅔ during the course of 1953, the ministers effectively acknowledged that no further quantitative increases could be expected. See Table 3.3 and Watson, *History of the JCS,* 297–300, 319. Instead, they once again emphasized improvements in the quality and effectiveness of these forces. The council's final communique is reprinted in *DSB* 30 (Jan. 4, 1954): 8–9.

42. See, for example, *FRUS, 1952–54,* 5: 449–54.

43. Ibid., 499–501.

44. A further reason for this decision was that the allies had put a good deal of pressure on the United States to discuss the problem of consultation and conditions governing the use of atomic weapons in the European area. See Memorandum, Arneson to MacArthur, Apr. 9, 1954, RG 59, 740.5.

45. "Statement by the Secretary of State to the North Atlantic Council Closed Ministerial Session, Paris, April 23, 1954," n.d., *FRUS, 1952–54,* 5: 509–14.

46. JP(54)76, "ACE 1957 Capabilities Study," Sept. 2, 1954, and JP(54)77, "The Most Effective Pattern of NATO Military Strength for the Next Few Years," Sept. 2, 1954, both in PRO, DEFE 6/26.

47. For British reactions, see JP(54)76 and JP(54)77. For the views of the State Department, see Memorandum, Wolf to Moore and Palmer, "The New Weapons Issue in the NATO Council," June 25, 1954, and Memorandum,

Wolf to Elbrick, Sept. 27, 1954, both in RG 59, 740.5. For the views of the JCS, see Watson, *History of the JCS*, 305–6.

48. Memorandum, Elbrick to the Secretary, Oct. 12, 1954, RG 59, 740.5.

49. JP(54)86, "The Most Effective Pattern of Military Strength for the Next Few Years," Oct. 21, 1954, and JP(54)99, "Briefs for the Military Committee Meeting," Dec. 2, 1954, both in PRO, DEFE 6/26; and Memorandum, Wolf to Elbrick, Sept. 27, 1954, RG 59, 740.5.

50. Watson, *History of the JCS*, 312–13; *FRUS, 1952–54*, 5: 527–32; Memorandum of Conversation, "NATO 'New Approach' Studies," Nov. 3, 1954, RG 59, 740.5; and Col. Andrew J. Goodpaster, Memorandum for the President, Nov. 16, 1954, DDEL, DDEP, AWF, Administrative Series, Box 27, "NATO."

51. AOC(54)9th Meeting, Dec. 6, 1954, and AOC(54)10th Meeting, Dec. 13, 1954, both in PRO, CAB 134/767; and Memorandum, Merchant to the Secretary of State, Nov. 16, 1954, RG 59, 740.5.

52. Watson, *History of the JCS*, 317; and *FRUS, 1952–54*, 5: 539–41, 547–48. For a complete summary of the tripartite meeting, see Memorandum of Conversation, Dec. 16, 1954, RG 59, 740.5. Dulles personally affirmed that the question of when nuclear weapons would be used must be left open. See Record of Background Press Conference by the Secretary of State, Dec. 16, 1954, *FRUS, 1952–54*, 5: 543–44.

53. For records of the NATO meeting, see *FRUS, 1952–54*, 5: 557–62.

54. JP(54)77; JP(54)86; JP(54)99; COS(55)58th Meeting, July 19, 1955, PRO, DEFE 4/78; and Watson, *History of the JCS*, 306, 312.

55. *FRUS, 1952–54*, 5: 530; JP(54)77; JP(54)99; ALO 264, Feb. 10, 1955, RG 218, CCS 092 Western Europe (3-12-48) (2), sec. 6; and USM-77-55, Nov. 21, 1955, same file, sec. 41.

56. JP(54)77; JP(54)99; Watson, *History of the JCS*, 306, 312; and CSAFM-243-54, "Military Comments on the 1954 Annual Review Report," Nov. 26, 1954, RG 218, CCS 092 Western Europe (3-12-48), sec. 314.

57. Statement by Gen. Alfred M. Gruenther, Mar. 26, 1955, Senate, Foreign Relations, *NATO and the Paris Accords*, 6. See also JCS 2073/644, Sept. 10, 1953, RG 218, CCS 092 Western Europe (3-12-48), sec. 231.

58. See esp. the statement by Gen. Lawton Collins, the U.S. representative to the NATO Standing Group, reported in Hughes to the Department of State, Sept. 16, 1954, *FRUS, 1952–54*, 5: 522–24.

59. "Gruenther's Address on the Defense of Europe," *DSB* 31 (Oct. 18, 1954): 562; Statement by Gen. Gruenther, Mar. 26, 1955, Senate, Foreign Relations, *NATO and the Paris Accords*, 5–6; *FRUS, 1952–54*, 5: 523; Watson, *History of the JCS*, 311, 320; JCS 2073/1139, Sept. 14, 1955, RG 218, CCS 092 Western Europe (3-12-48) (2), sec. 32; Memorandum by the Chief of Staff of the Army, "The Most Effective Pattern of Military Force—Report No. 2," Sept. 16, 1955, same file, sec. 33; and C(54)86, Mar. 5, 1954, PRO, CAB 129/66.

60. *FRUS, 1952–54*, 5: 533, 558; *DSB* 32 (Jan. 3, 1955): 10–12, cited in Watson, *History of the JCS*, 320; and JP(54)99. JP(54)77 suggests that the forward

defensive line was to be established much closer to the inter-German border along the Weser River.

61. See, for example, Dulles's remarks in *FRUS, 1952–54*, 5: 561; and Statement by Gen. Gruenther, Mar. 26, 1955, Senate, Foreign Relations, *NATO and the Paris Accords*, 6.

62. JP(54)76.

63. In this respect, the council's treatment of MC 48 was analogous to Eisenhower's qualified interpretation of the relevant passage of NSC 162/2: "The purpose . . . is primarily to permit the military to make plans on the basis of the availability of nuclear weapons. . . . [It] is not a decision in advance that atomic weapons will in fact be used in the event of any hostilities." Eisenhower reaffirmed this interpretation of U.S. policy the following year. See Lay Memorandum for the Secretary of State, et al., Jan. 4, 1954, *DDRS*, 1988, no. 2266; and Lay Memorandum for the Secretary of State, et al., Mar. 14, 1955, *DDRS*, 1989, no. 342.

64. Memorandum, Wolfe to Moore and Palmer, June 25, 1954, RG 59, 740.5. U.S. policy simply called for providing the allies with more information about the new weapons.

65. The Ridgway study assumed that NATO would face a total of 55 Soviet and 45 satellite divisions after 30 days of mobilization. (Figures for the Central Region alone are not available.) See COS(53)490. Between 1950 and 1954, moreover, the total number of Polish, Czech, and Hungarian divisions increased by nearly 50 percent, to 43, while the new East German army grew to 7 divisions, although doubts remained as to how useful these forces would be to the Soviet Union in a conflict. See JIC 607/10, "Soviet and Satellite Order of Battle," Nov. 24, 1953, RG 218, CCS 092 Western Europe (3-12-48); and the statement by Gen. Matthew Ridgway in House of Representatives, Foreign Affairs, *Selected Executive Session Hearings, 1951–56*, 10: 39–72.

66. The M-Day division breakdown by country was as follows: Belgium, 2; Netherlands, 2; Britain, 4; France, 5; United States, 5; Germany, 12; Canada, 1/3. See House of Representatives, Foreign Affairs, *Selected Executive Session Hearings, 1951–56*, 19: 28; ibid., 21: 135; "Interview with General Lauris Norstad," *NATO Letter* 4, no. 12 (Dec. 1956): 34; and Memorandum by Steel, Apr. 26, 1956, PRO, FO 371/123187.

67. COS(55)45th Meeting, June 15, 1955, PRO, DEFE 4/77; COS(55)34, Feb. 18, 1955, PRO, DEFE 5/56; ALO 264; and "Problems General Gruenther May Raise," Mar. 10, 1955, RG 218, CCS 092 Western Europe (3-12-48) (2), sec. 8. The British estimated that as many as 38 new airfields would have to be built and that the U.K. forces in Germany alone would have to be increased by as many as 43,000 men.

68. JP(54)77; JP(54)99; Secretary of State and Secretary of Defense to the President, Nov. 2[?], 1954, *FRUS, 1952–54*, 5: 529–32; and CSAFM-243-54, Nov. 11, 1954, RG 218, CCS 092 Western Europe (3-12-48), sec. 314. Even the NATO Council noted that some countries must "allocate resources for defense at a higher level than currently indicated." See Watson, *History of the JCS*, 319.

69. Memorandum for the Secretary of Defense, "Policy Aspects of the United States Reply to the NATO Annual Review Questionnaire," July 16, 1953, RG 218, CCS 092 Western Europe (3-12-48), sec. 227, and JCS 2073/596, July 6, 1953, same file; and JCS 2073/711, "Estimates of NATO Forces, 1954–57," Nov. 25, 1953, *Records of the JCS*, reel 9, 329–39.

70. Memorandum of Discussion at a Special Meeting of the National Security Council, Mar. 31, 1953, *FRUS, 1952–54*, 2: 278. For Eisenhower's and Dulles's views, see also ibid., 455–60. Dulles felt that even a reduction in the planned deployment of additional U.S. forces would add to the difficulty of maintaining European defense efforts at their existing levels and might well present another obstacle to the ratification of the EDC. See *FRUS, 1952–54*, 5: 435.

71. For Eisenhower's instructions, see Watson, *History of the JCS*, 16–17, who argues that these "left no doubt that he expected the new appointees to recommend a military strategy that could be implemented with smaller forces and would thus justify lower military budgets in the future."

72. The JCS report is summarized and excerpted in Watson, *History of the JCS*, 18–19. See also Snyder, " 'New Look,' " 414–15, who adds that the Chiefs concluded that the local defense of distant regions should henceforth be the responsibility of indigenous forces, although they conditioned their judgment of the feasibility of such an approach on the assumption that the build-up of Korean and German forces would proceed on schedule.

73. See *FRUS, 1952–54*, 2:443–54.

74. Report to the National Security Council by the NSC Planning Board, Sept. 30, 1953, ibid., 489–514, esp. 508–9.

75. Ibid., 526–29. For the final version of the text, see ibid., 593.

76. Dulles, "Evolution of Foreign Policy."

77. Dulles, "Policy for Security and Peace," 358. See also Department of State press release no. 728, Dec. 21, 1954. Ironically, Western Europe was perhaps the one place where the threat of massive retaliation had any credibility, according to critics of the doctrine, precisely because of the region's importance to the United States. See Kaufmann, "Requirements of Deterrence."

78. Snyder, " 'New Look,' " 436; *New York Times*, Oct. 20, 21, 29, 1953; and *FRUS, 1952–54*, 5: 447–48. For Dulles's reactions, see *FRUS, 1952–54*, 2: 549–50.

79. See *FRUS, 1952–54*, 5: 449–54; and a scathing correspondence between Wilson and Dulles dated Dec. 7, 1953, in RG 59, 740.5.

80. Watson, *History of the JCS*, 27–32; Memorandum for the Secretary of Defense, "Military Strategy and Posture," Dec. 9, 1953, and Enclosures, DDEL, DDEP, AWF, Administration Series, "Wilson, C. E. 1953 (2)"; and Snyder, " 'New Look,' " 451–56, 479–83. See also *FRUS, 1952–54*, 5: 482–84.

81. See, for example, *FRUS, 1952–54*, 5: 449–54, 532–33. The 1956 episode, associated with the leak of the so-called Radford Plan, is discussed in Chapter 4.

82. For British views, see COS(53)536, and AOC(52)106.

83. For U.S. statistics and attitudes, see Memorandum, Miller to Knight, Jan. 27, 1953, RG 59, 740.5; and *FRUS, 1952–54*, 6: 553. For the British perspective, see AOC(53)29.

84. Specifically, Eisenhower requested $1.0 billion in defense support funds for Europe and $1.6 billion for offshore procurement. Most of the remainder would be used to provide military equipment. Because of the backlog of unused appropriations from previous years, the administration nevertheless proposed to spend $6.5 billion in FY 1954 and $6.3 billion in FY 1955 on military assistance (*FRUS, 1952–54*, 2: 313–14).

85. *FRUS, 1952–54*, 6: 540–43, 566–71.

86. *FRUS, 1952–54*, 5: 853; and *FRUS, 1952–54*, 6: 566. State Department officials nevertheless estimated that the amendment, which did not prohibit the placement of procurement orders, would not have an appreciable effect for twelve to eighteen months, by which time they expected the EDC to be in operation (*FRUS, 1952–54*, 5: 796–97).

87. *FRUS, 1952–54*, 6: 614–16, 618–20.

88. Topol 425, Oct. 26, 1953, RG 59, 740.5; and *FRUS, 1952–54*, 6: 1422. See also Poole, *History of the JCS*, 299–302; and *FRUS, 1952–54*, 6: 599.

89. Polto 2006, June 4, 1954, RG 59, 740.5.

90. See JP(53)25, Feb. 6, 1953, PRO, DEFE 6/21; COS(53)108, "UK Force Contribution to NATO," Feb. 19, 1953, PRO, DEFE 5/45; and C(53)112, "German Financial Contribution to Defence in 1953–54," Mar. 23, 1953, PRO, CAB 129/60.

91. CC(53)29th Meeting, Minute 7, Apr. 28, 1953, PRO, CAB 128/26; and *FRUS, 1952–54*, 5: 371–73.

92. COS(53)328, "The Radical Review," July 8, 1953, PRO, DEFE 5/47.

93. CC(53)50th Meeting, Minute 2, Aug. 25, 1953, PRO, CAB 128/26.

94. See, for example, *FRUS, 1952–54*, 5: 473.

95. See, for example, AOC(54)10th Meeting.

96. COS(53)121st Meeting, Minute 1, and AOC(53)50. See also *FRUS, 1952–54*, 5: 447.

97. AOC(53)17th Meeting, "NATO: New Look and US Forces in Europe," Nov. 23, 1953, PRO, CAB 134/765; and AOC(53)50.

98. *FRUS, 1952–54*, 6: 1298–1300; and Memorandum, Vass to Moore, Aug. 4, 1953, RG 59, 740.5. For a complete breakdown of France's NATO assigned forces, see Message from USCINCEUR to the JCS, Sept. 10, 1955, RG 218, CCS 092 Western Europe (3-12-48) (2), sec. 32. At the end of 1954, there were fourteen French divisions in the NATO area, as called for by the revised force goals adopted in April 1953, but two of these were stationed in North Africa (*FRUS, 1952–54*, 6: 1385; and Polto 954, Dec. 13, 1953, RG 59, 740.5).

99. *FRUS, 1952–54*, 6: 1289, 1358–59.

100. Bipartite U.S.-French Conversations, First Session, Apr. 22, 1953, RG 59, 740.5. For an abbreviated summary of these talks, see *FRUS, 1952–54*, 6: 1347–49.

101. See, for example, *FRUS, 1952–54*, 6: 1299.

102. For U.S. estimates, see ibid., 1291; Mutual Security Agency, Title I (Europe) Defense Support, "Analysis of Effects of Assumed Expenditure Limitations, France," Mar. 14, 1953, *DDRS*, 1987, no. 2279; and Bipartite U.S.-French Conversations, First Session, Apr. 22, 1953, RG 59, 740.5.

103. "Analysis of Effects of Assumed Expenditure Limitations, France," Mar. 14, 1953, RG 59, 740.5; and Harrison, *Reluctant Ally*, 33, 38. Altogether, the United States contributed $785 million for Indochina in FY 1954, exclusive of end items (Topol 497, Nov. 6, 1953, RG 59, 740.5).

104. *FRUS, 1952–54*, 6: 1300; and Bipartite U.S.-French Conversations, First Session, Apr. 22, 1953, RG 59, 740.5.

105. The increased requirement was based on the expectation that military costs would rise sharply but there could be no comparable increase in French defense spending. See "Analysis of Effects of Assumed Expenditure Limitations, France," Mar. 14, 1953, RG 59, 740.5; *FRUS, 1952–54*, 6: 1318, 1344; and Bipartite U.S.-French Conversations, Apr. 22, 1953, RG 59, 740.5.

106. *FRUS, 1952–54*, 6: 1367–69; and Memorandum, Vass to Moore, Aug. 4, 1953, RG 59, 740.5.

107. Paris 2211, Dec. 8, 1953, RG 59, 740.5. For further figures on the French defense budget for 1953, see Memorandum, Labouisse to Stassen, Apr. 23, 1953, RG 59, 740.5; and *FRUS, 1952–54*, 6: 1359. The United States had hoped France would spend at least $400 million more. See Topol 497, Nov. 6, 1953, RG 59, 740.5.

108. JCS 2073/945, "Movement of French Forces to North Africa," Nov. 10, 1954, RG 218, CCS 092 Western Europe (3-12-48), sec. 311; JSPC 876/996, "Movement of French Forces to North Africa," July 7, 1955, RG 218, CCS 092 Western Europe (3-12-48) (2), sec. 22; USM-29-55, Aug. 30, 1955, same file, sec. 31; and *FRUS, 1952–54*, 6: 1495.

109. JSPC 876/996; DA IN 166010, Message from USCINCEUR to the JCS, Sept. 10, 1955, RG 218, CCS 092 Western Europe (3-12-48) (2), sec. 32; and "Problems General Gruenther May Raise," Mar. 10, 1955, same file, sec. 8.

110. C(54)86, Mar. 5, 1954, PRO, CAB 129/66; Memorandum, Martin to Moore, Mar. 24, 1954, RG 59, 740.5; and Statement by Gen. Gruenther, Mar. 26, 1955, Senate, Foreign Relations, *NATO and the Paris Accords*, 3.

111. "Problems General Gruenther May Raise," Mar. 10, 1955; and Memorandum by the Chief of Staff of the Army, "The Most Effective Pattern of Military Strength for the Next Few Years—Report No. 2," Sept. 16, 1955, RG 218, CCS 092 Western Europe (3-12-48) (2), sec. 33.

112. In mid-1954, West Germany was planning to raise an army of twelve divisions and 375,000 men, an air force of 1,326 aircraft and 45,000 men, and a navy of 20,000 men (COS[54]248, "Restrictions on German Rearmament," July 30, 1954, PRO, DEFE 5/53).

113. For a glimpse of Dulles's views at the time, see *FRUS, 1952–54*, 5: 711–17. See also Steininger, "John Foster Dulles."

114. See *FRUS, 1952–54*, 5: 708–9.

115. Fursdon, *European Defence Community*, 210.

116. For documentation, see *FRUS, 1952–54,* 5: 1548–82. See also Eisenhower's letter to Gruenther of Feb. 10, 1953, in DDEL, DDEP, AWF, Administration Series, "Gruenther, Gen. Alfred 1952–53 (3)."

117. Memorandum, Parsons to Bruce, Jan. 27, 1953, RG 59, 740.5; and *FRUS, 1952–54,* 5: 728–29, 749–50, 1554, 1565. Dulles was also motivated by the fear that Congress was likely to reduce substantially the appropriation for European assistance if there was no tangible evidence of progress toward ratification by April.

118. Dockrill, *Britain's Policy,* 114–15. For a detailed contemporary analysis of the sources of French opposition to the EDC, see Memorandum, Parsons to Bruce, Jan. 27, 1953.

119. See *FRUS, 1952–54,* 5: 1560–61; and Fursdon, *European Defence Community,* 207. The full texts of the protocols are in *FRUS, 1952–54,* 5: 719–26.

120. *FRUS, 1952–54,* 5: 719–21, 741–43, 758, 775; Fursdon, *European Defence Community,* 209, 217; and Dockrill, *Britain's Policy,* 115–16.

121. For the details of the French and British proposals, see *FRUS, 1952–54,* 5: 730–32, 745–48, 804–5. See also Fursdon, *European Defence Community,* 200; Dockrill, *Britain's Policy,* 116–19; and Melandri, "France," 278.

122. Dockrill, *Britain's Policy,* 119.

123. *FRUS, 1952–54,* 5: 768, 790–92; and Dockrill, *Britain's Policy,* 119–20.

124. Fursdon, *European Defence Community,* 219; and Dockrill, *Britain's Policy,* 124–26.

125. Dockrill, *Britain's Policy,* 126–27.

126. The proceedings of the Washington conference are described in *FRUS, 1952–54,* 5: 1607–95. See esp. Second Tripartite Foreign Ministers Meeting, July 11, 1953, ibid., 1621–31. See also Dockrill, *Britain's Policy,* 127–28. The text of the invitation to the Soviet Union is in *FRUS, 1952–54,* 5: 1701–2. U.S. officials nevertheless hoped that the summit would clearly demonstrate that Soviet intentions had not changed (ibid., 799).

127. *FRUS, 1952–54,* 5: 800, 807–9.

128. Kaplan, "NATO-Indochina Connection," 13–14. See also Melandri, "France," 276; and Fursdon, *European Defence Community,* 200.

129. Dockrill, *Britain's Policy,* 129–30; and Melandri, "France," 276.

130. *FRUS, 1952–54,* 5: 823.

131. *FRUS, 1952–54,* 7: 509–10; and Statement of Policy by the National Security Council, NSC 160/1, Aug. 17, 1953, ibid., 510–20. On the strength of U.S. support at the time, see *FRUS, 1952–54,* 5: 798–800.

132. *FRUS, 1952–54,* 5: 1769–86, 1794–1806. Some British officials considered this to be the main goal of the meeting. See AOC(53)17th Meeting.

133. Statement by the Secretary of State to the NAC, Dec. 14, 1953, *FRUS, 1952–54,* 5: 462–64. See also the account in the *New York Times,* Dec. 15, 1953, p. 14.

134. For further discussion of the Berlin summit, see Fursdon, *European Defence Community,* 234–42; and Watson, *History of the JCS,* 303–4.

135. *FRUS, 1952–54*, 5: 875–77, 879–82. The second condition was an outgrowth of the rumors of U.S. troop withdrawals that had followed the development of the New Look the previous fall.

136. Ibid., 901, n. 4, and 911–12.

137. Dockrill, *Britain's Policy*, 134–38; and Fursdon, *European Defence Community*, 253–54. At the same time, Eden announced that the United Kingdom would commit an armored division to the EDC once the organization was established.

138. Fursdon, *European Defence Community*, 256; and Dockrill, *Britain's Policy*, 138–39. For the text of the U.S. declaration and detailed State Department commentary, see *DSB* 30 (Apr. 26, 1954): 619–20; and *FRUS, 1952–54*, 5: 959–63.

139. Dockrill, *Britain's Policy*, 139; and *FRUS, 1952–54*, 5: 927. French opponents of the EDC argued that the U.S. assurances contained nothing new. See ibid., 960.

140. *FRUS, 1952–54*, 5: 940–41.

141. Fursdon, *European Defence Community*, 260–61; Kaplan, "NATO-Indochina Connection," 29; and Melandri, "France," 279.

142. The final vote was 319 to 264 with 12 abstentions. Fursdon, *European Defence Community*, chap. 8, offers a detailed account of these events.

143. Dockrill, *Britain's Policy*, 140–41; and *FRUS, 1952–54*, 5: 1122–25. For the work of the Anglo-American Study Group, see ibid., 988–89, 997–1018.

144. *FRUS, 1952–54*, 5: 975, 983–84, 988.

145. For Dulles's statement as well as the considerations going into its preparation, see *FRUS, 1952–54*, 5: 1114–22.

146. Dockrill, *Britain's Policy*, 142–43; and Fursdon, *European Defence Community*, 304–14.

147. See, for example, *FRUS, 1952–54*, 5: 1192–94.

148. Ibid., 1184–1205.

149. Ibid., 1216–17.

150. Dockrill, *Britain's Policy*, 145–46.

151. On Britain's previous aversion to making any commitment to station a fixed number of forces on the continent, see C(52)434, "European Defence Community and Alternative Plans," Dec. 10, 1952, PRO, CAB 129. Dockrill, in contrast, argues that the declaration did not mark a significant departure in British policy (*Britain's Policy*, 147–48). Some of the implications of this commitment for British conventional force levels on the continent are discussed in Chapter 4.

152. Dockrill, *Britain's Policy*, 143–45; and Fursdon, *European Defence Community*, 321–23. For the proceedings of the London conference as well as the resulting agreements, see *FRUS, 1952–54*, 5: 1294–1364. For a concise summary, see ibid., 1378–84.

153. *FRUS, 1952–54*, 5: 1331–32, 1369–70.

154. For the records of the Paris meetings, see ibid., 1404–63. For Dulles's report, see ibid., 1470–80.

155. See ibid., 1464–1542.

156. Statement by Gen. Gruenther, Mar. 26, 1955, Senate, Foreign Relations, *NATO and the Paris Accords*, 6.

157. See, for example, Office of the U.S. Special Representative in Europe, "Certain European Issues Affecting the United States," May 15, 1953, DDEL, DDEP, AWF, Administration Series, Box 13, "William H. Draper, Jr. (2)," esp. p. 28.

158. Ibid., 29.

159. Ibid., 30, 35; and Memorandum, Merchant to Nash, et al., Nov. 17, 1953, RG 59, 740.5, pp. 1, 5.

160. In addition to the sources cited above, see Memorandum, MacArthur to the Undersecretary of State, July 20, 1953, RG 59, 740.5; and Memorandum, JCS for the Secretary of Defense, July 16, 1953, RG 218, CCS 092 Western Europe (3-12-48), sec. 227, for further evidence of such views.

Chapter 4

1. For example, early tests of the U.S. Army's Pentomic division showed that it could not effectively wage two-sided nuclear operations. The army estimated that the new divisions would be rendered ineffective after being struck by as few as seven to fourteen nuclear weapons of 40 kilotons or less. See Midgley, *Deadly Illusions*, 73–74.

2. See, for example, William W. Kaufmann, "The Requirements of Deterrence," Memorandum No. 6, Center of International Studies, Princeton University (Princeton, 1954), which was subsequently reprinted as in Kaufmann, "Requirements of Deterrence," and Brodie, "Unlimited Weapons."

3. For detailed discussions of army attitudes during this period, see Midgley, *Deadly Illusions*, and Bacevich, *Pentomic Era*.

4. *FRUS, 1952–54*, 2: 773.

5. NSC 5501, "Basic National Security Policy," Jan. 7, 1955, and NSC 5602/1, "Basic National Security Policy," Mar. 15, 1956, both in RG 273. NSC 5501 formally replaced NSC 162/2, the basis of the Eisenhower administration's New Look, which had been adopted in October 1953.

6. See, for example, JSCP 877/299, Nov. 23, 1955, RG 218, CCS 381 (11-29-49), sec. 27.

7. In early 1956, for example, Robert R. Bowie, assistant secretary of state for policy planning and a close adviser of Dulles, called for "mobile forces capable of effective action with or without atomic weapons" to deter local aggression (Memorandum for the Secretary, "Sharing Control of Atomic Weapons," Jan. 4, 1956, Seely G. Mudd Library, JFDP, Subject Series, Box 4, "Papers on Nuclear Weapons (5)").

8. See CSAFM 87-53, Dec. 2, 1953, RG 218, 381 U.S. (1-31-50), sec. 32; and Watson, *History of the JCS*, 28.

9. In November 1954, for example, Wilson expressed the hope that NATO's New Approach study would allow the United States to withdraw two divisions in the near future. See *FRUS, 1952–54*, 5: 533.

10. Although not a leading advocate, Eisenhower was sympathetic toward this position. See Memorandum for Record, "Conference of Joint Chiefs of Staff with the President," Feb. 10, 1956, DDEL, WHO, OSS, Subject Series, DOD Subseries, Box 4, "JCS (2)," and Memorandum of Conference with the President, May 24, 1956, DDEL, DDEP, AWF, DDE Diaries, Box 15, "May 1956 Goodpaster."

11. Memorandum for the Chairman of the Joint Chiefs of Staff from the Secretary of Defense, "Military Strategy and Posture," Jan. 27, 1956, RG 218, CJCS Radford Files, 381 (Military Strategy & Posture), Box 47.

12. For a brief description of the process of developing force requirements during this period, see Taylor, *Uncertain Trumpet*, 35, 85–87.

13. CM-264-56, Mar. 28, 1956, RG 218, CJCS Radford Files, 381 (Military Strategy & Posture), Box 47; Memorandum for the Chairman of the Joint Chiefs of Staff from B. L. Austin, Apr. 17, 1956, same file; and CM-266-56, Apr. 3, 1956, RG 218, CJCS Radford Files, "Chairman's Messages—1956," Box 65.

14. JCS 2143/56, Apr. 12, 1956, RG 218, CCS 381 (11-29-49), sec. 30.

15. Memorandum for the Record by Captain Richard H. Phillips, May 22, 1956, RG 218, CCS 381 (11-29-49), sec. 30.

16. SM-423-56, May 23, 1956, RG 218, CCS 381 (11-29-49), sec. 30.

17. Enclosure "A" to SM-506-56, June 18, 1956, RG 218, CCS 381 (11-29-49), sec. 31.

18. Memo for Admiral Radford from Bird, June 1, 1956, CM-311-56, June 5, 1956, and CM-316-56, June 11, 1956, all in RG 218, CJCS Radford Files, 381 (Military Strategy & Posture), Box 47.

19. "JSOP Military Strategy and Force Structure," n.d., DDEL, WHO, OSS, Briefing Series, DOD Subseries, Box 2, "Budget & Program, Defense FY 1958 (2)." For additional background on the Radford plan, see *New York Times*, July 13, 1956, p. 1; and Taylor, *Uncertain Trumpet*, 38–42.

20. Draft Memorandum for the Secretary of Defense, n.d., RG 218, CJCS Radford Files, JSOP Military Strategy and Force Structure (Investigation) (1956), Box 63.

21. See the page 1 stories in *New York Times* on July 13, 14, 15, and 17, 1956. The leak occurred even though only six copies of the original document were made. Radford subsequently conducted an internal investigation to determine the source of the leak, which was never identified.

22. Wilson had stated as recently as May 16 that the United States planned to keep its present military force of 2.9 million men for the "foreseeable future." See *New York Times*, May 17, 1956, p. 16.

23. See, for example, *DSB* 35 (July 30, 1956): 181; *DSB* 35 (Aug. 13, 1956): 263; and Stebbins, *United States in World Affairs, 1956*, 217–18.

24. *New York Times*, July 13, 1956.

25. See Chapter 3.

26. Statement by Gen. Alfred Gruenther, Mar. 26, 1955, Senate, Foreign Relations, *NATO and the Paris Accords*.

27. See Chapter 3.

28. JP(55)94, "The Most Effective Pattern of NATO Military Strength for the Next Few Years—Report No. 2," Sept. 9, 1955, PRO, DEFE 6/31; JP(55)146, "Military Comments on the 1955 Annual Review," Nov. 23, 1955, PRO, DEFE, 6/32; and USM-77-55, Nov. 21, 1955, RG 218, CCS 092 Western Europe (3-12-48) (2), sec. 41.

29. In 1955, the Soviets announced a planned cut of 640,000 men, and in May 1956, they declared that an additional reduction of 1.2 million troops, including 63 divisions, would be effected within twelve months. This amounted to an overall decline in the size of the Soviet armed forces from 5.7 to 3.9 million in just two years. See Stebbins, *United States in World Affairs, 1956,* 22, 166; and Wolfe, *Soviet Power,* 164.

30. See, for example, *FRUS, 1955–57,* 4: 25, 571–72.

31. USM-77-55; *FRUS, 1955–57,* 4: 24–26; and DEF 991588, Nov. 4, 1955, RG 218, CCS 092 Western Europe (3-12-48) (2), sec. 40.

32. JP(55)146; and DC(55)40, "UK Position in the NATO 1955 Annual Review," Sept. 27, 1955, PRO, CAB 131/16.

33. See *FRUS, 1955–57,* 4: 81, 571.

34. See Chapter 3 and USM-77-55.

35. It was not yet clear that the United States would have enough warheads to meet its own requirements. In 1952, the JCS had assumed that only 80 atomic warheads—about 8 percent of the total U.S. stockpile at the time—would be available for use in Western Europe (SM-271-52, Memorandum for General Eisenhower, "Planning Assumptions," Jan. 28, 1952, RG 218, CCS 471.6 [8-15-45], sec. 7). This number probably began to grow significantly only in 1956, when the overall size of the arsenal increased sharply. The growth of the U.S. atomic stockpile is charted in Cochran, Arkin, and Hoenig, *Nuclear Weapons Databook,* Table 1.6, p. 15.

36. *FRUS, 1955–57,* 4: 18, n. 5.

37. USM-77-55.

38. USM-5-56, Jan. 12, 1956, enc. to JCS 2073/1219, Jan. 17, 1956, RG 218, CCS 092 Western Europe (3-12-48) (2), sec. 48; and USM-13-56, Jan. 27, 1956, same file, sec. 49.

39. DC(55)13th Meeting, Minute 1, Nov. 4, 1955, PRO, CAB 131/16. See also Bartlett, *Long Retreat,* 105.

40. COS(56)282, "NATO Annual Review—UK Submission," July 24, 1956, PRO, DEFE 5/70; Bartlett, *Long Retreat,* 105–6, 109; and Pierre, *Nuclear Politics,* 97–98.

41. *FRUS, 1955–57,* 4: 32.

42. COS(55)176, "Long-Term Defence Programme (Strategic Factors)," July 25, 1955, PRO, DEFE 5/59; DC(55)8th Meeting, Minute 3, Aug. 26, 1955, PRO, CAB 131/16; and COS(56)219, "Long-Term Defence Review," June 7, 1956, PRO, DEFE 5/68. On the financial impact of the British nuclear program, especially the decision to develop the hydrogen bomb, see Pierre, *Nuclear Politics,* 196; and Bartlett, *Long Retreat,* 133.

43. DC(55)43, "UK Defence Programme," Oct. 14, 1955, PRO, CAB 131/16; and Eden to the Lord President et al., May 31, 1956, PRO, FO 371/123187.

44. DC(56)15, "Costs of British Forces in Germany," June 11, 1956, PRO, CAB 131/17; COS(56)282; COS(56)234, "Policy Review—Assessment of Further Studies," June 14, 1956, PRO, DEFE 5/69; ZP5/56G, July 19, 1956, PRO, FO 371/123188; and Meeting of the Committee on Service Costs in Germany, June 5, 1956, PRO, FO 371/124622.

45. The Shield forces consisted of the conventional forces in Europe along with their associated tactical nuclear capabilities.

46. *FRUS, 1955–57*, 4: 84–87, 89–92; CM(56)44th Meeting, Minute 10, June 19, 1956, PRO, CAB 128/30, Part 1; and JP(56)120, "NATO Strategy and Level of Forces—Military Brief," June 27, 1956, PRO, DEFE 6/36.

47. JP(56)120; COS(56)234; and COS(56)282.

48. *FRUS, 1955–57*, 4: 84–86; JP(56)120; COS(56)266, "Planning for War in Northwest Europe," July 11, 1956, PRO, DEFE 5/69; and JP(56)162, "SACEUR's Force Requirements 1960/62," Nov. 16, 1956, PRO, DEFE 6/36. As an afterthought, the British acknowledged that NATO's conventional forces should be able to impose some delay on the progress of a Soviet invasion until the full impact of NATO's nuclear retaliation was felt. See *FRUS, 1955–57*, 4: 91–92.

49. At one point, the British floated the idea of a fifteen-division force, including only two British, two to three U.S., and eight German divisions. See Memorandum, Timmons to Dulles, July 12, 1956, RG 59, 740.5.

50. Memorandum, Elbrick to Dulles, "U.K. Proposal for NATO Review of Strategy," June 23, 1956, and Tel. 7716, June 22, 1956, both in RG 59, 740.5; *FRUS, 1955–57*, 4: 84–87; and CM(56)44th Meeting, Minute 10.

51. *FRUS, 1955–57*, 4: 89–90; and Memorandum of Conversation, July 2, 1956, and MacArthur, Memorandum for the Record, Aug. 28, 1956, both in RG 59, 740.5.

52. Memorandum of Conversation, "British Proposal for Review of NATO Strategy, Looking Toward a Reduction in Conventional Force Levels," Aug. 29, 1956, RG 59, 740.5; COS(56)89th Meeting, Minute 4, "Draft Revised Directive to the NATO Military Authorities," PRO, DEFE 4/90; Minister of Defence to the Chairman, Chiefs of Staff Committee, Sept. 3, 1956, and "NATO Reappraisal," Oct. 8, 1956, both in PRO, AIR 8/2065; and *FRUS, 1952–54*, 4: 102.

53. Topol 599, Oct. 13, 1956, RG 59, 740.5; and *FRUS, 1955–57*, 4: 102.

54. MD 53, Oct. 24, 1956, PRO, AIR 8/2065.

55. JSPC 876/1019, Aug. 24, 1955, RG 218, CCS 092 Western Europe (3-12-48) (2), sec. 29, and USM-11-56, Jan. 27, 1956, same file, sec. 50.

56. JP(56)132, "Overall Strategic Concept for the Defence of the North Atlantic Area," Aug. 10, 1956, PRO, DEFE 6/37; JP(56)150, "Overall Strategic Concept for the Defence of the North Atlantic Area," Sept. 25, 1956, PRO, DEFE 6/37; COS(56)377, "Briefs for the 14th Session of the NATO Military Committee," Oct. 11, 1956, PRO, DEFE 5/71; "NATO Reappraisal," Oct. 8, 1956, PRO, AIR 8/2065; and USM-68-56, July 13, 1956, enc. to JCS 2073/1280, July 16, 1956, RG 218, CCS 092 Western Europe (3-12-48) (2), sec. 60.

57. Specifically, the role of the Shield forces was to be limited to dealing

with local infiltrations and incursions, enabling aggressive intentions to be identified as such, providing a shield against a satellite attack, and holding any Soviet attack until the strategic counteroffensive became effective. See Annex to CP(56)269, "UK Forces in the FRG," Nov. 28, 1956, PRO, CAB 129/84.

58. *FRUS, 1955–57*, 4: 96–98, 114–15; and Memorandum, MacArthur to Dulles, "Your Meeting with Selwyn Lloyd on NATO Today," Oct. 7, 1956, and Topol 592, Oct. 12, 1956, both in RG 59, 740.5.

59. Memorandum, Wolfe to Elbrick, et al., "Review of NATO Strategy and Force Levels," RG 59, 740.5. By mid-1956, even SACEUR Alfred Gruenther had acknowledged the possibility of "small, localized actions" in the event of nuclear parity. See *DSB* 35 (July 12, 1956): 112.

60. The mid-1955 NATO air maneuver, Carte Blanche, for example, had suggested that a nuclear war in Europe would quickly result in millions of civilian casualties. See Speier, *German Rearmament*, 144–47, 182–93; and Kelleher, *Germany*, 34–43.

61. Bonn 329, July 24, 1956, RG 59, 762A.00; Memorandum of Conversation, "Implications of Possible Reductions in Force Levels," July 26, 1956, RG 59, 740.5; Memorandum, MacArthur to Dulles, "Review of NATO Military Strategy," July 27, 1956, RG 59, 740.5; Memorandum of Conversation between Dulles and Krekeler, July 30, 1956, RG 59, 740.5; COS(57)107, "Conversation Between Prime Minister and Heusinger," May 10, 1957, PRO, DEFE 5/75; Craig, "Germany and NATO," 240; and Cioc, *Pax Atomica*, 33. West Germany had not participated in the formulation of MC 48, having been admitted to NATO only in 1955.

62. Bonn to Foreign Office, Tel. 402, Nov. 29, 1956, PRO, FO 371/124609; CP(56)269; JP(57)28, "CPX 7—Brief for Chairman of the Chiefs of Staff," Mar. 22, 1957, PRO, DEFE 6/40; and *FRUS, 1955–57*, 4: 146, 157.

63. COS(56)97th Meeting, Minute 1, "Overall Strategic Concept," Oct. 9, 1956, PRO, DEFE 4/90.

64. See Letter, Martin to Perkins, Sept. 25, 1956, RG 59, 740.5/9-2756.

65. NATO Heads of Government Meeting, "NATO Defense Policy and Strategy" (Background Paper), Dec. 4, 1957, DDEL, DDEP, WHCF, Subject Series, Box 46, "NATO (5)"; and JP(57)15, "Review of the World Situation," Feb. 21, 1957, PRO, DEFE 4/95.

66. "NATO Defense Policy and Strategy," Dec. 4, 1957, DDEL, DDEP, WHCF, Subject Series, Box 46, "NATO (5)"; JP(57)15; COS(57)244, "NATO Minimum Forces Studies," Nov. 14, 1957, PRO, DEFE 5/79; and COS(57)280, "MC 70," Dec. 23, 1957, PRO, DEFE 5/80.

67. JP(57)30, "Overall Strategic Concept," Mar. 22, 1957, PRO, DEFE 6/40.

68. JP(57)11; and COS(57)24th Meeting, 1st Minute, "Overall Strategic Concept," Mar. 28, 1957, PRO, DEFE 4/96.

69. COS(57)74. Indeed, the JCS had recommended that there be no explicit upper limit on the duration of the initial phase. See SM-111-57, Feb. 9, 1957, RG 218, CCS 092 Western Europe (3-12-48) (2), sec. 73.

70. JP(57)11; COS(57)24th Meeting, Minutes 1 and 2, Mar. 28, 1957, PRO,

DEFE 4/96; JP(57)32, "Measures to Implement the Strategic Concept," Mar. 27, 1957, PRO, DEFE 6/40; and COS(57)75, "Measures to Implement the Strategic Concept," Mar. 28, 1957, PRO, DEFE 5/74.

71. JP(57)30; COS(57)74; COS(57)30th Meeting, Minute 1, "Meeting with Admiral Denny on Military Committee Meeting," Apr. 12, 1957, PRO, DEFE 4/96; and SM-111-57.

72. COS(57)11th, Minute 3, Feb. 8, 1957, PRO, DEFE 4/95; JP(57)11, "NATO—Overall Strategic Concept," Feb. 5, 1957, PRO, DEFE 6/40; and COS(57)74, "Overall Strategic Concept," Mar. 28, 1957, PRO, DEFE 5/74.

73. JP(57)30; and JP(57)129, "ACE Minimum Forces Study," Nov. 6, 1957, PRO, DEFE 6/43.

74. JP(57)30; and COS(57)74.

75. SM-111-57; JP(57)30; and COS(57)74. The Germans in particular remained concerned lest any small incident trigger the use of tactical nuclear weapons on their territory. See COS (57)107, "Conversation Between the Prime Minister and Heusinger on May 8, 1957," May 10, 1957, PRO, DEFE 5/75; and JP(57)63, "Brief for Meeting of British and West German Ministers of Defence," May 21, 1957, PRO, DEFE 6/41.

76. For details, see JP(57)129, "ACE Minimum Forces Study, 1958–62," Nov. 6, 1957, PRO, DEFE 6/43; and COS(57)244, "NATO Minimum Forces Studies," Nov. 14, 1957, PRO, DEFE 5/79. An extensive extract of the text is printed in Wampler, "NATO Strategic Planning," 40–41.

77. SM-111-57; and "A Review of North Atlantic Problems for the Future," Mar. 1961, JFKL, NSF, Box 220, "NATO Acheson Report, 3/61," 31.

78. Memorandum for the Secretary of Defense, "Understanding of Certain Terms," Apr. 9, 1958, RG 218, CCS 092 Western Europe (3-12-48) (2), sec. 104; and JP(57)129.

79. COS(57)74, "Overall Strategic Concept," Mar. 28, 1957, PRO, DEFE 5/74; and COS(57)244, "NATO Minimum Forces Studies," Nov. 14, 1957, PRO, DEFE 5/79.

80. "Understanding of Certain Terms," Apr. 9, 1958; JP(57)129; and COS(57)244.

81. "SHAPE Statement on Need for 'Shield' Forces," Aug. 2, 1956, RG 59, 740.5; Norstad, "NATO: Deterrent and Shield"; and Statement by Gen. Lauris Norstad, June 11, 1957, House of Representatives, Foreign Affairs, *Selected Executive Session Hearings, 1957–60*, 19: 25–26. See also JP(56)162, "SACEUR's Force Requirements 1960/62," Nov. 16, 1956, PRO, DEFE 6/37.

82. *FRUS, 1955–57*, 4: 170–71; JP(57)129; COS(57)244; Norstad, "Text"; Statement by Gen. Lauris Norstad, Mar. 25, 1958, House of Representatives, Foreign Affairs, *Selected Executive Session Hearings, 1957–60*, 19: 487–94; and Statement of Gen. Lauris Norstad, Mar. 26, 1958, Senate, Foreign Relations, *Mutual Security Act of 1958*, 187, 200–201.

83. COS(57)244; and JP(57)129.

84. COS(58)70, "MC 70," Mar. 11, 1958, PRO, DEFE 5/82; and JP(58)12, "Minimum Essential Force Requirements, 1958–63," Feb. 17, 1958, PRO, DEFE 6/49.

85. "NATO Defense Policy and Strategy," Dec. 4, 1957, and NATO Heads of Government Meeting, "U.S. Defense Policy and Force Posture in NATO" (Background Paper), Dec. 4, 1957, both in DDEL, DDEP, WHCF, Subject Series, Box 46, "NATO (5)."

86. Although Norstad denied admitting the possibility of limited war in the NATO area, he nevertheless reiterated the importance of having the means to deal decisively with any situation less than general war. He now described the deterrent function of the Shield forces as imposing a "pause," which might last as long as weeks or months and during which the Soviet Union would presumably reconsider the potential consequences of its actions and decide to break off hostilities. See Statement of Gen. Lauris Norstad, Apr. 9, 1959, House of Representatives, Foreign Affairs, *Mutual Security Act of 1959*, 445–47, 466–67.

87. "Address by General Lauris Norstad," n.d., JFKL, NSF, Box 220, "NATO, 11/21/60–2/15/61"; "A Review of North Atlantic Problems for the Future," Mar. 1961, JFKL, NSF, Box 220, "NATO Acheson Report, 3/61," 33; and Statement by Gen. Lauris Norstad, Mar. 10, 1960, House of Representatives, Foreign Affairs, *Selected Executive Session Hearings, 1957–60*, 21: 133–35. See also Norstad, "NATO: Strength and Spirit," 9.

88. *FRUS, 1955–57*, 4: 129.

89. The difference consisted of one French and one British division. The independent U.S. brigades were presumably not counted. For the MC 70 requirements, see Statement by Gen. Norstad, Apr. 9, 1959, House of Representatives, Foreign Affairs, *Mutual Security Act of 1959*, p. 449; Statement by Gen. Norstad, Mar. 10, 1960, House of Representatives, Foreign Affairs, *Selected Executive Session Hearings, 1957–60*, 21: 135–36; and DOD, "Remarks by Secretary McNamara."

90. JP(57)129; and COS(57)244.

91. *FRUS, 1955–57*, 4: 92–93, n. 7. Eisenhower agreed with Dulles's observation that the Germans would view any reduction as tantamount to U.S. abandonment. See ibid., 94, n. 4.

92. *FRUS, 1955–57*, 4: 94.

93. Letter, Martin to Perkins, Sept. 27, 1956, RG 59, 740.5; and *FRUS, 1955–57*, 4: 96, n. 1.

94. Memorandum, Elbrick and MacArthur to Dulles, "Review of NATO Strategy and Withdrawal of Forces," Sept. 27, 1956, RG 59, 740.5.

95. *FRUS, 1955–57*, 4: 99–102. Eisenhower had expressed this view to Dulles in mid-September. See Letter, Dulles to Robertson, Sept. 26, 1956, RG 59, 740.5.

96. Message, Norstad to the Secretary of Defense, DA IN 71824, Nov. 18, 1957, RG 218, CCS 092 Western Europe (3-12-48) (2), sec. 91.

97. For a detailed discussion of the Pentomic division, see Midgley, *Deadly Illusions*, chap. 2.

98. DOD News Release No. 399-59 (Apr. 16, 1959); and Midgley, *Deadly Illusions*, 69–71, 92–93. In artillery, the new divisions were limited to 30 105mm, 12 155mm, and 4 8-inch guns, in contrast to Korean War divisions,

which had 54 105mm and 18 155mm guns. In an effort to correct this weakness, the Pentomic division was modified in 1960 to have 30 155mm guns.

99. Submission by the United States of America in Response to the Questionnaire for Annual Review 1958, RG 218, CCS 092 Western Europe (3-12-48) (2), BP 21, AF-III. Another source states that these aircraft were all nuclear-capable by the end of 1956. See *FRUS, 1955–57,* 4: 159.

100. *New York Times,* July 17, 1957, p. 1; Memorandum of Conference with the President, Aug. 16, 1957, DDEL, DDEP, WHO, OSS, Subject Series, DOD Subseries, Box 1, "DOD II(1)"; and Memorandum from the Secretary of Defense to the Secretary of the Army, the Secretary of the Navy, and the Secretary of the Air Force, Sept. 19, 1957, RG 218, CCS 092 Western Europe (3-12-48) (2), sec. 87. See also *New York Times,* Aug. 12, 1957, p. 1, and Aug. 17, 1957, p. 7.

101. JCS Memorandum to the Secretary of Defense, Apr. 26, 1957, RG 218, CCS 092 Germany (5-4-49), sec. 33.

102. Wilson did say, however, that the United States might reduce these forces by 10 percent over the next two years through further streamlining. He hoped eventually to withdraw an additional 35,000 men from Europe in this way. See Memorandum of Conference with the President, Aug. 16, 1957, DDEL, WHO, OSS, Subject Series, DOD Subseries, Box 1, "DOD II(1)," and "The Review of US and UK Forces," enc. to Memorandum for the President, Mar. 22, 1957, *DDRS,* 1987, no. 149.

103. Taylor, *Uncertain Trumpet,* 52.

104. JCS 931016, Oct. 11, 1957, RG 218, CCS 092 Western Europe (3-12-48) (2), sec. 89, and DA IN 63492, Oct. 18, 1957, same file. "If the United States . . . now reduces its own commitments," Norstad warned, "the example may be irresistible."

105. Memorandum for the Secretary of Defense, "Force Levels for FY 1959 Budget," Nov. 1, 1957, RG 218, CCS 370 (8-19-45), sec. 60; and Taylor, *Uncertain Trumpet,* 54. See also "Memorandum of Discussion at the 346th Meeting of the National Security Council, November 22, 1957," Nov. 25, 1957, *DDRS,* 1987, no. 1690.

106. See, for example, "Memorandum of Conference with the President, November 28, 1958," Dec. 9, 1958, DDEL, DDEP, AWF, DDE Diaries, Box 37, "Staff Notes, November 1958"; "Memorandum of Conference with the President, December 12, 1958," Dec. 15, 1958, DDEL, DDEP, AWF, DDE Diaries, Box 37, "Staff Notes, December 1958"; "Memorandum of Conference with the President, August 24, 1959," Aug. 25, 1959, *DDRS,* 1987, no. 1713; Memorandum, Weiss to Ernst, Aug. 11, 1959, *DDRS,* 1987, no. 289; and "Issues of U.S. Policy Regarding the Defense Posture of NATO," enc. to "Memorandum for the National Security Council," Nov. 10, 1959, *DDRS,* 1986, no. 1717.

107. See, for example, Speier, *German Rearmament.*

108. See Kelleher, *Germany,* 38–41; Craig, "Germany," 242; and Speier, *German Rearmament.*

109. Kelleher, *Germany,* 38–41, 54–55; and Craig, "Germany," 242.

110. On the impact of the Radford plan, see Kelleher, *Germany,* 43–49.

111. See, for example, Bonn 329, July 24, 1956, RG 59, 762A.oo.

112. *New York Times*, July 21, 1956, p. 1, July 26, 1956, p. 4, and July 31, 1956, p. 1.

113. *New York Times*, July 21, 1956, p. 1; Kelleher, *Germany*, 43–47; Craig, "Germany," 243; and Bonn 329.

114. Memorandum, Murphy to the Secretary, "Your appointment with Ambassador Krekeler," July 18, 1956; and Bonn 1456, 16 Oct. 1956, both in RG 59, 740.5. Despite official denials to the contrary, a steady stream of stories in the press suggested that U.S. policy was shifting toward ever greater emphasis on nuclear weapons at the expense of cuts in conventional manpower and that American officials questioned the need for twelve German divisions. See, for example, *New York Times*, July 17, 1956, p. 1, Aug. 3, 1956, p. 9, Aug. 5, 1956, p. 34, and Sept. 27, 1956, p. 19.

115. Memorandum of Conversation, "German Concern About Press Reports of American Troop Reductions," Aug. 24, 1956, RG 59, 740.5; Memorandum, Murphy to Dulles, July 18, 1956; and Craig, "Germany," 241. See also Speier, *German Rearmament*, 207.

116. Just days before the crisis broke, and only shortly after a trip to the United States, Adenauer had insisted that U.S. and NATO authorities continue to stress the absolute need for a German force of 500,000 men. See Kelleher, *Germany*, 46.

117. See Bonn 1456.

118. Memorandum, Elbrick to Murphy, "Your meeting with Ambassador Krekeler," July 17, 1956, and Memorandum of Conversation, "Press Speculation Regarding U.S. and NATO Force Reductions," July 19, 1956, both in RG 59, 740.5; and *New York Times*, July 26, 1956, p. 4.

119. *New York Times*, July 26, 1956, p. 4, and Sept. 28, 1956, p. 1; and Kelleher, *Germany*, 48. In an unusually blunt statement to the press, the German government blamed the reduced term of service squarely on the disclosure of the Radford plan.

120. JCS 1868/601, Sept. 27, 1956, RG 218, CCS 092 Western Europe (3-12-48) (2), sec. 64; Secretary of the Army to the Secretary of State, Oct. 12, 1956, RG 218, CJCS Radford files, CJCS 092.2 NATO (1956), Box 22; and *New York Times*, Sept. 28, 1956, p. 3. For a more detailed discussion, see Kelleher, *Germany*, 49–56.

121. On Strauss's views, see Kelleher, *Germany*, 64–74.

122. On October 17, Strauss announced that German plans now called for only 70,000 men in uniform at the end of 1956, instead of the original goal of 96,000, and only 120,000 men at the end of 1957, instead of 270,000. At the same time, he stated that the Bundeswehr should be equipped with nuclear weapons if the other Europeans were so equipped. And two days later, it was reported that he would seek a 300,000-man force equipped with atomic weapons. See *New York Times*, Oct. 18, 1956, p. 3, and Oct. 20, 1957, p. 3; USM-109-56, Nov. 5, 1956, enc. to JCS 2124/178, Nov. 5, 1956, RG 218, 092 Germany (5-4-49), sec. 32; and Bonn to Foreign Office, Tel. 402, Nov. 28, 1956, PRO, FO 371/124609.

123. See *New York Times*, July 2, 1957, p. 1, July 14, 1957, p. iv: 4, July 30, 1957, p. 3, and Oct. 25, 1957, p. 14; Bonn to Foreign Office, Tel. 321, Nov. 16, 1957, PRO, FO 371/130777; and IISS, *Military Balance, 1960–1961*.

124. Bonn to Foreign Office, Tel. 402, Nov. 28, 1956; COS(57)41st Meeting, Minute 1, May 24, 1957, PRO, DEFE 4/97; and PRS D-6/17, NATO Heads of Government Meeting, "German NATO Contribution" (Bilateral Position Paper), Dec. 4, 1957, *DDRS*, 1986, no. 2326.

125. "German NATO Contribution," Dec. 4, 1957, *DDRS*, 1986, no. 2326. In April 1957, it had been reported that the German armed forces would have a strength of only 200,000 by the end of 1961. See *New York Times*, Apr. 16, 1957, p. 17.

126. ZP5/56G; ZP5/61G, Memorandum by G. L. McDermott, July 25, 1956, and ZP5/68G, Aug. 8, 1956, both in PRO, FO 371/123188; COS(56)234; COS(56)282, "NATO Annual Review for 1956—UK Submission," July 24, 1956, PRO, DEFE 5/70; DC(56)15; and AUS(A) to Secretary of State for Air, Dec. 18, 1956, PRO, AIR 8/2046.

127. Memorandum, Norman Brook to the Minister of Defence, Dec. 5, 1956, PRO, FO 371/123189; CM(56)97th Meeting, Minute 3, Dec. 7, 1956, PRO, CAB 128/30; and CP(56)269, Memorandum by the Minister of Defence, "UK Forces in Germany," Nov. 28, 1956, PRO, CAB 129/84.

128. COS(56)97th Meeting, Minute 1, Oct. 9, 1956, PRO, DEFE 4/90; MD 53, Denny to the Chiefs of Staff, Oct. 24, 1956, PRO, AIR 8/2065; CP(56)269; COS(56)125th Meeting, Minutes 2 and 4, Nov. 27, 1956, PRO, DEFE 4/92; and COS(56)124th Meeting, Minute 2, Nov. 26, 1956, PRO, DEFE 4/92.

129. Memorandum, Clinton-Thomas to the Foreign Secretary, Dec. 18, 1956, PRO, FO 371/123189; Gen. 564/1st Meeting, Dec. 18, 1956, PRO, CAB 130/122; Private Secretary to Chief of the Air Staff, Dec. 21, 1956, PRO, AIR 8/2046; COS(57)34, "Long-Term Defence Programme, Memorandum by the Chiefs of Staff," Feb. 5, 1957, PRO, DEFE 5/73; and Misc/M(56)195, Dec. 18, 1956, PRO, AIR 8/2046.

130. Chiefs of Staff to Denny, Jan. 27, 1957, PRO, AIR 8/2066; and Brook to the Minister of Defence, Dec. 5, 1956, PRO, FO 371/123189.

131. Clinton-Thomas to the Foreign Secretary, Dec. 18, 1956, PRO, FO 371/123189; COS(56)97th Meeting, Minute 1; COS(56)229, "Role and Composition in 2TAF," June 11, 1956, PRO, DEFE 5/68; COS(56)282; and CP(56)269. Initially, the British talked of maintaining four divisions in Germany, which would be greatly reduced in size, from 18,500 men to approximately 10,000 men each. Subsequently, however, they decided to reorganize the remaining forces into six brigade groups assigned to three divisional headquarters. A fourth division and two additional brigades would be stationed in Britain. See Clinton-Thomas to the Foreign Secretary, Dec. 18, 1956, PRO, FO 371/123189; *FRUS, 1955–57*, 4: 123–33; and GEN. 570/1st Meeting, "British Forces in Germany," Feb. 8, 1957, PRO, CAB 130/122.

132. CM(56)97th Meeting, Minute 3.

133. CP(56)269; Brook to the Minister of Defence, Dec. 5, 1956, PRO, FO

371/123189; and Clinton-Thomas to the Foreign Secretary, Dec. 18, 1956, PRO, FO 371/123189.

134. Clinton-Thomas to the Foreign Secretary, Dec. 18, 1956, PRO, FO 371/123189; *FRUS, 1955–57*, 4: 123–33; Telegram, Minnich to Goodpaster, n.d., *DDRS*, 1986, no. 3263; GEN. 564/1st Meeting; and Chiefs of Staff to Denny, Jan. 2, 1957. On the U.S. attitude toward the British cuts, see also *FRUS, 1955–57*, 4: 165–66.

135. COS(57)11th Meeting, Minutes 1 and 3, Feb. 8, 1957, and COS(57) 12th Meeting, Minute 1, Feb. 12, 1957, both in PRO, DEFE 4/95; and GEN. 570/1st Meeting. For Norstad's views, see also *New York Times*, Feb. 1, 1957, p. 2, and Feb. 15, 1957, p. 3.

136. COS(57)47, Feb. 22, 1957, PRO, DEFE 5/74; and D.(57)2nd Meeting, Feb. 27, 1957, PRO, CAB 131/18.

137. CC(57)8th Conclusions, Minute 4, Feb. 5, 1957, and CC(57)12th Conclusions, Minute 5, Feb. 19, 1957, both in PRO, CAB 128/31; COS(57)47; Bermuda Meeting, March 21–23, 1957, "Reduction of British Forces in Germany," n.d., *DDRS*, 1987, no. 771; Stebbins, *United States in World Affairs, 1957*, 57, 98–99; and *New York Times*, Feb. 15, 1957, p. 3, and Mar. 11, 1957, p. 1.

138. JP(57)16th Meeting, Minute 1, Feb. 21, 1957, PRO, DEFE 4/95; D.(57) 2nd Meeting; and CC(57)14th Conclusions, Minute 1, Feb. 28, 1957, and CC(57)17th Conclusions, Minute 5, Mar. 12, 1957, both in PRO, CAB 128/31. The United States also pressed Britain to accept Norstad's proposals. See "Reduction of British Forces in Germany," n.d., *DDRS*, 1987, no. 771.

139. COS(57)197, Sept. 3, 1957, PRO, DEFE 5/77; COS(57)103, May 8, 1957, PRO, DEFE 5/75; "Reduction of British Forces in Germany," n.d., *DDRS*, 1987, no. 771; and *New York Times*, Mar. 19, 1957, p. 1, and Mar. 20, 1957, p. 20.

140. See *New York Times*, Apr. 5, 1957, p. 4.

141. COS(57)197; COS(57)35th Meeting, Minute 1, May 10, 1957, PRO, DEFE 4/97; JP(57)63, May 21, 1957, PRO, DEFE 6/41; COS(57)30th Meeting, Minute 1, Apr. 12, 1957, PRO, DEFE 4/96; and D.(57)13th Meeting, July 2, 1957, PRO, CAB 131/18.

142. C(58)5, Jan. 8, 1958, PRO, CAB 129/91; GEN. 638/2, "Withdrawal of Forces from Germany," Feb. 25, 1958, PRO, CAB 130/145; *New York Times*, Oct. 15, 1957, p. 16, and Dec. 5, 1957, p. 8; Pierre, *Nuclear Politics*, 189; and Beach, "British Forces," 164.

143. C(58)5; GEN. 638/2; COS(58)145, May 30, 1958, PRO, DEFE 5/84; GEN. 638/2nd Meeting, "British Forces in Germany," July 8, 1958, PRO, CAB 130/145; Macmillan Talks, Washington, June 9–11, 1958, "Support Costs for U.K. Forces," June 1, 1958, *DDRS*, 1987, no. 214; Macmillan Talks, Washington, June 9–11, 1958, "NATO Strategy," June 1, 1958, *DDRS*, 1987, no. 215; and Beach, "British Forces," 165–66.

144. Memorandum for Chairman, Joint Chiefs of Staff, "Withdrawal of French Army Forces from NATO," Aug. 14, 1956, RG 218, CJSC Radford

Files, CJCS 092.2 NATO; and Statement by Gen. Norstad, June 11, 1957, House of Representatives, Foreign Affairs, *Selected Executive Session Hearings, 1957–60,* 19: 29. See also *New York Times,* Mar. 16, 1956, p. 3, and Mar. 18, 1956, p. 1.

145. *FRUS, 1955–57,* 4: 176; and Statement by Gen. Norstad, Apr. 9, 1959, House of Representatives, Foreign Affairs, *Mutual Security Act of 1959,* 449. In 1959, the United States estimated that France would have only two and two-thirds divisions "capable of opposing aggression effectively during the first month of a war." See NSC 5910/1, "U.S. Policy on France," Nov. 4, 1959, RG 273, p. 6.

146. Polto 96, July 12, 1956; Polto 501, Sept. 8, 1956; and Brussels 369, Oct. 9, 1956, all in RG 59, 740.5; and *New York Times,* June 16, 1957, p. 20, and July 30, 1957, p. 3.

147. *FRUS, 1955–57,* 4: 156; *NATO Letter* 5 (Mar. 1, 1957): 13; Statement by Gen. Norstad, June 11, 1957, 38; and Statement by Gen. Norstad, Mar. 25, 1958, House of Representatives, Foreign Affairs, *Selected Executive Session Hearings, 1957–60,* 19: 487.

148. Memorandum, Chief of Naval Operations to the Joint Chiefs of Staff, n.d., RG 218, CCS 092 Western Europe (3-12-48) (2), sec. 43; *New York Times,* Mar. 20, 1956, p. 2; *FRUS, 1955–57,* 4: 160–61; and DA IN 30706, June 20, 1957, RG 218, CCS 092 Western Europe (3-12-48) (2), sec. 81.

149. Memorandum from the Chairman, Joint Chiefs of Staff, "SHAPE Planning Guidance 1957," June 28, 1957, enc. to JCS 2073/1418, July 3, 1957, RG 218, CCS 092 Western Europe (3-12-48) (2), sec. 82. On Norstad's views, see "Interview with General Norstad," *NATO Letter* 4 (Dec. 1956): 37, and *NATO Letter* 5 (May 1957): 13.

150. *FRUS, 1955–57,* 4: 123–33, 150, 157, 160; and *New York Times,* Dec. 14, 1956, p. 1, and Dec. 15, 1956, p. 12.

151. The British had suggested two general alternatives: allied divisions might have U.S. atomic forces brigaded in their support, or there might be some mechanism for providing nuclear weapons to the allies in an emergency. See *FRUS, 1955–57,* 4: 128, 131.

152. *New York Times,* Feb. 16, 1957, p. 1.

153. Memorandum for the Secretary of Defense, "Use of Atomic Weapons," Aug. 28, 1957, RG 218, CCS 471.6 (8-15-45), sec. 95.

154. Memorandum of Conference with the President, Oct. 22, 1957, DDEL, WHO, OSS, Subject Series, DOS Subseries, Box 2, "DOS 1957 (August–October) (2)"; DA IN 64141, Oct. 21, 1957, RG 218, CCS 471.6 (8-15-45), sec. 102; Memorandum for the Secretary, JCS, "Proposed NATO Stockpile System," Oct. 30, 1957, same file, sec. 103; and *FRUS, 1955–57,* 4: 209.

155. *DSB* 38 (Jan. 6, 1958): 8–9; and *New York Times,* Dec. 11, 1957, p. 1, and Dec. 17, 1957, p. 1.

156. On the NATO decision, see the NATO Communique of December 19, 1957, in *DSB* 38 (Jan. 6, 1958): 14. For a glimpse of SHAPE planning, see CINCEUR to CINCUSAREUR, Apr. 23, 1958, RG 218, CCS 092 Western Europe (3-12-48) (2), sec. 105.

157. The training of Belgians to use the NIKE air defense missile, for example, did not begin before October 1957 (DA IN 30706, June 20, 1957, RG 218, CCS 092 Western Europe [3-12-48] [2], sec. 81). The delayed arrival of nuclear-capable weapons for non-U.S. forces is described in Message from USRO to the Secretary of Defense, Mar. 27, 1958, RG 218, CCS 092 Western Europe (3-12-48) (2), sec. 102. The first two British regiments to receive the Corporal missile did not become operational until 1959, and Britain did not receive warheads for either its Honest John or Corporal missiles until 1960. See Bartlett, *Long Retreat*, 137, and Pierre, *Nuclear Politics*, 167.

158. General Gruenther had pointed out as early as August 1956 that "some countries will not be able to buy, produce, or maintain complicated and expensive new equipment in any quantity for many years in the future." See Gruenther letter to the JCS, "Planning for New Weapons Under MDAP for NATO Countries," Aug. 2, 1956, RG 218, CJCS Radford Files, CJCS 092 NATO (1956), Box 22.

159. Statement by Gen. Norstad, Mar. 10, 1960, House of Representatives, Foreign Affairs, *Selected Executive Session Hearings, 1957–60*, 21: 118. Other sources, however, suggest that the requirements were even higher. See COS(58)145, "UK Reply to the NATO Annual Review," May 30, 1958, PRO, DEFE 5/84.

160. By December 1957, the United States had programmed approximately $750 million in military assistance for the purchase of modern weapons, mostly short-range missiles and aircraft, for allied forces. See *DSB* 38 (Jan. 6, 1958): 8.

161. Pierre, *Nuclear Politics*, 167.

162. Kelleher, *Germany*, 89–110.

163. NSC 5910/1, Nov. 4, 1959, RG 273, p. 8.

164. See, for example, the similar conclusions reached by Dulles and Eden in Memorandum of Conversation, "The Present Political Situation," Oct. 23, 1957, DDEL, DDEP, AWF, International Series, Box 20, "Macmillan October 23–25, 1957 (1)." For an especially bleak description of the situation, see DA IN 154643, Sept. 18, 1958, RG 218, CCS 381 (11-15-48), sec. 20.

165. Norstad provided the following breakdown: five U.S., five German, three British, two French, two Belgian, one Dutch, and one-third Canadian divisions. See Statement by Gen. Norstad, Mar. 26, 1958, Senate, Foreign Relations, *Executive Sessions*, 10: 158.

166. DOD, "Remarks by Secretary McNamara," 9.

167. Kelleher, *Germany*, 89.

168. *FRUS, 1955–57*, 4: 79.

169. See Dulles's remarks in *DSB* 36 (Jan. 7, 1957): 3–4.

170. Further evidence of the strength of these institutional influences is the relatively small size of the reduction in the BAOR (30 percent) in comparison with the substantial cuts the British government decided to make in the overall size of the armed forces (50 percent).

171. Previously, the Eisenhower administration had limited itself to declaring that the United States would "continue to maintain in Europe, in-

cluding Germany, such units of its armed forces as may be necessary and appropriate to contribute its fair share of the forces needed for the joint defense of the North Atlantic area while a threat to that area exists." See *DSB* 30 (Apr. 26, 1954): 619–20.

Chapter 5

1. "The North Atlantic Nations: Tasks for the 1960's," Aug. 1960, DDEL, WHO, Office of the Special Assistant for Disarmament (Stassen), Box 9, "Bowie Report (1)." Bowie felt that such an objective was well within NATO's reach, especially once the German buildup was completed. Noting that various factors would sharply limit the size of the forces that the Soviet Union could bring to bear on the central front, he concluded that a force of 30 divisions in the region would be adequate, although these would have to be better trained and equipped and have larger reserves than they presently did. See also "Memorandum of Conference with the President, August 16, 1960," Aug. 19, 1960, *DDRS*, 1987, no. 1139.

2. "NATO in the 1960's: U.S. Policy Considerations," Sept. 9, 1950, DDEL, WHO, OSS, International Trips and Meetings, Box 5, "NATO (5)."

3. NSC 6017, "NATO in the 1960's," Nov. 8, 1960, RG 273.

4. "Excerpt from Statement by Secretary of State Herter at NATO Meeting, Paris, December 1960," n.d., *DDRS*, 1986, no. 2332; and "Statement by Secretary Gates to the NATO Ministerial Meeting—December 1960," n.d., *DDRS*, 1987, no. 1141. Emphasis added.

5. For a glimpse of the wide range of views prevailing in the Eisenhower administration at this time, see "Memorandum of Meeting with the President, November 29, 1960," Dec. 1, 1960, DDEL, WHO, OSANSA, Special Assistant Series, Presidential Subseries, Box 5, "1960 Meetings with the President II (3)."

6. For a useful summary of Kennedy's personal views, see Stromseth, *Origins of Flexible Response*, 26–28. See also Kennedy, *Strategy of Peace*, 183–85.

7. See, for example, Rusk Memorandum to Bell, "Foreign Policy Considerations Bearing on U.S. Defense Posture," Feb. 4, 1961, JFKL, NSF, Departments and Agencies, Box 273, "Department of Defense, 2/61"; Memorandum for the President, "Message to the Governments of All NATO Countries," Feb. 4, 1961, *DDRS*, 1986, no. 2333; and Bundy Memorandum for Sorenson, "Defense Message," Mar. 13, 1961, JFKL, NSF, Departments and Agencies, Box 273, "Department of Defense, 3/61."

8. "A Review of North Atlantic Problems for the Future," Mar. 1961, JFKL, NSF, Box 220, "NATO Acheson Report, 3/61." Other members of the group were Paul Nitze, Henry Rowen, and Albert Wohlstetter (Stromseth, *Origins of Flexible Response*, 30).

9. Such an approach, the report noted, would be consistent with Norstad's elaboration of the NATO strategic concept in 1960. See Chapter 4.

10. Deputy National Security Adviser Robert Komer described Acheson's

thinking as similar to that in the Bowie report (Komer Memorandum to Kennedy, "March 7 Meeting with Acheson," Mar. 6, 1961, JFKL, NSF, Meetings and Memoranda, Box 321, "Staff Memoranda, Robert W. Komer, 1/1/61–3/14/61"). According to Douglas G. Brinkley, Acheson explicitly used the earlier analysis as his point of departure and incorporated many of Bowie's recommendations into his own report (Brinkley, "Dean Acheson," 148).

11. "Review of North Atlantic Problems." In reviewing the Acheson report, the Joint Chiefs of Staff agreed that NATO strategy, together with Norstad's interpretation, adequately contemplated a flexible response to the various forms of aggression that might take place in the NATO area. See CSAFM-99-61, Mar. 28, 1961, RG 218, 9050/3070 NATO (Mar. 16, 1961), sec. 2.

12. National Security Memorandum No. 40, "Policy Directive Regarding NATO and the Atlantic Nations," Apr. 24, 1961, LBJL, VPSF, Box 4, "NSC—1961."

13. "President's Trip to Ottawa, May 16–18, 1961, NATO Developments, Text of U.S. Presentation to NATO Council on April 26, 1961," JFKL, POF, Box 13, "Canada-Security, JFK Trip to Ottawa, 5/61 (B)."

14. Paris 4522, Apr. 20, 1961, JFKL, NSF, Countries, Box 70, "France—General, 3/16/61–4/21/61."

15. Paris 4522; and "Memorandum of Conversation with Ambassador Alphand, Mr. Nitze, and Mr. Palen, French Counselor," May 8, 1961, Attachment to Bundy Memorandum to Battle, May 9, 1961, JFKL, NSF, Countries, Box 70, "France—General, 5/1/61–5/10/61."

16. Brussels 1991, May 29, 1961, JFKL, NSF, Countries, Box 70, "France—General, 5/19/61–5/29/61." It was also reported that de Gaulle feared excessive emphasis on a NATO non-nuclear buildup would detract from the credibility of the nuclear deterrent in defending Berlin ("De Gaulle's Views on Conventional Forces and Berlin," May 26, 1961, *DDRS*, 1986, no. 2928).

17. "De Gaulle's Program for the French Armed Services," n.d., JFKL, NSF, Countries, Box 70, "France—General, CIA Briefing Packet, 5/18/61."

18. On German attitudes up to the Berlin crisis, see Kelleher, *Germany*, 156–65; and Bonn 1367, Mar. 16, 1961, JFKL, POF, Countries, Box 17, "German Security, Adenauer Meeting, General, 4/61 (C)."

19. For Strauss's views, see Current Intelligence Weekly Summary, "FRG Defense Minister Strauss," n.d., JFKL, POF, Countries, Box 17, "Germany Security, 7/61"; *New York Times*, May 11, 1961, p. 4; and Stromseth, *Origins of Flexible Response*, 129–31.

20. USM-160-61, Memorandum for the Joint Chiefs of Staff, "NATO Force Requirements for 1966," June 9, 1961, RG 218, CCS 9050/3410 NATO (Apr. 29, 1961), sec. 3. The Acheson report stated that NATO's nuclear capability should be strengthened only in ways that would not divert needed resources from non-nuclear tasks.

21. See, for example, Memorandum for the Record, "Discussion of NATO Strategy," May 25, 1961, RG 218, CCS 9050/3070 NATO (Apr. 19, 1961), sec. 1.

22. See, for example, "President's Visit to de Gaulle, Talking Points Paper, Berlin and Germany," n.d., JFKL, POF, Countries, Box 16A, "France, JFK Visit to de Gaulle, 5/61–6/61 (D)." Even the Acheson report suggested that the president should explicitly state that an effective nuclear capability would be maintained in Europe.

23. "De Gaulle's Views on Conventional Forces and Berlin," May 26, 1961, *DDRS*, 1986, no. 2928.

24. See, for example, "President's Visit to de Gaulle, Talking Points Paper, Berlin and Germany," n.d., JFKL, POF, Countries, Box 16A, "France, JFK Visit to de Gaulle, 5/61–6/61 (D)"; and the report of a speech by Deputy Secretary of Defense Roswell Gilpatrick, *New York Times*, June 7, 1961, p. 9.

25. Decision on JCS 2305/475, Report by the J-5 on "NATO Requirements Study (Project 106C)," May 5, 1961, RG 218, CCS 9050/3410 NATO (Apr. 29, 1961), sec. 1; and McNamara Memorandum for the Chairman of the Joint Chiefs of Staff, "NATO Planning," enc. to JCS 2305/565, June 2, 1961, RG 218, CCS 9050/9105 NATO (June 2, 1961), sec. 3.

26. The following discussion draws heavily on Schlesinger, *Thousand Days*, chap. 15; and Sorensen, *Kennedy*, chap. 21, which are two of the most detailed accounts of the Kennedy administration's handling of the crisis. Also useful is Trachtenberg, *History and Strategy*, chap. 5, which is based on a careful examination of the available primary source material.

27. At Vienna, Khrushchev warned Kennedy that the Soviet Union would use force if the borders of East Germany were violated. For descriptions of the meeting, see Schlesinger, *Thousand Days*, 370–74; and Sorensen, *Kennedy*, 584–86.

28. Kennedy's views are described in Schlesinger, *Thousand Days*, 374, 379; and Sorensen, *Kennedy*, 586; and *New York Times*, June 6, 1961, p. 1.

29. Schlesinger, *Thousand Days*, 388; Sorensen, *Kennedy*, 587–88; and Nitze, *From Hiroshima to Glasnost*, 197. Franz-Josef Strauss calculated that NATO could afford no more than a single brigade for the purposes of a probe ("Strauss-McNamara Conversation," July 14, 1961, RG 218, CCS 9165/5420 Germany [West] [June 16, 1961], Box 176).

30. Sorensen, *Kennedy*, 587–89; Schlesinger, *Thousand Days*, 380–82, 389; and Trachtenberg, *History and Strategy*, 217. On the use of a conventional buildup to demonstrate resolve, see also Owen Memorandum to Bundy, May 17, 1961, JFKL, NSF, Countries, Box 81, "Germany–Berlin, General,"; and the Sept. 20, 1961, testimony of Rusk in Senate, Foreign Relations, *Executive Sessions*, 13, pt. 2: 614.

31. As late as June 24, McNamara had stated that the United States still had no plans for a military buildup over the Berlin crisis (*New York Times*, June 24, 1961, p. 4). The planning process may have been given impetus by Khrushchev's speech of July 8, in which he announced an end to planned Soviet troop reductions and an increase in defense spending, citing rising NATO military budgets (*New York Times*, June 9, 1961, p. 1).

32. "Military Choices in Berlin Planning," July 13, 1961, JFKL, NSF, Coun-

tries, Box 81, "Germany–Berlin, General, 7/13/61." The most ambitious proposal, which had been put forward by the JCS, was designed to achieve a substantial increase in U.S. conventional military strength by the beginning of the following year. It called for mobilizing 4 National Guard divisions and 21 Air National Guard fighter squadrons and increasing the number of U.S. military personnel by 450,000. It would have involved sending 64,000 troops to Europe to round out existing units, while allowing the United States to deploy an additional six divisions by March 1962 ("Department of Defense Recommended Program Force Increases and Related Actions," n.d., JFKL, NSF, Departments and Agencies, Box 273, "Department of Defense, 2/61").

33. "Memorandum of Discussion in the National Security Council on July 19, 1961," July 24, 1961, JFKL, NSF, Meetings and Memoranda, Box 328–330, "NSAM 62"; "The Military Build-up: 'Low Key' or 'High Pressure'?" July 17, 1961, JFKL, POF, Departments and Agencies, Box 77, "Defense, 7/61–8/61"; and "Suggested Comments for Meeting with the JCS, 18 July 1961," n.d., JFKL, NSF, Departments and Agencies, Box 273, "Department of Defense, 6/61–7/61." See also Sorensen, *Kennedy*, 589–90.

34. These decisions were circulated on July 24 as NSAM 62. See "National Security Action Memorandum 62," July 24, 1961, JFKL, NSF, Meetings and Memoranda, Box 328–330, "NSAM 62." For accounts of the meetings at which these decisions were taken, see Bundy Memorandum for the President, "This Afternoon's Meetings," July 19, 1961, JFKL, NSF, Countries, Box 81, "Germany–Berlin, General, 7/19/61–7/22/61"; "Memorandum of Discussion in the National Security Council on July 19, 1961"; "Memorandum of Minutes of the National Security Council, July 20 [*sic*], 1961," July 25, 1961, *DDRS*, 1986, no. 2842.

35. Kennedy's letter to Macmillan is contained in State 10195, July 20, 1961, JFKL, POF, Countries, Box 27, "UK–General, 1/61–5/61." His letter to de Gaulle was transmitted in State 10196, July 20, 1961, *DDRS*, 1986, no. 1866.

36. McNamara Memorandum for the Chairman of the Joint Chiefs of Staff, "Allied Military Planning on Berlin," July 27, 1961, JFKL, NSF, Countries, Box 82, "Germany–Berlin, General, 7/27/61." The United States reportedly proposed inter alia that Britain station a fourth division on the continent and that Germany form three new divisions as quickly as possible. See *New York Times*, July 22, 1961, p. 1, and July 23, 1961, p. 1.

37. *New York Times*, July 23, 1961, p. 27, July 24, 1961, p. 1, and July 25, 1961, p. 8.

38. *New York Times*, Aug. 3, 1961, p. 2, Aug. 6, 1961, p. 1, Aug. 8, 1961, p. 1, Aug. 9, 1961, p. 1; and Nitze Memorandum for the Vice President, "Briefing Notes for Your Paris Discussions," Sept. 27, 1961, LBJL, VPSF, Box 2, "Paris Talks–September 30, 1961."

39. On August 17, France declared that it would reinforce its military units in West Germany and northeast France, and the British announced that they would form a new reserve division to be based in the United Kingdom.

On August 18, Adenauer announced that the Federal Republic would increase its troop strength (*New York Times*, Aug. 18, 1961, pp. 1–2, and Aug. 19, 1961, p. 1).

40. Nitze Memorandum for the Chairman of the Joint Chiefs of Staff, "NATO Military Build-up," Aug. 22, 1961, RG 218, CCS 9172 Berlin/3100 Germany (East) (Aug. 9, 1961) (1), Box 178; and USM-225-61, Memorandum for the Joint Chiefs of Staff, "NATO Military Build-Up," Sept. 8, 1961, RG 218, CCS 9050/3410 NATO (Sept. 8, 1961), Box 141.

41. Kennedy Memorandum for the Secretary of Defense, Aug. 14, 1961, JFKL, POF, Box 77, "Defense, 7/61–8/61." The president added that the United States should proceed to raise its long-term readiness to include the capability to send a six-division force to Europe by the end of the year.

42. Schlesinger, *Thousand Days*, 395; Kaufmann, *McNamara Strategy*, 258; and *New York Times*, Sept. 9, 1961, p. 1.

43. DOD, "Remarks by Secretary McNamara," 18–19.

44. In mid-October, McNamara ordered that the heavy equipment and supplies of the Fourth Infantry Division and the Second Armored Division be stationed in Europe. At the same time, he requested that the military develop airlift movement plans that would enable the divisional personnel to be deployed and combat ready in Europe ten to fourteen days after a decision to send them was made. See Memorandum for the Chief of Staff, US Army, "Movement Plans—USCINCEUR Augmentation," Oct. 18, 1961, RG 218, CCS 9172 Berlin/3100 (Aug. 9, 1961), sec. 1; and JCS 1874, Message from the JCS to the Chief of Staff of the Army, et al., Oct. 13, 1961, RG 218, CCS 9172 Berlin/3100 (Aug. 9, 1961), sec. 3.

Overall, the U.S. Army expanded from eleven to sixteen divisions as three training divisions were made combat ready and two National Guard divisions were ordered to active duty, while the number of army personnel grew from 860,000 to over a million. In addition, the U.S. tactical air force was increased from 16 to 21 wings and from 1,200 to 1,800 aircraft through the call-up of reserves. See Taylor Memorandum for the President, "Effect of Army Strength of 960,000," Dec. 9, 1961, JFKL, NSF, Departments and Agencies, Box 275, "DOD, Defense Budget FY 1963, 11/61–12/61"; "President's European Trip, June 1963, NATO Developments," June 13, 1963, LBJL, VPSF, Box 3, "Briefing Book, June 1963 (I)"; and DOD, *Annual Report for FY 1962*, 17.

45. See, for example, "Memorandum of Conversation between Vice President Lyndon B. Johnson, Ambassador Gavin, Ambassador Finletter and General Norstad, at the United States Embassy, Paris, France, 30 Sept. 1961," n.d., LBJL, VPSF, Box 2, "Paris Talks—September 30, 1961."

46. Memorandum for Colonel Howard Burris, "NATO Country Progress Toward Berlin Buildup," Sept. 27, 1961, LBJL, VPSF, Box 2, "Defense Briefing for Paris Discussions"; *New York Times*, Oct. 13, 1961, p. 1, Nov. 24, 1961, p. 3, Dec. 9, 1961, p. 3, and Dec. 31, 1961, p. 22; and Kelleher, *Germany*, 168.

47. Memorandum for Burris, "NATO Country Progress Toward Berlin Buildup," Sept. 27, 1961, LBJL, VPSF, Box 2, "Defense Briefing for Paris Dis-

cussions"; and Paris 1015, Aug. 24, 1961, JFKL, NSF, Countries, Box 70, "France—General, 8/21/61–8/31/61."

48. Even Norstad, who was presumably more in tune with allied sensibilities, had told the British on August 7 that their proposed effort represented little new or significant (Paris 2111, Oct. 18, 1961, JFKL, NSF, Countries, Box 170, "UK—General, 10/1/61–10/31/61").

49. Memorandum for Burris, "NATO Country Progress Toward Berlin Buildup," Sept. 27, 1961, LBJL, VPSF, Box 2, "Defense Briefing for Paris Discussions"; and Paris 2388, Nov. 2, 1961, JFKL, NSF, Countries, Box 17, "UK—General, 11/61." The BAOR was expected to remain at a level of no more than 51,000 men, well short of the 62,000 needed to achieve Norstad's goal.

50. Polto 28, Oct. 23, 1961, JFKL, POF, Countries, Box 17, "Germany Security, 8/61–12/61."

51. As McNamara told Congress in February 1962, "The events of the past year convinced us that NATO forces in Europe must be greatly strengthened. We want to have a choice other than doing nothing or deliberately initiating a general nuclear war" (Kaufmann, *McNamara Strategy*, 112).

52. DMP, "Recommended Department of Defense FY 1963 Budget and 1963–67 Program," Oct. 6, 1961, JFKL, POF, Departments and Agencies, Box 77, "Defense, 9/61–12/61"; and "Points for Discussion with Vice President Johnson," n.d., LBJL, VPSF, Box 9, "Germany and Berlin—1961." McNamara later told Kennedy that the U.S. Air Force would have been unable to provide any meaningful tactical air support for the army in the event of a non-nuclear limited war over Berlin in 1961 for lack of non-nuclear ordnance (McNamara Memorandum for the President, "Could the Defense Budget Be Cut to $43 Billion Without Weakening the Security of the United States?" Apr. 17, 1963, JFKL, NSF, Departments and Agencies, Box 274, "Department of Defense, 1/63–3/63").

53. Kaplan, *Wizards of Armageddon*, 301–3.

54. Sorensen, *Kennedy*, 629.

55. Enthoven and Smith, *How Much Is Enough?*, 120, 132–33; and Wolfe, *Soviet Power*, 166–67. For an analysis of the qualifications that attended some of these assessments, see Duffield, "Soviet Military Threat."

56. Sorensen, *Kennedy*, 627; and Kaufmann, *McNamara Strategy*, 83.

57. The evolution of the administration's conclusions can be followed in DOD, "Statement of Secretary McNamara on the FY 1964–68 Defense Program," and DOD, DMP, "Recommended FY 1964–68 General Purpose Forces." See also Kaufmann, *McNamara Strategy*, 113; and Enthoven and Smith, *How Much Is Enough?*, 136–42.

58. Wolfe, *Soviet Power*, 171–72.

59. "Basic National Security Policy," Mar. 25, 1963, JFKL, NSF, Subjects, Box 294, "Basic National Security Policy"; DOD, DMP, "Recommended FY 1964–68 General Purpose Forces"; Kaufmann, *McNamara Strategy*, 120–21; and DOD, "Statement of Secretary McNamara on the FY 1965–69 Defense Program." MC 26/4, which had superseded MC 70 in 1961, extended NATO force requirements through 1966. See Table 5.1.

60. On McNamara's December statement, see DOD, "Remarks by Secretary McNamara," 1, 16; and *New York Times*, Dec. 15, 1961, p. 6.

61. McNamara delivered an unclassified version of the Athens speech at Ann Arbor, Michigan, on June 16, 1962, which is reprinted in *Survival* 4 (Sept.–Oct. 1962): 194–96.

62. McNamara nevertheless insisted that equal attention be given to the quality of the forces provided, citing deficiencies in manning levels, support units, and stocks of ammunition and spare parts.

63. See *New York Times*, Dec. 15, 1962, p. 1; and Kelleher, *Germany*, 211.

64. Paris 3530, Jan. 21, 1962, JFKL, NSF, Countries, Box 70, "France—General, 1/1/62–1/31/62"; Legere Memorandum for Bundy, "French NATO Deficiencies," Oct. 9, 1962, JFKL, NSF, Countries, Box 71A, "France—General, De Murville Talks, 10/9/62"; Johnson Memorandum for Bundy, "Analysis of French and Belgian Force Changes in NATO," July 28, 1963, JFKL, NSF, Countries, Box 72, "France—General, Withdrawal of Naval Forces, 5/63–9/63"; and Burris Memorandum to the Vice President, "French Armed Forces," May 18, 1962, LBJL, VPSF, Box 5, "Memos to the Vice President from Colonel Burris, Jan. 61–Jan. 62."

65. IISS, *Military Balance* (various years).

66. For a more detailed discussion of the positions of the European allies, see Stromseth, *Origins of Flexible Response*, chaps. 6–8. For insightful contemporary accounts of the allied views, see Kaufmann, *McNamara Strategy*, 103–4; and Brodie, "Conventional Capabilities," 150–51. Kelleher, *Germany*, chap. 6, provides a valuable analysis of the German position.

67. As French foreign minister Maurice Couve de Murville flatly told Rusk, the Europeans believed that unless the West made it clear it was prepared to respond to any aggression with nuclear weapons, aggression would occur (Memorandum of Conversation, "Relations with Germany," Oct. 8, 1963, JFKL, NSF, Countries, Box 73, "France—General, 10/12/63–10/17/63").

68. See, for example, Brodie, "Conventional Capabilities," 151.

69. German military planners estimated that NATO would need 35 to 40 divisions in the Central Region to wage a sustained conventional defense (Stromseth, *Origins of Flexible Response*, 209–10).

70. "NATO Ministerial Meeting, Paris, December 16–18, 1963: Scope Paper," Nov. 30, 1963, LBJL, NSF, IMTF, Box 30-34, "NATO Defense Policy Conference, 12/2/63."

71. Kelleher, *Germany*, 169, 208. It is interesting to note the shift that had taken place since the mid-1950s in the German position, which had previously distinguished between limited forms of conventional aggression and a tactical nuclear war in Europe and had sought to ensure that NATO not overreact to the former. See Chapter 4.

72. Kelleher, *Germany*, 199, 208, 212. To defuse American criticism, German leaders denied that the use of tactical nuclear weapons would automatically result in escalation to the strategic level (ibid., 175, 213). This contention, however, ran counter to the argument that the threat to use such weapons made the risk of aggression incalculable.

73. Hilsman Memorandum to the Secretary, "West Germany: Political and Economic Prospects," Aug. 3, 1962, JFKL, NSF, Countries, Box 75, "Germany—General, 8/3/62."

74. The best general discussion of French policy toward NATO during this period is Harrison, *Reluctant Ally*. On the French nuclear program, see also Kohl, *French Nuclear Diplomacy*, and Mendl, *Deterrence and Persuasion*.

75. For Kennedy's frustration with the French rumor campaign, see Memorandum of Conversation, "Franco-American Relations and Europe," Oct. 7, 1961, JFKL, NSF, Countries, Box 73, "France—General, 10/8/63–10/11/63."

76. See, for example, *The Times* (London), Sept. 9, 1963.

77. The British regularly threatened to reduce the size of the BAOR unless the German government agreed to defray additional support costs. See, for example, *New York Times*, Jan. 6, 1962, p. 1, and Feb. 21, 1962, p. 3.

78. The total number went from approximately 2,500 to some 4,000. See Kaufmann, *McNamara's Strategy*, 310.

79. "President's European Trip, June 1963, Background Paper: Germany," June 11, 1963, LBJL, VPSF, Box 3, "President's European Trip Briefing Book, June 1963 (I)." A reinforcing factor was the slowness with which the U.S. Air Force in Europe reoriented its nuclear-capable aircraft for non-nuclear missions (Stromseth, *Origins of Flexible Response*, 71).

80. Stromseth, *Origins of Flexible Response*, 147–48. For further discussion of the MLF, see Schwartz, *NATO's Nuclear Dilemmas*, chap. 5.

81. Kugler, *Great Strategy Debate*, 104–5; and DOD, *Annual Report for FY 1963*, 21. Important equipment modernization measures included the replacement of the M-48 tank with the M-60, the introduction of more armored personnel carriers, and the substitution of 155mm self-propelled artillery for older towed pieces of lower caliber. ROAD stood for Reorganized Objective Army Division.

82. McNamara proposed that the active-duty strength of the tactical air force be increased from 1,200 to nearly 1,600 aircraft, or 250 more than proposed by the air force itself. Of this total, 400 planes would be dedicated to close air support of the ground forces. Indeed, McNamara regarded the provision of greater air support as the principal objective of the tactical air force programs. See DMP, "Recommended DOD FY 1963 Budget and 1963–1967 Program," Oct. 6, 1961, JFKL, POF, Departments and Agencies, Box 77, "Defense, 9/61–12/61"; DOD, "Statement of Secretary McNamara on the FY 1963–67 Defense Program"; and DOD, *Annual Report for FY 1963*, 25.

83. By late 1963, the Kennedy administration had increased U.S. airlift capacity by 75 percent and had plans for a 400 percent increase over 1961 levels by 1967 (Kaufmann, *McNamara Strategy*, 306). As a result of these efforts, the United States would have been able to deploy two additional combat-ready divisions in Europe in 10 days and two more divisions in 30 days after a deployment decision had been made (DOD, DMP, "Recommended FY 1964–1968 General Purpose Forces").

84. The administration set an interim procurement objective of sufficient stocks for 88 division-months of combat (DOD, "Statement of Secretary McNamara on the FY 1964–68 Defense Program"). Its long-term goal was for

the United States to be able to maintain 22 divisions in combat for the entire period between D-Day and the time when U.S. production lines would be able to catch up with the rate of combat consumption (DOD, *Annual Report for FY 1962*, 23).

85. DOD, DMP, "Recommended FY 1964–1968 General Purpose Forces."

86. In 1963, for example, the Defense Department calculated that a total additional expenditure by the Europeans of only $1.75 billion per year over the next five years would overcome the major deficiencies between their present forces and their MC 26/4 goals (DOD, "Statement of Secretary McNamara on the FY 1964–68 Defense Program").

87. See, for example, *New York Times*, Jan. 28, 1963, p. 1, May 12, 1963, p. 1, and June 14, 1963, p. 1.

88. See Treverton, *"Dollar Drain."*

89. An offset agreement was concluded with the Germans in 1961 and renewed in 1963 (ibid., 32–34). The administration sought to work out similar arrangements with the other countries in which U.S. troops were stationed. For an example of U.S. pressure on the French, see Memorandum of Conversation, Mar. 3, 1962, JFKL, NSF, Countries, Box 70, "France—General, 3/11/62–3/30/62."

90. By July 1962, the United States was bringing back the first contingent of 10,000 men (McNamara Memorandum for the President, July 16, 1962, JFKL, POF, Departments and Agencies, Box 77, "Defense, 7/62–12/62"; see also *New York Times*, May 5, 1962, p. 12, June 28, 1962, p. 1, and July 1, 1962, p. 26).

91. *New York Times*, Feb. 15, 1963, p. 1, Aug. 6, 1963, p. 9, Sept. 13, 1963, p. 34, and Oct. 31, 1963, p. 1.

92. These culminated in a special presidential message to Congress on the balance-of-payments problem in July 1963, in which Kennedy announced that the Defense Department would reduce the annual rate of military expenditures abroad by $300 million by 1965.

93. For details, see DOD, *Annual Report for FY 1964*, 17–18; *New York Times*, Sept. 24, 1963, p. 1; and Raj, *American Military*, 98.

94. See *New York Times*, Oct. 20, 1963, p. 66, and Oct. 23, 1963, p. 17; and "Vice President's Visit to the Benelux Countries, November 3–10, 1963, Review of Outstanding World Problems, NATO," Oct. 29, 1963, LBJL, VPSF, Box 3, "Vice President's Visit to the Benelux Countries, Nov. 4–10, 1963." In addition, former president Eisenhower published a widely read article in which he argued that the time had come to reduce substantially the American military presence in Europe (Eisenhower, "Let's Be Honest").

95. The original plan, which was approved by Kennedy on October 24, included the return of six combat units—one armored cavalry regiment, two tank battalions, and three artillery battalions—that had been sent to Europe during the Berlin crisis in the first half of 1964; the reorganization of the army's logistics forces in Europe to allow a reduction of 30,000 men by 1966; and the withdrawal of ten fighter squadrons—three from France and seven from the United Kingdom—by mid-1966 (NSAM 270, "10/24 Meeting with

the President Regarding the Possible Redeployment of U.S. Military Forces from Europe," Oct. 29, 1963, JFKL, NSF, Meetings and Memoranda, Box 342, "NSAM 270").

96. *New York Times*, Oct. 31, 1963, p. 1; and Note to Holder of NSAM 270, Nov. 5, 1963, JFKL, NSF, Meetings and Memoranda, Box 342, "NSAM 270." For administration denials that any such action was contemplated, see *New York Times*, Oct. 26, 1963, p. 6, and Nov. 1, 1963, p. 1.

97. See, for example, McNamara's address to the Economic Club of New York on November 18, 1963, in Kaufmann, *McNamara Strategy*, 309.

98. In March 1963, Paul Nitze described the most likely contingencies as "Communist probes, exploitations of weak spots, military actions at the lower end of the spectrum, and efforts to divide the West." In contrast, he regarded the risk of a massive conventional attack as small (ibid., 130).

99. Polto 715, Nov. 26, 1963, LBJL, NSF, IMTF, Box 30-34, "NATO Defense Policy Conference, 12/2/63."

100. "Chronology of Actions in NATO on NFP Exercise," n.d., LBJL, NSF, IMTF, Box 30-34, "NATO Defense Policy Conference, 12/2/63."

101. The official title of MC 100/1 was "Appreciation of the Military Situation as It Affects NATO up to 1970" (ibid.; and "Chronology of Actions by Military Committee/Standing Group Concerning NATO Long Term and Force Planning," n.d., LBJL, NSF, IMTF, Box 30-34, "NATO Defense Policy Conference, 12/2/63." See also Beer, *Integration and Disintegration*, 115).

102. "Chronology of Actions in NATO," "Chronology of Actions by Military Committee/Standing Group," and "NATO Force Planning," n.d., all in LBJL, NSF, IMTF, Box 30-34, "NATO Defense Policy Conference, 12/2/63."

103. "Proposed Defense Position: Troop Levels and Balance of Payments," Dec. 2, 1963; "State Paper," n.d.; and "NATO Ministerial Meeting, Paris, December 16–18, 1963, Scope Paper," Nov. 30, 1963, all in LBJL, NSF, IMTF, Box 30-34, "NATO Defense Policy Conference, 12/2/63."

104. "State Paper" and "NATO Ministerial Meeting, Paris, December 16–18, 1963, Scope Paper," both in LBJL, NSF, IMTF, Box 30-34, "NATO Defense Policy Conference, 12/2/63"; and Bundy Memorandum to the President, Dec. 2, 1963, LBJL, NSF, Aides File, Box 1, "Bundy—Memos for the President, Vol. 1."

105. "Draft Outline of McNamara's Remarks," Nov. 29, 1963, LBJL, NSF, IMTF, Box 30-34, "NATO Defense Policy Conference, 12/2/63."

106. See, for example, Stromseth, *Origins of Flexible Response*; Schwartz, *NATO's Nuclear Dilemmas*; and Legge, *Theater Nuclear Weapons*.

107. See McNamara's testimony in January 1963 to the House Armed Services Committee cited in Kaufmann, *McNamara Strategy*, 127–28, and in Brodie, "Conventional Capabilities," 148.

108. Stromseth, *Origins of Flexible Response*, 55; and Kugler, *Great Strategy Debate*, 63–64.

109. Before the December 1961 NATO meeting, for example, the JCS pointedly recommended that McNamara explicitly deny any intention of handling an all-out Soviet conventional attack with non-nuclear means alone

(Decision on JCS 2305/677, "Draft Remarks to Be Made by the Secretary of Defense at the NATO Ministerial Meeting, 14 December 1961," Dec. 7, 1961, RG 218, 9050/5410 NATO [Nov. 3, 1961], sec. 1).

110. As one close observer notes, "The joint assessment that was conducted during the trialogue provided the basis for an extensive revision of NATO's strategic concept. . . . It is doubtful whether the Allies would have reached an agreement as easily on revising NATO's strategy had it not been for the intensive interaction which established a solid consensus among the three central states" (Whitt, "Quarrel on the Rhine," 464, 468).

111. Von Hassel, "Detente Through Firmness."

112. Kelleher, *Germany*, 215; Stromseth, *Origins of Flexible Response*, 139–41; and von Hassel, "Organizing Western Defense." See also the communique issued after a meeting between McNamara and von Hassel in *New York Times*, Nov. 15, 1964. These adjustments in the German position did not eliminate all differences with the United States. For example, the Germans continued to insist that limited nuclear means be used selectively at an early stage.

113. Kelleher, *Germany*, 204–5; Stromseth, *Origins of Flexible Response*, 136–41; and Kugler, *Great Strategy Debate*, 69–72, 112.

114. See the report on the 1965 British White Paper on Defence in *New York Times*, Feb. 24, 1965; Hughes Memorandum to the Secretary, "Analysis of the 1965 British Defense White Paper," Mar. 5, 1965, LBJL, NSF, Country File, Box 206-207, "UK Cables, Vol. III, 2/65–4/65"; and Stromseth, *Origins of Flexible Response*, 164–66. See also Minister of Defense Denis Healey's remarks in the House of Commons, *New York Times*, Mar. 4, 1965; and London 4248, Mar. 4, 1965, LBJL, NSF, Country File, Box 206-207, "UK Cables, Vol. III, 2/65–4/65."

115. DOD, DMP, "Role of Tactical Nuclear Forces," 31–32; and "The Role of Tactical Nuclear Forces in NATO Strategy," Defense Background Brief, n.d., LBJL, NSF, Committee File, CNP, Box 4, p. 3. See also McNamara's December 1964 remarks to the NATO defense ministers in "The Troop Problem and Burden Sharing," LBJL, SDAH, vol. 1, chap. 3.

116. For a description of U.S. strategic war plans during this period, see Sagan, *Moving Targets*, 24–30.

117. Private communications with John Bellinger, Aug. 5, 1988, Mar. 21, 1989, and Apr. 4, 1989. Bellinger was the U.S. Army action officer responsible for the issue at the time.

118. "Role of Tactical Nuclear Forces," LBJL, NSF, Committee File, CNP, Box 4, pp. 1–2; and DOD, DMP, "Role of Tactical Nuclear Forces," 6–21.

119. "Role of Tactical Nuclear Forces," LBJL, NSF, Committee File, CNP, Box 4, pp. 1–2; and DOD, DMP, "Role of Tactical Nuclear Forces," 1, 19, 21, 41, 46.

120. "Remarks by Secretary McNamara, Defense Ministers' Meeting, Paris, France," May 31, 1965, LBJL, NSF, Agency File, Box 11, "DOD 6/65, Vol. III"; Polto 395, Sept. 29, 1965, LBJL, NSF, Country File, Box 171, "France, Cables, Vol. VII, 6/65–8/65"; "Troop Problem and Burden Sharing," LBJL,

SDAH, vol. 1, chap. 3; *New York Times*, Dec. 17, 1965, p. 9; and interview with Timothy W. Stanley, Mar. 20, 1989, Washington, D.C. Stanley was a high-level official in the U.S. mission to NATO during this period.

At the end of the first year of the process, NATO ministers would adopt "force goals" for the period under consideration. These would be derived from "force proposals" prepared by the NATO military authorities (NMA), which would be based in turn on "political guidance" issued by the ministers. During the second year, countries would submit plans covering the forces they actually planned to provide, which would be reviewed in an attempt to reconcile them with the agreed force goals and then combined and adopted by the ministers as a "NATO Force Plan" at the end of the second year. At the same time, the ministers would approve new political guidance to the NMAs, starting the cycle all over again.

121. For further discussion, see Harrison, *Reluctant Ally*, 134–63.

122. "Troop Problem and Burden-Sharing," LBJL, SDAH, vol. 1, chap. 3; "France/NATO," Dec. 3, 1966, LBJL, NSF, IMTF, Box 35-36, "NATO, Ministerial Meeting, Paris, December, 1966"; and Harrison, *Reluctant Ally*, 160–61.

123. Another factor was the establishment of the Special Committee of Defense Ministers, the forerunner of the Nuclear Planning Group, and especially its working group on nuclear planning, which met four times in 1966. This forum allowed the United States to educate some of the allies, including West Germany, about the tremendous difficulties and disadvantages inherent in relying on nuclear weapons and may have made them more amenable to a revision of the strategic concept (Stanley interview. See also Schwartz, *NATO's Nuclear Dilemmas*, 182–87).

124. DOD, DMP, "NATO Strategy and Force Structure," Sept. 21, 1966.

125. "Background Paper, BAOR and Offset Agreement," July 27, 1966, LBJL, NSF, Country File, Box 215-216, "UK, PM Wilson Visit Briefing Book, 7/29/66"; Treverton, "*Dollar Drain*," 34, 141–42; "Trilateral Talks, Background Paper, UK/German Military Offset Relationship," Nov. 5, 1966, DDRS, 1986, no. 2669; and *New York Times*, July 21, 22, 26, 30, 1966, and Aug. 20, 1966.

126. Treverton, "*Dollar Drain*," 35; and *New York Times*, June 11, 12, 1966.

127. Williams, *Senate and U.S. Troops*, 139–48; Raj, *American Military*, 145–60; and *New York Times*, Sept. 1, 1966.

128. Francis M. Bator, Briefing Memorandum for the President, Aug. 23, 1966, DDRS, 1986, no. 2935; "Results of the Meeting with the President on August 23, 1966," n.d., DDRS, 1987, no. 1730; "Message from President to Prime Minister," Aug. 26, 1966, DDRS, 1985, no. 649; and Treverton, "*Dollar Drain*," 142.

129. State 41140, Sept. 3, 1966, DDRS, 1985, no. 1008; Memorandum for the President, "Your 10:00 AM Meeting on the Erhard Visit," Sept. 21, 1966, DDRS, 1986, no. 2937; Memorandum for the President, "Erhard Visit, September 26–27, 1966," Sept. 25, 1966, DDRS, 1986, no. 2936; and State 59292,

Oct. 4, 1966, *DDRS*, 1985, no. 185. On the British reaction, see also *New York Times*, Sept. 2, 1966. The mandate of the talks is mentioned in the communique issued at the end of Erhard's September meeting with Johnson, which is reprinted in *New York Times*, Sept. 28, 1966.

130. State 83128, Nov. 10, 1966, LBJL, NSF, National Security Council Histories, Box 50, "Trilateral Negotiations and NATO, Book 1, Tabs 26–44." See also "NATO Strategy and Forces," enc. to Letter, John J. McCloy to the President, Nov. 21, 1966, *DDRS*, 1986, no. 1017.

131. State 65019, Oct. 13, 1966, *DDRS*, 1985, no. 1011; Paris 6006, Oct. 22, 1966, *DDRS*, 1985, no. 1013; and "NATO Force Planning," Dec. 9, 1966, and "Annotated Agenda," Dec. 9, 1966, both in LBJL, NSF, IMTF, Box 35-36, "NATO, Ministerial Meeting, Paris, December, 1966." It is important to distinguish between the NATO force planning process, which went into operation in 1966 and continues, in somewhat modified form, to this day, and the NATO Force Planning Exercise. The former grew out of the latter, but the Force Planning Exercise was largely ad hoc and contained no requirement to provide political guidance to the NATO military authorities, the element of the planning process that was critical for the development of MC 14/3.

132. Background Paper, "Trilateral Talks and NATO Force Planning," June 5, 1967, LBJL, NSF, IMTF, Box 35-36, "NATO Ministerial Meeting, Luxembourg, June 13–14, 1967." See also *New York Times*, May 10, 1967, p. 9.

133. Detailed analysis is provided in Kugler, *Great Strategy Debate*, chap. 8. See also Stromseth, *Origins of Flexible Response*, 175–78, and Legge, *Theater Nuclear Weapons*, 9–10.

134. DOD, DMP, "NATO Strategy and Force Structure," Jan. 16, 1968, p. 4. Now even SACEUR discounted the possibility of a massive, premeditated Soviet assault. See "Major Issues," Feb. 25, 1967[?], *DDRS*, 1986, no. 75.

135. Facer, *Conventional Forces*, 5. In hammering out the political guidance on which the strategy was based, allied officials struck an explicit bargain whereby the United States agreed to the inclusion of the concept of deliberate escalation in return for European assent to greater emphasis on conventional forces and the concept of flexibility (Interview with Timothy W. Stanley).

136. DOD, DMP, "NATO Strategy and Force Structure," Jan. 7, 1969, pp. 2–3.

137. DOD, DMP, "NATO Strategy and Force Structure," Jan. 16, 1968, pp. 3–5.

138. This point is made in Baldauf, "Implementing Flexible Response," 39.

139. "After years of effort," McNamara lamented, "this is the most ambitious strategy we have been able to convince our Allies to accept" (DOD, DMP, "NATO Strategy and Force Structure," Jan. 16, 1968, pp. 3–4).

140. For detailed treatments of the ambiguities and compromises embodied in MC 14/3, see Stromseth, *Origins of Flexible Response*, chap. 9; and Kugler, *Great Strategy Debate*, chap. 9.

141. In addition to Stromseth, *Origins of Flexible Response*, 176, see Daalder, "NATO Nuclear Targetting," 266.

142. "Trilateral Talks and NATO Force Planning," June 5, 1967, LBJL, NSF, IMTF, Box 35-36, "NATO, Ministerial Meeting, Luxembourg, June 13–14, 1967"; DOD, DMP, "NATO Strategy and Force Structure," Jan. 16, 1968, p. 4; DOD, "Statement of Secretary McNamara on the Fiscal Year 1969–73 Defense Program," 81; Cleveland, *NATO*, 81; and Whitt, "Quarrel on the Rhine," 403, 458–62. For U.S. interpretations of the concept, see State 176210, Apr. 15, 1967, *DDRS*, 1986, no. 223.

143. Whitt, "Quarrel on the Rhine."

144. DOD, "Statement of Secretary McNamara on the FY 1965–69 Defense Program"; "Troop Problem and Burden-Sharing," LBJL, SDAH, vol. 1, chap. 3; "Remarks by Secretary McNamara," May 31, 1965, LBJL, NSF, Agency File, Box 11, "DOD 6/65, Vol. III"; and DOD, DMP, "NATO Strategy and Force Structure," Sept. 21, 1966.

145. Stromseth, *Origins of Flexible Response*, 142; Kugler, *Great Strategy Debate*, 110; and *New York Times*, Apr. 11, 1965.

146. DOD, *Annual Report for FY 1964*, 16, 21; and DOD, *Annual Report for FY 1965*, 26.

147. Most of the remaining Berlin round-out units were removed in the spring and summer of 1964 (*New York Times*, Apr. 11, May 16, 1964). Between 1963 and 1966, the United States also withdrew approximately 8,000 logistics troops (Raj, *American Military*, 99). A 1964 proposal to homebase ten additional squadrons from Britain and France, however, encountered resistance in the State Department and was put on hold ("U.S. Troop Presence in Germany," Nov. 19, 1964, LBJL, NSF, Country File, Box 191, "Schroder Visit").

148. *New York Times*, Dec. 16, 1965; and Raj, *American Military*, 104.

149. *New York Times*, Apr. 8, 1966; and Raj, *American Military*, 105–7. On allied concerns, see Bonn 3314, Apr. 16, 1966, LBJL, NSF, Subject File, Box 20-22, "[NATO] McCloy Talks."

150. Kugler, *Great Strategy Debate*, 103.

151. For overviews of French policy between 1958 and 1965, see Harrison, *Reluctant Ally*, 134–40; and Kaplan and Kellner, "Lemnitzer," 103–4. On the deployment of the remaining French forces in Germany, see Hunt, "NATO Without France," 7.

152. These forces amounted to two divisions and four squadrons of tactical aircraft. See Harrison, *Reluctant Ally*, 153.

153. Harrison, *Reluctant Ally*, 153–56; and Kaplan and Kellner, "Lemnitzer," 112–13. The text of the 1954 convention is printed in *FRUS, 1952–54*, 5: 1437–39.

154. Harrison, *Reluctant Ally*, 157–58. On SACEUR's concerns, see Kaplan and Kellner, "Lemnitzer," 121. For U.S. views, see DOD, DMP, "NATO Strategy and Force Structure," Jan. 16, 1969, p. 15.

155. See Harrison, *Reluctant Ally*, 148–50; Raj, *American Military*, 116–19; and DOD, *Annual Report for FY 1967*, 46.

156. Harrison, *Reluctant Ally*, 150–51; and Hunt, "NATO Without France."

157. Morton H. Halperin, Memorandum for the Record, "Trilateral Meeting of 20–21 March 1967," Mar. 22, 1967, and attachment, *DDRS*, 1986, no. 73; John J. McCloy, Letter to the President, Mar. 22, 1967, *DDRS*, 1986, no. 1025; and "Trilateral Talks, Agreed Minute on Procedure in NATO and WEU," Tab E to John J. McCloy, Letter to the President, May 17, 1967, *DDRS*, 1986, no. 1026.

158. "Trilateral Talks and NATO Force Planning," June 5, 1967, LBJL, NSF, IMTF, Box 35-36, "NATO, Ministerial Meeting, Luxembourg, June 13–14, 1967"; *New York Times*, May 3, 1967; and Whitt, "Quarrel on the Rhine," iv, 458–64.

159. U.S. officials estimated the foreign exchange costs of the British deployment at $110 million per year ("UK/German Military Offset Relationship," attachment to Trilateral Talks, Background Paper, Nov. 5, 1966, *DDRS*, 1986, no. 2669). British leaders had reportedly decided to reduce the BAOR by as many as 20,000 men. See *New York Times*, Oct. 21, 1966, p. 3; and *Congressional Record*, 90th Cong., 1st sess., vol. 13, pt. 1, p. 1001.

160. Francis M. Bator, Memorandum for the President, "Agenda for Meeting on European Policy," Aug. 23, 1966, *DDRS*, 1987, no. 3517; "UK/German Military Offset Relationship," *DDRS*, 1986, no. 2669; and *New York Times*, Oct. 18, 1966.

161. On the events surrounding the fall of Erhard's government, see Treverton, *"Dollar Drain,"* 74–76.

162. Cable, Walt Rostow to the President, Nov. 22, 1966, *DDRS*, 1988, no. 1122; and *New York Times*, Dec. 10, 1966.

163. JCSM-60-67, Memorandum for the Secretary of Defense, "Military Redeployments from Europe," Feb. 2, 1967, *DDRS*, 1986, no. 2571; Memorandum for the President, "U.S. Position in the Trilateral Negotiations," Feb. 23, 1967, *DDRS*, 1986, no. 2941; DOD, DMP, "NATO Strategy and Force Structure," Sept. 21, 1966; and Raj, *American Military*, 121.

164. The JCS position is outlined in JCSM-693-66, Memorandum for the Secretary of Defense, "Withdrawal of US Forces from Europe," Oct. 27, 1966, *DDRS*, 1986, no. 2570; JCSM-46-67, Memorandum for the Secretary of Defense, "Redeployment of US Forces Withdrawn from Europe," Jan. 28, 1967, *DDRS*, 1988, no. 756; JCSM-60-67; and Raj, *American Military*, 119–21.

165. McCloy's views are detailed in the enclosures to Letter, McCloy to the President, Nov. 21, 1966, *DDRS*, 1986, nos. 1017–23. For JCS criticism of McNamara's assumptions about redeployment times, see JCSM-60-67.

166. "Results of the Meeting with the President on February 25, 1967," n.d., *DDRS*, 1987, no. 1091; and "Record of the President's February 27 Meeting with the Congressional Leadership," n.d., *DDRS*, 1987, no. 1735.

167. *New York Times*, Jan. 20, 1967; Raj, *American Military*, 173; and Whitt, "Quarrel on the Rhine," 289.

168. McCloy conditioned his approval of even this limited rotation plan on allied approval, the demonstration of an effective U.S. capability to re-

inforce its forces in Europe as advertised, and a U.S. commitment not to engage in any further withdrawals unless they were justified by reciprocal Soviet reductions or other major positive changes in the security situation. On McCloy's views, see Memorandum for the President, "Force Levels in Europe," Feb. 23, 1967, *DDRS*, 1986, no. 1024. For the State Department position, see Eugene V. Rostow, Memorandum for the Secretary, "Force Levels in Europe," Jan. 30, 1967, *DDRS*, 1987, no. 1767; and "U.S. Position in the Trilateral Negotiations," Feb. 23, 1967, *DDRS*, 1986, no. 2941.

169. "Record of President's March 9 Meeting with Messrs. Rusk, McNamara, Fowler, McCloy, Walt Rostow, and Bator," n.d., *DDRS*, 1987, no. 1092. See also *New York Times*, Mar. 4, 1966, and Treverton, *"Dollar Drain,"* 38–39.

170. "Record of President's March 9 Meeting," n.d., *DDRS*, 1987, no. 1092; Francis M. Bator, Memorandum for the Record, "Trilaterals—Status Report," Mar. 17, 1967, *DDRS*, 1987, no. 1093; and Treverton, *"Dollar Drain,"* 92, 131–32, 151–52.

171. See Robert S. McNamara, Memorandum for the Chairman, Joint Chiefs of Staff, "Redeployment of US Forces from Germany," Mar. 23, 1967, *DDRS*, 1986, no. 74; JSCM-180-67, Memorandum for the Secretary of Defense, "Redeployment of US Forces from Germany," Mar. 30, 1967, *DDRS*, 1988, no. 729; DOD, *Annual Report for FY 1967*, 45; and DOD, *Annual Report for FY 1968*, 26, 36. The annual exercise became known as REFORGER (REturn of FORces to GERmany).

172. Bonn 12730, Apr. 24, 1967, *DDRS*, 1985, no. 1021; Francis M. Bator, Letter to the President, Apr. 27, 1967, *DDRS*, 1987, no. 1094; State 183936, Apr. 27, 1967, *DDRS*, 1985, no. 191; Treverton, *"Dollar Drain,"* 132–33; and *New York Times*, Apr. 23, 1966.

173. *New York Times*, Oct. 1, 1966; Senate, Combined Subcommittee of Foreign Relations and Armed Services, *United States Troops in Europe*, 6.

174. Dean, *Watershed in Europe*, 100; *New York Times*, July 7, 8, 1967; and Raj, *American Military*, 185–87.

175. *New York Times*, Apr. 14, 1968; and Senate, Combined Subcommittee of Foreign Relations and Armed Services, *United States Troops in Europe*, 6.

176. Williams, *Senate and U.S. Troops*, 155–56.

177. Ibid.; *New York Times*, Jan. 11, 12, 1968; and Raj, *American Military*, 189–90.

178. *New York Times*, June 20, Aug. 19, 1968; and Raj, *American Military*, 193–94.

179. See, for example, the views expressed in *New York Times*, Sept. 5, 1968.

180. *New York Times*, Sept. 5, Oct. 17, 29, 1968. This conclusion was not universally accepted. Even an internal Pentagon study emphasized the decreased dependability of the Czech forces and the possibility that other Warsaw Pact forces would have to be used to neutralize them (Raj, *American Military*, 136–38).

181. Cleveland, *NATO*, 119–20; IISS, *Strategic Survey, 1968*, 12. For an account of NATO's response to Soviet preparations for the invasion, see Betts, *Surprise Attack*, 81–86.

182. Raj, *American Military*, 195. See also *New York Times*, Aug. 22, Sept. 5, 1968.

183. Cleveland, *NATO*, 122. This move did not, however, prevent the United States from completing the withdrawals resulting from the trilateral talks as scheduled (Raj, *American Military*, 129).

184. *New York Times*, Oct. 19, 25, 1968.

185. Cleveland, *NATO*, 122–24. For press reports of U.S. appeals at a special November meeting of the NATO defense ministers, see *New York Times*, Nov. 15, 16, 17, 1968.

186. See also the analysis of Stromseth, *Origins of Flexible Response*, 190–91.

187. Ibid., 71, 135–36.

Chapter 6

1. For an excellent analysis of the debate over NATO's theater nuclear posture and employment doctrine under flexible response, see Daalder, *Nature and Practice of Flexible Response*.

2. Specifically, Canada planned to pull back its single brigade group from NATO's forward line of defense and to reorganize it as a light "combat group." Canada's six air squadrons would be similarly reduced. Overall, the Canadian presence would shrink from more than 10,000 to at most 3,500 military personnel. See Rempel, "Canada's Troop Deployment," 220; and *New York Times*, Apr. 4, 1969, p. 1, Apr. 28, 1969, p. 11, and May 29, 1969, p. 1.

3. In the face of considerable allied pressure, Canada ultimately limited its reduction to approximately 50 percent, retaining one-half of the brigade group, three squadrons, and approximately 5,000 military personnel in Germany. See Rempel, "Canada's Troop Deployment," 220; Senate, Foreign Relations, *United States Forces in Europe*, 2141–43; and *New York Times*, Sept. 20, 1969, p. 1.

4. Raj, *American Military*, 139; Treverton, *"Dollar Drain,"* 41–43; Williams, *Senate and U.S. Troops*, 160–61; and *New York Times*, Jan. 19, 1969, p. 46, and Jan. 26, 1969, p. 7. In the end, Nixon limited the reduction in support personnel to 6,000. See Senate, Foreign Relations, *United States Forces in Europe*, 2062.

5. IISS, *Strategic Survey, 1969*, 13; and *New York Times*, Dec. 7, 1969, p. 18. The choice of mid-1971 may have been influenced by the fact that the two-year offset agreement signed with the Federal Republic in July 1969 was scheduled to expire at that time. See Williams, *Senate and U.S. Troops*, 161–62.

6. On the Mansfield resolution, see *New York Times*, Dec. 2, 1969, p. 2; and Williams, *Senate and U.S. Troops*, 162–64. For Symington's views, see Senate, Foreign Relations, *United States Forces in Europe*.

7. For more detailed discussions, see Newhouse, *U.S. Troops in Europe*,

esp. chap. 1; Williams, *Senate and U.S. Troops*, 160–234; and Yochelson, "American Military Presence," 799–802.

8. Senate, Foreign Relations, *United States Forces in Europe*, 2077; Williams, *Senate and U.S. Troops*, 164; Treverton, *"Dollar Drain,"* 44; and Cromwell, *Eurogroup*, 14. Any lingering doubts the allies may have held were dispelled at a meeting of the NATO defense ministers in early June 1970, when Secretary of Defense Melvin Laird offered his colleagues a sobering evaluation of the situation in the United States. He enumerated "four realities" that pointed toward reduced U.S. force levels in Europe. These were the need for the United States to devote its resources to maintaining at least strategic parity with the Soviet Union; the growing congressional criticism of maintaining so many U.S. troops in Europe; the difficulty of finding sufficient manpower; and growing fiscal constraints. See House of Representatives, Armed Services, Special Subcommittee on NATO Commitments, Hearings, 12700, 12784–86.

9. Federal Minister of Defense, *White Paper 1970*, 32; Senate, Foreign Relations, *United States Forces in Europe*, 2142, 2251; Beach, "British Forces," 168; and *New York Times*, Mar. 5, 1970, p. 1. Later in the decade, the British established a further infantry brigade in Germany as a bargaining counter in the negotiations on mutual and balanced force reductions. See Beach, "British Forces," 169.

10. Cromwell, *Eurogroup*, 1. For additional background, see also NIS, *Eurogroup*.

11. Cromwell, *Eurogroup*, 15–16; House of Representatives, Armed Services, Special Subcommittee on NATO Commitments, Hearings, 13093; Treverton, *"Dollar Drain,"* 44; Williams, *Senate and U.S. Troops*, 165; NIS, *Eurogroup*; and *New York Times*, June 12, 1970, p. 3, Aug. 20, 1970, p. 14, and Oct. 9, 1970, p. 3.

12. NIS, *NFC, 1949–74*, 253; NIS, *Eurogroup*, 11–12; Carrington, "Eurogroup," 11; and *New York Times*, Dec. 2, 1970, p. 1.

13. The Eurogroup's communique indicated that the EDIP was being undertaken "on the basis that the United States . . . would for its part maintain [its] forces at substantially current levels" (Cromwell, *Eurogroup*, 16).

14. Laird had favored a modest cut of 20,000 to 40,000 support troops, but this option had been opposed by the Department of State and the Joint Chiefs of Staff as well as the president. See Yochelson, "American Military Presence," 784–85, 793–94; Kissinger, *White House Years*, 401–2; Williams, *Senate and U.S. Troops*, 166; Treverton, *"Dollar Drain,"* 45; NIS, *NFC, 1949–74*, 243; and *New York Times*, Nov. 9, 1970, p. 12, Dec. 1, 1970, p. 9, and Dec. 4, 1970, p. 1.

15. House of Representatives, Armed Services, *American Commitment to NATO*; and Cromwell, *Eurogroup*, 16–17. As one contemporary analysis put it, "the ten countries, after long and arduous consultation, brought forth a mouse" (Newhouse, *U.S. Troops in Europe*, 17).

16. The most detailed discussion of AD-70 is provided by Baldauf, "Implementing Flexible Response," esp. chap. 4.

17. House of Representatives, Armed Services, Special Subcommittee on

NATO Commitments, Hearings, 12569; House of Representatives, Armed Services, *American Commitment to NATO*, 63; Sorley, "Goodpaster," 129–31; and interview with General Andrew Goodpaster, Washington, D.C., Sept. 9, 1987. Goodpaster was SACEUR from July 1969 until December 1974. The decision of the Canadian government to reduce its 100,000-man force by only half and to reequip its remaining armored battalion was due in no small measure to Goodpaster's actions.

18. Goodpaster interview.

19. IISS, *Strategic Survey, 1970*, 18; House of Representatives, Armed Services, Special Subcommittee on NATO Commitments, Hearings, 12693, 12721, 13058; NIS, *NFC, 1949–74*, 240; and *New York Times*, Apr. 12, 1970, p. 2. For the text of the president's speech, see Nixon, *U.S. Foreign Policy*. Brosio nevertheless cautioned against reopening the question of NATO strategy, which was considered to be potentially too divisive. See Baldauf, "Implementing Flexible Response," 51.

20. NIS, *NFC, 1949–74*, 247–53; House of Representatives, Armed Services, Special Subcommittee on NATO Commitments, Hearings, 12783, 13058; Facer, *Conventional Forces*, 33–34; and *New York Times*, Dec. 3, 1970, p. 18. The eight areas were armor/antiarmor capabilities, aircraft protection, maritime capabilities, maldeployment of ground forces in the Central Region, defense of the flanks, mobilization and reinforcement capabilities, communications, and war reserve stocks. A ninth area, electronic warfare, was added in 1971.

21. Baldauf, "Implementing Flexible Response," 181, 189; and Facer, *Conventional Forces*, 34.

22. NIS, *NFC, 1949–74*, 256, 271–72; and DOD, *Annual Report FY 1973*, 109. Among the measures recommended for early action in December 1971 were the purchase of additional antitank weapons and tanks, electronic warfare equipment, improved aircraft, improved air defenses and aircraft protection, and larger ammunition stocks.

23. Cromwell, *Eurogroup*, 17; and House of Representatives, Armed Services, Special Subcommittee on NATO Commitments, Hearings, 13061.

24. Cromwell, *Eurogroup*, 18; and House of Representatives, Armed Services, Special Subcommittee on NATO Commitments, Hearings, 12695.

25. Williams, *Senate and U.S. Troops*, 169–74; Yochelson, "American Military Presence," 802–4; and *New York Times*, Dec. 2, 1970, p. 9, May 12, 1971, p. 9, and May 13, 1971, p. 8.

26. For accounts of the administration's campaign to defeat the 1971 Mansfield amendment, see Williams, *Senate and U.S. Troops*, 174–86; Kissinger, *White House Years*, 740–47; and Hodgson, "Establishment," 29–32. See also *New York Times*, May 14, 1971, p. 1, and May 16, 1971, p. 1.

27. See Keliher, *Negotiations*, chap. 2, and Garthoff, *Detente and Confrontation*, 110–17.

28. Williams, *Senate and U.S. Troops*, 186–92; Garthoff, *Detente and Confrontation*, 115–16; Kissinger, *White House Years*, 946–47; and *New York Times*, May 15, 1971, p. 1. The reasons for the change in Soviet policy have

been the subject of debate. Garthoff argues that Soviet leaders were concerned about the possible consequences of large-scale U.S. withdrawals, which might have precipitated a compensatory European military buildup and even German acquisition of nuclear weapons. They may have also sought to induce a more favorable Western attitude toward the Soviet proposal for a European security conference. See, for example, Gellner, "Mansfield Proposals." Brezhnev had indicated his interest in conventional arms control talks on March 30, but this signal had not elicited much of a response in the United States. Williams suggests that it was only "the introduction of the amendment [that] precipitated an unprecedented burst of enthusiasm for mutual force reductions by the Nixon administration" (*Senate and U.S. Troops*, 187).

29. Williams, *Senate and U.S. Troops*, 192–204; and *New York Times*, May 20, 1971, p. 1, Nov. 19, 1971, p. 1, and Nov. 24, 1971, p. 1. The November amendment was initially approved by the Senate Appropriations Committee but then defeated in the full Senate by the narrower margin of 54 to 39.

30. In 1972, the overall U.S. balance-of-payments deficit had reached $9.2 billion while the net military balance-of-payments deficit with respect to NATO Europe was between $1.31 billion and $1.47 billion, twice its 1969 level. See Cromwell, *Eurogroup*, 35; and Senate, Foreign Relations, *U.S. Security Issues*, 3.

31. For a more detailed discussion of the context surrounding the 1973 debate, see Williams, *Senate and U.S. Troops*, 205–11. Even the administration acknowledged a need to review and to reinvigorate the overall relationship between the United States and Western Europe, as evidenced by its decision to make 1973 the "Year of Europe."

32. Williams, *Senate and U.S. Troops*, 208–11. According to Williams, "It was the attitudes and policies of President Nixon . . . which transformed a probable confrontation into an inevitable one" (p. 209).

33. Williams, *Senate and U.S. Troops*, 211–19. See also DOD, *Annual Report FY 1974*, 5. The administration's argument was not without merit. Even Senate staff members recognized that to the extent the United States emphasized the possibility of unilateral cuts in talks with its allies, it weakened NATO's position in East-West arms reduction negotiations. See Senate, Foreign Relations, *U.S. Security Issues*, 25.

34. One reason is that there was much more sustained pressure for reductions in the House of Representatives in 1973 than in any previous year. See Williams, *Senate and U.S. Troops*, 214.

35. Senate, Foreign Relations, *U.S. Security Issues*, 1–2. For the text of Schlesinger's presentation, see DOD, "Principal Remarks by Secretary Schlesinger." Schlesinger's demarche was followed by similar requests by Secretary of State Henry Kissinger before the North Atlantic Council later that month and by the U.S. ambassador to NATO in August.

36. Senate, Foreign Relations, *U.S. Security Issues*, 3, 7.

37. Williams, *Senate and U.S. Troops*, 219–20; and Cromwell, *Eurogroup*, 36.

38. The Cranston amendment first passed by a narrow margin only to be defeated six hours later by a vote of 51 to 44. See Williams, *Senate and U.S. Troops*, 220–21.

39. On the conduct of the MBFR talks, see Keliher, *Negotiations*; and Dean, *Watershed in Europe*, esp. chap. 7.

40. Williams, *Senate and U.S. Troops*, 235–43, 248. For a comprehensive discussion of the impact of the war on U.S.-Soviet relations, see Garthoff, *Detente and Confrontation*, chap. 11.

41. Cromwell, *Eurogroup*, 39; and Williams, *Senate and U.S. Troops*, 247–49.

42. For a comparison of numbers of U.S. and Soviet strategic launchers during this period, see IISS, *Military Balance, 1977–78*, 80. The SALT I "Interim Agreement" limiting strategic offensive arms is printed in ACDA, *Arms Control and Disarmament Agreements*, 148–59.

43. For a forceful presentation of this view to the allies, see DOD, "Principal Remarks by Secretary Schlesinger."

44. The total number of Soviet military personnel in Eastern Europe, including the air forces, had increased by as much as 150,000. For details, see Senate, Armed Services, *NATO and the New Soviet Threat*, 4; Lawrence and Record, *U.S. Force Structure*, 13; Blechman et al., *Soviet Military Buildup*, 9; Aspin, "Surprise Attack," 7; DOD, *Annual Report FY 1979*, 74–75; Federal Minister of Defense, *White Paper 1975/1976*, 35; and CBO, *Strengthening NATO*, 12–14.

45. Senate, Armed Services, *NATO and the New Soviet Threat*, 4–5; DOD, *Annual Report FY 1977*, 97; Blechman et al., *Soviet Military Buildup*, 9; Aspin, "Surprise Attack," 7–8; DOD, *Annual Report FY 1979*, 74–75; Federal Minister of Defense, *White Paper 1975/1976*, 34–35; CBO, *Strengthening NATO*, 13–14; *Washington Post*, Feb. 24, 1977; and Baldauf, "Implementing Flexible Response," 107.

46. On the capabilities of Soviet tactical aircraft in 1968, see Enthoven and Smith, *How Much Is Enough?* 154–56.

47. Senate, Armed Services, *NATO and the New Soviet Threat*, 4–6; DOD, *Annual Report FY 1977*, 96, 98; Blechman et al., *Soviet Military Buildup*, 12; Aspin, "Surprise Attack," 8; and *Washington Post*, Feb. 24, 1977. See also Berman, *Soviet Air Power*. According to a Pentagon analysis, the Warsaw Pact modernized 36 percent of its tactical aircraft between 1974 and 1978 as against only 5 percent for NATO. See DOD, "NATO Center Region Military Balance Study," II-7.

48. The earliest proponents of this view were Canby, "NATO Muscle"; Komer, "Treating NATO's Self-Inflicted Wound"; and Lawrence and Record, *U.S Force Structure*, who argued (p. 17) that "the size, disposition, and structure of Soviet forces in Eastern Europe reveal a military establishment designed for decisive offensive operations in a short war." In early 1975, even the U.S. Department of Defense acknowledged that Warsaw Pact forces appeared to be geared for a short, intense war. See DOD, *Annual Report FY 1976*, III-41.

49. Lawrence and Record, *U.S. Force Structure*, 9–10, 14–15; DOD, *Annual Report FY 1976*, I-19; and Canby, "NATO Muscle," 42.

50. Canby maintained that less than 30 percent of the cost of the U.S. forces in Europe was related to fighting a short war in the Central Region while almost 50 percent of the cost was associated with maintaining a long war capability. See "NATO Muscle," 44, 46. Overall, NATO's posture was based on the assumption that a conflict would last from one to six months (Nunn, "Address by Senator Nunn," 31–32). This was not to say that NATO forces were well-prepared for a long war, however. Many of the allies lacked the war reserve stocks needed to fight for more than a brief period.

51. Senate, Armed Services, *Policy, Troops, and the NATO Alliance*, 3, 6. Nevertheless, the U.S. government continued to estimate that NATO forces could hold for as long as 90 days.

52. Aspin, "Surprise Attack," 7. See also Senate, Armed Services, *NATO and the New Soviet Threat*, 18.

53. A 1973 Defense Intelligence Agency study concluded that the Warsaw Pact could mobilize in one week the same attacking force that was previously thought to require 30 days of preparation. Subsequently, Pentagon analysts agreed that the Pact could be ready for war in only 14 days. Consequently, U.S. reinforcements would be needed by M+10, assuming that NATO mobilization lagged by only 4 days. See Baldauf, "Implementing Flexible Response," 136–37.

54. Nunn, "Address by Senator Nunn," 32; Senate, Armed Services, *NATO and the New Soviet Threat*, 4, 6, 9; DOD, *Annual Report FY 1978*, 91–94; and Aspin, "Surprise Attack," 7. Many experts and officials disputed the more extreme versions of the surprise attack scenario. See Aspin, "Surprise Attack"; Enthoven, "U.S. Forces in Europe"; and *Washington Post*, Feb. 24, 1977.

55. See DOD, "Principal Remarks by Secretary Schlesinger."

56. Ibid. For U.S. assessments of the conventional balance and the feasibility of conventional defense at this time, see also *Washington Post*, June 7, 1973; Baldauf, "Implementing Flexible Response," 99–135; and DOD, *Annual Report FY 1975*, 87–91.

57. DOD, *Annual Report FY 1975*, 88; and NIS, *NFC, 1949–74*, 303. The additional measures were electronic warfare, mobile air defense, and modern air-delivered munitions. See Facer, "Conventional Forces," 34.

58. Facer, "Conventional Forces," 35–36; and Baldauf, "Implementing Flexible Response," 26.

59. Baldauf, "Implementing Flexible Response," 72, 132, 135.

60. For discussions of the differences between U.S., West German, and NATO threat assessments and planning factors, see Baldauf, "Implementing Flexible Response," 106–31, 154–56; and Senate, Armed Services, *Policy, Troops, and the NATO Alliance*, 5, 8.

61. See esp. Canby, "NATO Muscle"; Komer, "Treating NATO's Self-Inflicted Wound"; and Lawrence and Record, *U.S. Force Structure*. The Soviet ratio of combat to support troops was estimated to be 75:25, while that of

Notes to Pages 210–12

the United States was no better than 60 : 40 (Lawrence and Record, *U.S. Force Structure*, 41–42).

62. Like the "basic issues" initiative, these proposals were predicated on the assumption that NATO could probably count on a reasonable amount of time to mobilize and deploy its forces before a Warsaw Pact attack began. See esp. Komer, "Treating NATO's Self-Inflicted Wound," 38, 44.

63. Senate, Armed Services, *Policy, Troops, and the NATO Alliance*, 9; and CBO, *Strengthening NATO*, 5. Nunn had originally estimated that the number of U.S. Army personnel in Europe could be reduced by as many as 29,000 to 60,000—from a total of about 200,000—without any loss in combat strength.

64. DOD, *Annual Report FY 1976*, III-29, III-40–41; and DOD, *Annual Report FY 1977*, 103. The two brigades represented the forward elements of the two U.S.-based divisions with prepositioned equipment in Europe. The United States also added at least two artillery battalions. See DOD, *Annual Report FY 1979*, 238; and DOD, *Annual Report FY 1980*, 47.

65. See Lawrence and Record, *U.S. Force Structure*, 70; and DOD, *Annual Report FY 1978*, 154.

66. Baldauf, "Implementing Flexible Response," 154–55.

67. Federal Minister of Defense, *White Paper 1975/1976*, 109–10, 113; and Federal Minister of Defense, *White Paper 1979*, 150–54. One Home Defense Brigade would be manned at 85 percent of its wartime strength, two at 65 percent, and three at 52 percent. Unlike regular army units, however, they would not necessarily be assigned to NATO.

68. Enthoven, "U.S. Forces in Europe," 521, citing then SACEUR General Andrew Goodpaster; and Komer, "Ten Suggestions," 67, citing T. A. Callaghan, *US/European Economic Cooperation in Military and Civil Technology* (Arlington, Va., 1974). Komer put the matter bluntly: "There is really no such thing as a NATO defence posture, only a collection of heterogeneous national postures, which differ far more in their equipment, organization and procedures than do their Warsaw Pact counterparts" (p. 68).

69. See NIS, *NFC, 1949–74*, 290–93, 300–303, 310–13, 323–26; and NIS, *NFC, 1975–80*, 34–37, 48–51.

70. NIS, *NFC, 1975–80*, 36; GAO, *NATO's New Defense Program*, 17; and Haig, "NATO and the Security of the West," 9.

71. DOD, *Rationalization/Standardization Within NATO: Seventh Annual Report*, 7.

72. Senate, Armed Services, *NATO and the New Soviet Threat*, 17.

73. As the principal author of the LTDP put it, "The substantive content of the LTDP as it has emerged is not all that new. For the most part it simply builds on many previous proposals long since seen as essential to cope with the evolving threat. What the LTDP concept really adds is a coherent management framework for pulling together key NATO requirements—already largely analyzed and accepted—so that all the allies can see clearly what is needed, and commit themselves to respond." See Komer, "Origins and the Objectives," 12.

74. Komer et al., *Alliance Defense in the Eighties.* Some of the report's details remain classified, but much of the content can be gleaned from Komer, "Ten Suggestions," 67–72, and Komer, "Origins and the Objectives," 9–12.

75. These measures included restructuring NATO ground forces and increasing their readiness, expediting the arrival of reinforcements, generating reserve forces more quickly, and putting more emphasis on the ground attack role of tactical aircraft. See Komer, "Ten Suggestions," 69.

76. Some 40 percent of the 1,100 force goals adopted in 1976 had been assigned top priority. See GAO, *NATO's New Defense Program,* 18–19.

77. Specifically, the report proposed extending the existing five-year force plan to ten years and establishing a ten-year armaments plan.

78. Komer, "Origins and the Objectives," 10.

79. Wendt and Brown, *Improving the NATO Force Planning Process,* 16.

80. The president also suggested that NATO undertake high-priority improvements in its conventional forces over the next year and that it improve cooperation in the development, production, and procurement of defense equipment. See Carter, "Carter Remarks."

81. NIS, *NFC, 1975–80,* 66.

82. For the communiques of both meetings as well as the text of the ministerial guidance, see NIS, *NFC, 1975–80,* 56–57, 68–74.

83. Specifically, the guidance indicated the need for action to improve the readiness and combat capability of in-place forces—with special emphasis on defense against armor and air attack—and to enable NATO to reinforce and augment its forces quickly. See Mumford, "Foreword," 4.

84. This measure had been under consideration since the previous year and was not explicitly linked to the LTDP or any other specific military requirements. See House of Representatives, Armed Services, *NATO Standardization, Interoperability, and Readiness,* 33. For further discussion of the 3 percent pledge, see Foreign Policy Research Institute, *Three Percent Solution.*

85. NIS, *NFC, 1975–1980,* 68; and Gundersen, "Military Perspective," 5–9. The nine areas were readiness; reinforcement; reserve mobilization; maritime posture; air defense; command, control, and communications; electronic warfare; rationalization; and consumer logistics. A tenth area, theater nuclear forces, was assigned to the Nuclear Planning Group and handled separately. The defense ministers also agreed to initiate a program of short-term force improvement measures, which included antiarmor capabilities, war reserve munitions, and readiness and reinforcement. See Facer, *Conventional Forces,* 41.

86. Mumford, "Foreword," 4; Gundersen, "Military Perspective," 6; Quinlan, "LTDP," 14; and interview with Robert Fiss, Falls Church, Va., Aug. 21, 1987. Fiss was a Pentagon official responsible for NATO affairs. The task forces nevertheless received considerable input from the United States. In many cases, the task force director or an influential member was an American, and many of the task forces required analytic support that only the Department of Defense could provide. This influence was also the result of a

conscious effort on the part of Komer and other U.S. officials, who sought to make the LTDP as ambitious as possible. Consequently, they left very little to chance. Komer set up a special group within the Pentagon, consisting of representatives from the Joint Staff, the military services, and the Office of the Secretary of Defense, to review the work of the task forces and to provide U.S. input. More generally, he seized every opportunity to inject U.S. positions into the task force reports.

87. Quinlan, "LTDP," 14; and Fiss interview. In Quinlan's view, "It wasn't possible for leaders to give a clear view on every single issue or to undertake specific commitments about all of them in May."

88. Quinlan, "LTDP," 15; and Fiss interview.

89. NIS, *NFC, 1975–1980,* 87–88, 95; and "The Long-Term Defence Program: A Summary," *NATO Review* 26 (Aug. 1978): 29–30.

90. Mumford, "Foreword," 4; GAO, *NATO's New Defense Program,* 6.

91. Areas that required further study as late as mid-1979 included logistical arrangements, war reserves, the air defense program, crisis management, and improvements in European facilities to receive reinforcements. See GAO, *NATO's New Defense Program,* 9, 11, 14.

92. Ibid., 5, 12–13; Fiss interview; House of Representatives, Armed Services, *NATO Standardization, Interoperability, and Readiness.*

93. House of Representatives, Government Operations, *Implementation,* 8.

94. As one British official put it, "Special arrangements cutting across normal patterns can be very helpful . . . but if they are prolonged too far the law of diminishing returns and increasing dissipation of effort may set in." See Quinlan, "LTDP," 15.

95. GAO, *NATO's New Defense Program,* 14–15; DOD, *Rationalization/ Standardization Within NATO: Fifth Annual Report* (Washington, D.C., 1979), cited in House of Representatives, Foreign Affairs, *NATO and Western Security,* 62–63.

96. House of Representatives, Foreign Affairs, *NATO and Western Security,* 63; and DOD, *Rationalization/Standardization Within NATO: Sixth Annual Report,* 6.

97. Fiss interview.

98. House of Representatives, Government Operations, *Implementation,* 2, 5; and Fiss interview.

99. House of Representatives, Government Operations, *Implementation,* 2; and DOD, *Rationalization/Standardization Within NATO: Sixth Annual Report,* 6.

100. NIS, *NFC, 1975–80,* 119; and House of Representatives, Government Operations, *Implementation,* 2.

101. Nicholls, "Long Term Defence Planning." Although not technically a part of the LTDP, the new planning procedure shared "the goal of achieving closer coordination at both the national and international level in setting Alliance objectives and in allocating resources for defence." See NIS, *NFC, 1975–80,* 133.

102. See Schwartz, *NATO's Nuclear Dilemmas*, chap. 7.

103. NIS, *NFC, 1975–80*, 131–32; House of Representatives, Government Operations, *Implementation*, 4, 8–9; and DOD, *Annual Report FY 1982*, 207–8. The special post-Afghanistan measures included war reserve stocks, electronic warfare, air defense, and protection against chemical, biological, and nuclear attacks. For further discussion, see Baldauf, "Implementing Flexible Response," 248–52.

104. The United States proposed inter alia that its partners increase the number of reserve units that they could make available. Although the Europeans agreed in 1980 to commit some 55 long-range transport aircraft for the purpose of ferrying U.S. reinforcements, the allies could agree to little more than reemphasizing the importance of implementing a number of existing force goals and LTDP measures. See Kupchan, "NATO and the Persian Gulf"; DOD, *Annual Report FY 1982*, 204; and Baldauf, "Implementing Flexible Response," 250–51.

105. Komer, "Is Conventional Defense of Europe Feasible?" 85.

106. NIS, *NFC, 1975–80*, 149; House of Representatives, Government Operations, *Implementation*, 2; DOD, *Rationalization/Standardization Within NATO: Seventh Annual Report*; and DOD, *Annual Report FY 1982*, vi. House of Representatives, Government Operations, *Implementation*, offers a detailed discussion of the progress attained in three of the nine program areas through early 1981.

107. DOD, "NATO Center Region Military Balance Study," 8.

108. Komer, "Is Conventional Defense of Europe Feasible?" 83; DOD, "NATO Center Region Military Balance Study," 8; and CBO, *Strengthening NATO*, 29.

109. Gundersen, "Military Perspective," 6.

110. Interview with a Pentagon official, Washington, D.C., Feb. 3, 1987.

111. As a result of the restructuring that followed the 1974 Nunn amendment, each division now maintained a forward deployed element of one brigade in Europe. Equipment was stored for the remaining six brigades and one armored cavalry regiment, as well as divisional support units, or the equivalent of approximately two and one-third divisions. See DOD, *Annual Report FY 1980*, 196.

112. Ibid., 7, 227.

113. DOD, *Annual Report FY 1979*, 7, 140; DOD, *Annual Report FY 1980*, 47, 199, 201, 204; DOD, *Annual Report FY 1982*, 197; and CBO, *Strengthening NATO*, xiv, 5.

114. CBO, "Equipping the Total Army," iii–v; and Lowe, "US Mobilization," 104.

115. DOD, *Annual Report FY 1979*, 138, 238; DOD, *Annual Report FY 1980*, 47; and DOD, "NATO Central Region Military Balance Study," 5–6.

116. DOD, *Annual Report FY 1980*, 47; and DOD, "NATO Central Region Military Balance Study," 7, II-7.

117. DOD, "NATO Central Region Military Balance Study," 5–6, 9, I-13.

118. Symbolic of this shift is that the separate treatment NATO-related

programs had received in the annual reports of the Defense Department during the Carter administration was eliminated in the reports of the Reagan administration. And such issues are not mentioned at all in Caspar W. Weinberger's memoir of his tenure as Reagan's secretary of defense, *Fighting for Peace.*

119. Kugler, *Commitment to Purpose,* 363–64, 392–93.

120. Fiss interview; Braband, "Long Term Defence Programme," 46–52; and Baldauf, "Implementing Flexible Response," 27, 261.

121. GAO, *U.S.-NATO Burden Sharing,* 31.

122. See Braband, "Long Term Defence Programme," 52, which offers a relatively positive assessment.

123. In 1982, SACEUR Bernard Rogers called upon the NATO countries to achieve 4 percent increases, but this objective was widely regarded as politically infeasible and was never adopted. For Rogers's proposal, see *New York Times,* Feb. 15, 1982, p. 4, and Rogers, "Atlantic Alliance."

124. See Kugler, *Commitment to Purpose,* 394, who also discusses some of the allies' reasons for not spending more.

125. GAO, *U.S.-NATO Burden Sharing,* 29–30.

126. GAO, *U.S.-German Wartime Host Nation Support Agreement.*

127. Kugler, *Commitment to Purpose,* 366, 393. For details, see also the annual reports to Congress by the secretary of defense.

128. Bundy et al., "Nuclear Weapons." The debate was further fueled the following year when McNamara revealed that he had counseled Presidents Kennedy and Johnson never to initiate, under any circumstances, the use of nuclear weapons. See McNamara, "Military Roles of Nuclear Weapons."

129. In addition to Bundy et al., "Nuclear Weapons," see Bundy et al., "Back from the Brink."

130. For discussions of the implications of PGMs, see Mearsheimer, "Precision-Guided Munitions"; Kennedy, "Precision ATGMs"; and Walker, "Precision-Guided Weapons."

131. See Steering Group of the European Security Study, *Strengthening Conventional Deterrence;* Sutton et al., "Deep Attack Concepts"; Rogers, "Follow-on Forces Attack"; OTA, *New Technology for NATO.*

132. See esp. Kaiser et al., "Nuclear Weapons." For a sophisticated American critique of NFU, see Mearsheimer, "Nuclear Weapons."

133. A further familiar argument was that the threat of nuclear retaliation would cause an attacker to disperse its forces to reduce their vulnerability, thereby degrading their effectiveness in conventional operations. See Steering Group of the European Security Study, *Strengthening Conventional Deterrence,* 29.

134. For a more detailed discussion of congressional attitudes, see Williams, "Nunn Amendment."

135. See, for example, Senate, Foreign Relations, *Crisis in the Atlantic Alliance.*

136. Williams, "Nunn Amendment," 4; *Aviation Week and Space Technology,* Dec. 6, 1982, 22–24; and IISS, *Strategic Survey, 1982–1983,* 39.

137. The text of the Nunn amendment is printed in *Congressional Record*, June 18, 1984, S7721. For further analysis, see Williams, "Nunn Amendment."

138. Williams, "Nunn Amendment," 5. See also Abshire, *Preventing World War III*, 72.

139. Williams, "Nunn Amendment," 6.

140. In late 1987, for example, the House Armed Services Committee created a special panel on defense burden-sharing. Although the work of the panel grabbed headlines during the following months and many of its members held highly critical views of allied military efforts, the panel stopped short of proposing any dramatic action to bring about a more equitable distribution of the defense burden. See House of Representatives, Armed Services, *Report of the Defense Burdensharing Panel.*

141. For example, the 1983 German defense White Paper maintained that the Warsaw Pact's advantage in most major weapons systems had grown since 1970 in both relative and absolute terms. While NATO was concentrating on qualitative improvements, moreover, the Soviet Union had added tank, motor rifle, and artillery battalions to its divisions. And the Warsaw Pact began to deploy a new generation of combat aircraft, greatly complicating NATO's efforts to regain the technological lead in that area. See Federal Minister of Defense, *White Paper 1983*, 86–92, 103–7.

142. Rogers, "Atlantic Alliance"; and Abshire, "NATO's Conventional Defense Improvement Effort," esp. 49–50.

143. DOD, "Improving NATO's Conventional Capabilities."

144. Interviews with Pentagon officials, July and August 1987, Washington, D.C. German motives are discussed in Kugler, *Commitment to Purpose*, 397–98.

145. NATO Press Service, Press Communique M-DPC-2(84)27, Dec. 5, 1984. The ministers also agreed to expand and accelerate programs to increase stocks of certain critical munitions and to double the budget for the common infrastructure program, which would provide for most of the remaining support requirements, such as hardened aircraft shelters, for U.S. tactical air reinforcements.

146. The most thorough description of the CDI is in Abshire, "NATO's Conventional Defense Improvement Effort." Abshire was the U.S. ambassador to NATO from 1983 until 1986. See also Stewart, "Conventional Defence Improvements"; and Rogers, "NATO's Conventional Defense Improvement Initiative."

147. These were in-place ready forces, mobilizable reserve forces, follow-on force attack capabilities, offensive counter air operations, support for early tactical air reinforcements, a NATO aircraft identification system, military assistance to NATO's poorer southern flank members, maritime posture, and sustainability. See Abshire, "NATO's Conventional Defense Improvement Effort," 54.

148. GAO, *NATO Burden-Sharing*, 31; and Stewart, "Conventional Defence Improvements."

149. Kugler, *Commitment to Purpose*, 410–11.

150. Bell, "Flexible Response," 15; and Steinberg, "Rethinking the Debate on Burden-Sharing," 66.

151. GAO, *NATO Burden-Sharing*, 32. See also Moodie and Fischmann, "Action Plan for NATO."

152. Garthoff, *Detente and Confrontation*, 476.

153. For further analysis, see Senate, Foreign Relations, *Crisis in the Atlantic Alliance*, 9–25; and Baldauf, "Implementing Flexible Response," 32–34. See also Daalder, *Nature and Practice of Flexible Response*, esp. pp. 1–39.

154. On U.S.-West German differences, see Hanrieder, *Germany, America, Europe*, 211–17, 366–67.

155. This situation represented a reversal of the usual geographical logic, whereby the alliance members that are closest to a common potential adversary will perceive a greater threat and will thus be the leading proponents of defensive preparations.

156. Baldauf, "Implementing Flexible Response," 169, 254, 507.

157. See Kohl, "Nixon-Kissinger Foreign Policy System," 29–30.

Chapter 7

1. Similarly, the number of tactical aircraft squadrons in the Central Region has been highly stable since the mid-1960s. See Bowie, Lorell, and Lund, *Trends in NATO Central Region Tactical Fighter Inventories*.

2. Of note, however, is the role that NATO military authorities played in shaping the details and interpretations of MC 48 and MC 14/2.

3. A related purpose of the U.S. political commitment and military presence was to reassure the other countries of Western Europe with respect to the possibility of a resurgent Germany so as to facilitate Germany's political, economic, and—eventually—military integration into the West. See Ireland, *Creating the Entangling Alliance*; and Leffler, *Preponderance of Power*.

4. The new emphasis on tactical nuclear weapons was also necessitated by the belief that the Soviet Union would acquire tactical nuclear weapons in the near future.

5. As U.S. officials acknowledged in late 1963, "Although motivations differ somewhat, to the extent that they have been articulated, French, UK, and German views on strategy are closer together than to the US view. Each has a common strong 'trip-wire' flavor, depreciates the value of non-nuclear forces, the value of air superiority and the importance of the contingency of limited conflicts affecting NATO. There is a good chance that an explicit UK, French and German consensus on strategy, very different from our own view, will emerge in the coming months." See "NATO Force Planning," n.d., LBJL, NSF, IMTF, Box 30-34, "NATO Defense Policy Conference, 12/2/63."

6. Snyder, "Alliance Theory," 118–20.

7. In contrast, Kenneth Waltz argues that the contributions of the European NATO allies have been fairly unimportant so the United States could have disregarded their views (*Theory of International Politics*, 169).

8. Harrison, *Reluctant Ally*, 24.

9. Record, *U.S. Nuclear Weapons;* and Leitenberg, "Background Information," 77–80. See also DOD, DMP, "NATO Strategy and Force Structure," Jan. 7, 1969, p. 3.

10. It could be argued that these opposing pressures tended to cancel each other out. It seems highly unlikely, however, that they would have done so quite so precisely.

11. The cases of Britain and France seem especially puzzling because beginning in the 1960s, the disengagement of both countries from overseas commitments would have allowed them to devote a greater percentage of their defense resources to NATO. For a more detailed analysis, see Duffield, "International Regimes," 831–32.

12. Charles Kupchan argues that European force levels have been determined primarily by domestic factors ("NATO and the Persian Gulf," 317–46).

13. See the discussion of German force planning in the 1970s in Baldauf, "Implementing Flexible Response," 507–8.

14. Detailed evidence of the influence of regime considerations in the calculations of decision makers and of the internalization of behavior in accordance with regime injunctions, especially in the United States and Britain, was presented in Chapters 2 through 6 and is not repeated here. See also Duffield, "International Regimes."

15. The alliance adopted a measure in 1955 requiring members to inform the North Atlantic Council of any important changes in their force contributions that were being considered. See DC(55)40, "UK Position in the 1955 NATO Annual Review, Memorandum by the Minister of Defence," Sept. 27, 1955, PRO, CAB 131/16.

16. National goals for ready forces in the Central Region during this period, like actual force contributions during the subsequent three decades, remained relatively constant. In most cases, they varied by no more than one division. The main exception was the goal for French forces, which declined sharply in the late 1950s, from seven to only four ready divisions, following the withdrawal of most NATO-assigned French forces for duty in North Africa in the mid-1950s.

17. Once again, the only significant difference lay in the French forces. At the time, the gap between German requirements and actual force levels was expected to be closed within several years with the completion of the West German military buildup.

18. Young, *International Cooperation*, 81–83; and Schelling, *Strategy of Conflict*, 67–70, 111–13.

19. See, for example, Buchan, "United States," 311.

20. See Simon, "Human Nature."

21. For an illustration of such thinking, see the testimony of Generals David A. Burchinal and Andrew J. Goodpaster in Senate, Foreign Relations, *United States Forces in Europe.*

22. Many U.S. policy makers subscribed to the questionable belief that in

the absence of a firm U.S. commitment, the European countries would readily "bandwagon" with the Soviet Union. See Lepgold, *Declining Hegemon,* 86–92. For evidence that bandwagoning behavior in international relations is the exception rather than the rule, see Walt, *Origins of Alliances.*

23. On the distinction between general and specific beliefs, see Odell, *U.S. Monetary Policy,* 62–64.

24. The importance of shared assumptions or mind-sets is stressed in Kohl, "Nixon-Kissinger Foreign Policy System," 29–30. For further examples of the views of high-level U.S. officials and leading American defense intellectuals, see Raj, *American Military,* 147, 182, 189.

25. On the concept of organizational "essence," see Halperin, *Bureaucratic Politics,* 28–40.

26. Yochelson, "American Military Presence," 788. For analogous examples of State Department opposition to reductions in the U.S. nuclear stockpile in Europe, see Daalder, *Nature and Practice of Flexible Response,* 154.

Epilogue

1. The causes and content of the new thinking are discussed in Snyder, "Gorbachev Revolution"; Garthoff, "New Thinking"; Meyer, "Sources and Prospects"; Holloway, "Gorbachev's New Thinking"; Legvold, "Revolution in Soviet Foreign Policy"; and Warner, "New Thinking."

2. For details, see Dean, *Meeting Gorbachev's Challenge,* esp. chaps. 1 and 4.

3. See esp. Meyer, "Sources and Prospects"; and Warner, "New Thinking."

4. See, for example, *Washington Post,* Mar. 21, 1987, p. A23; and *New York Times,* July 9, 1988, p. 1.

5. Subsequently, Soviet officials revealed that four of the six divisions would be withdrawn from East Germany, one from Czechoslovakia, and one from Hungary. For details, see Dean, *Meeting Gorbachev's Challenge,* 54–55, 132–33, and app. 5; and "Documentation: Conventional Arms Control," *Survival* 31 (May–June 1989): 269–73. The text of Gorbachev's address is reprinted in *New York Times,* Dec. 8, 1988, p. A6.

6. Holloway, "Gorbachev's New Thinking," 75; and Dean, *Meeting Gorbachev's Challenge,* 54, 132.

7. For example, many Western officials pointed out that the Soviet bloc would retain a large numerical advantage over NATO in important measures of conventional strength. See *New York Times,* Dec. 8, 1988, p. A1, and Dec. 10, 1988, p. 5.

8. The mandate is reprinted in Dean, *Meeting Gorbachev's Challenge,* app. 6.

9. The Warsaw Pact position paper is reprinted, ibid., 375–77.

10. Important disagreements concerned the types of weapons to include, how weapons were to be defined, and the desired extent of the reductions. See Ghebali, "Putting Europe Back Together"; and *New York Times,* Mar. 9, 1989, p. A12.

11. See, for example, *New York Times*, Mar. 7, 1989, p. A1, and May 25, 1989, p. A9.

12. For a brief summary of the negotiations, see Daalder, *CFE Treaty*, 5–9.

13. For further analysis, see ibid., esp. 25–29.

14. See, for example, *Washington Post*, May 31, 1990, p. A27.

15. For details, see IISS, *Strategic Survey, 1989–90*, 38–51.

16. *New York Times*, Dec. 20, 1989, p. A16; and *Washington Post*, Feb. 4, 1990, p. C2.

17. See, for example, *Washington Post*, Apr. 8, 1988, p. B5.

18. *Washington Post*, Feb. 27, 1990, p. A12, and Mar. 11, 1990, p. A32. For more detailed analysis, see Feinstein, "Soviet Cutbacks."

19. *New York Times*, June 7, 1990, p. A8, and June 8, 1990, p. A1.

20. *Washington Post*, July 2, 1991, p. A11; *Non-Offensive Defence and Conversion*, no. 19 (Aug. 1991): 14; and ibid., no. 20 (Sept. 1991): 17.

21. "Speech by Chancellor Helmut Kohl in the German Parliamentary Budget Debate: Policy Towards Germany," Nov. 28, 1989, reprinted in *Survival* 32 (Jan.–Feb. 1990): 86–87.

22. Useful accounts include Kaiser, "Germany's Unification"; Pond, "Wall Destroyed"; and Pond, *After the Wall*.

23. The quotation is from Van Orden, "Bundeswehr in Transition," 361.

24. "Final Settlement with Respect to Germany," reprinted in *Arms Control Today* 20 (Oct. 1990): 33–34.

25. *Washington Post*, Mar. 17, 1992, p. A10, and May 23, 1992, p. A24. At the end of 1993, Russian leaders indicated that Russia would maintain 2.1 million men under arms, but even this higher figure still represented a substantial reduction in the potential threat. See *Jane's Defense Weekly*, Jan. 15, 1994, p. 8.

26. Some might question whether it is possible to speak of an alliance as having a grand strategy. Because interests are unlikely to be identical, an alliance grand strategy will promote only a fraction of each member's overall objectives. As long as members have some security interests in common, however, one can imagine them employing a joint strategy to achieve their shared goals.

27. See also Schmidt, *Grand Strategy for the West*.

28. The report is reprinted in NIS, *NATO: Facts and Figures*, 289–91.

29. NATO's new grand strategy was nowhere spelled out as such. Nevertheless, the principal elements were articulated in a number of statements and documents. See, in particular, "London Declaration on a Transformed North Atlantic Alliance," reprinted in *NATO Review* 38 (Aug. 1990): 32–33; "North Atlantic Council Ministerial Communique, December 1990," reprinted in *NATO Review* 38 (Dec. 1990): 22–24; "Partnership with the Countries of Central and Eastern Europe" and "NATO's Core Security Functions in the New Europe," reprinted in *NATO Review* 39 (June 1991): 28–31; and "Rome Declaration on Peace and Cooperation" and "The Alliance's New Strategic Concept," reprinted in *NATO Review* 39 (Dec. 1991): 19–24 and 25–32.

30. For further details, see Flanagan, "NATO and Central and Eastern Europe"; Watt, "Hand of Friendship"; and "Partnership for Peace: Invitation" and "Partnership for Peace: Framework Document," both reprinted in *NATO Review* 42 (Feb. 1994): 28–30.

31. Some of the reasons are discussed in Taylor, "NATO and Central Europe."

32. This statement, which was repeated in the November "Rome Declaration on Peace and Cooperation," was described at the time as the "closest thing NATO could do to offering a security guarantee." See *New York Times*, June 7, 1991, p. A1.

33. "The Situation in the Soviet Union," reprinted in *NATO Review* 39 (Aug. 1991): 8–9.

34. See, for example, *Washington Post*, May 24, 1990, pp. A33 and A36, and May 25, 1990, p. A34.

35. *Washington Post*, Feb. 4, 1990, p. A27.

36. "London Declaration."

37. "Defence Planning Committee and Nuclear Planning Group Communique," reprinted in *NATO Review* 39 (June 1991): 33–35.

38. The following analysis is based primarily on the text of "The Alliance's New Strategic Concept."

39. "London Declaration." For a discussion of the role of nuclear weapons in the new strategy, see also Daalder, *Nature and Practice of Flexible Response*, 293.

40. In September 1991, President George Bush announced that the United States would withdraw all of its nuclear artillery shells and all nuclear warheads for short-range ballistic missiles from Europe. The only U.S. nuclear weapons remaining on the continent would be air-delivered. For the text of Bush's speech, see *Arms Control Today* 21 (Oct. 1991): 3–5. On the basis of this and other initiatives, NATO planned to reduce its stockpile of substrategic nuclear weapons in Europe by roughly 80 percent. See *NATO Review* 39 (Dec. 1991): 33.

41. *NATO Review* 40 (June 1992): 31, and (Dec. 1992): 28–31. See also *New York Times*, June 5, 1992, p. A1; and *Washington Post*, June 5, 1992, p. A41.

42. For a useful discussion, see Lowe and Young, "Multinational Corps."

43. On this last point, see also "London Declaration" and "Defence Planning Committee and Nuclear Planning Group Communique."

44. The actual cuts were somewhat larger because NATO holdings at the time the treaty was signed included equipment that had belonged to East Germany. See Daalder, *CFE Treaty*, 24–25.

45. For details, see Van Orden, "Bundeswehr in Transition," 366 and n. 37; and *Non-Offensive Defence*, no. 18 (Mar. 1991): 23–24. In contrast, the Federal Republic had previously fielded a regular army of 12 divisions and 36 brigades and a territorial army of 6 Home Defense Brigades with manning levels of 50 percent or above.

46. *Non-Offensive Defence and Conversion*, no. 25 (Feb. 1993): 13.

47. See, for example, *Washington Post*, Feb. 7, 1993, p. A26.

48. *Jane's Defense Weekly*, Mar. 23, 1991, p. 424, cited in *Non-Offensive Defence and Conversion*, no. 19 (Aug. 1991): 10; and *Armed Forces Journal International* 129 (Dec. 1991): 18.

49. See, for example, Sam Nunn, "A New Europe—A New Military Strategy," *Washington Post*, Apr. 24, 1990, p. A25; and the remarks of Senator William Cohen reported in *Armed Forces Journal International* 129 (Apr. 1992): 14.

50. *Washington Post*, June 9, 1993, p. A12. Both House and Senate authorization committee members endorsed a similar measure in late 1991. See *Armed Forces Journal International* 129 (Dec. 1991): 18.

51. *New York Times*, Mar. 30, 1993, p. A5.

52. *Washington Post*, July 26, 1990, pp. A29 and A32; *Jane's Defense Weekly*, Aug. 4, 1990, pp. 152–53; IISS, *Military Balance, 1991–92*, 50; IISS, *Military Balance, 1992–93*, 34; and Mager, "Continental Commitment."

53. *Jane's Defense Weekly*, Sept. 1, 1990, p. 301, and Nov. 24, 1990, p. 1011; *Washington Post*, May 17, 1992, p. A31; IISS, *Military Balance, 1992–93*, 32; and Carton, "French Policy."

54. *Jane's Defense Weekly*, Dec. 22, 1990, p. 1266; De Vos, "Scutum Belgarum"; IISS, *Military Balance, 1992–93*, 34; and pers. comm., Embassy of Belgium, Washington, D.C., Sept. 20, 1993.

55. *Armed Forces Journal International* 128 (May 1991): 25; IISS, *Military Balance, 1991–92*, 49; Siccama, "Enduring Interests"; and pers. comm., Royal Netherlands Embassy, Washington, D.C., Sept. 14, 1993.

56. *Washington Post*, Mar. 11, 1992, p. A16; and IISS, *Military Balance, 1992–93*, 33–34.

57. The airmobile multinational division in the Central Region would consist of one brigade each from Germany, Britain, Belgium, and the Netherlands. In addition, the United States might assign one of its two divisions based in Germany. For details, see *Armed Forces Journal International* 129 (Dec. 1991): 31; *New York Times*, May 29, 1991, p. A1; and IISS, *Military Balance, 1992–93*, 30–31.

58. Germany would contribute forces to all of the multinational corps. Altogether, the main forces would include approximately seven and two-thirds German divisions, two U.S. divisons, two (later reduced to one) Dutch divisions, one Danish division, and two Belgian brigades. See *Armed Forces Journal International* 129 (Dec. 1991): 31. Under the terms of the agreement on German unification, foreign forces cannot be stationed on the territory of the former German Democratic Republic. See "Treaty on the Final Settlement with Respect to Germany," Art. 5.

59. See esp. Krasner, "Regimes and the Limits of Realism," 355–58. For analogies with domestic institutions, see Krasner, "Approaches to the State," and Ikenberry, "Conclusion."

60. Krasner borrows the concept "punctuated equilibrium" from evolutionary theory to describe the relationship between institutions and their environments. See "Approaches to the State," 242–43.

61. Walt, *Origins of Alliances.*

62. For general discussions of how NATO helped to solve the German problem, see Ireland, *Creating the Entangling Alliance,* and Joffe, *Limited Partnership.*

63. Ikenberry, "Conclusion," 224. See also Krasner, "Approaches to the State," 240–41.

64. For a similar conceptualization that is couched in terms of physical and behavioral laws on the one hand and rules on the other, see Ostrom, "Agenda for the Study of Institutions."

65. On the indeterminacy of structural factors, see, for example, Waltz, *Theory of International Politics,* 68–73; and Waltz, "Reflections on *Theory of International Politics,*" 343–44.

Bibliography

MANUSCRIPT COLLECTIONS AND ARCHIVES

Dwight D. Eisenhower Library, Abilene, Kans.
  John Foster Dulles Papers, 1951–59
  Dwight D. Eisenhower Papers
    Papers as President, 1953–61 (Ann Whitman File)
    Records as President, White House Central Files, 1953–61
  White House Office
    Office of the Special Assistant for Disarmament (Harold Stassen): Records, 1955–58
    Office of the Special Assistant for National Security Affairs: Records, 1952–61
    Office of the Special Assistant for Science and Technology: Records, 1957–61
    Office of the Staff Secretary: Records, 1952–61
Lyndon Baines Johnson Library, Austin, Tex.
  Lyndon Baines Johnson Papers
    National Security File
    State Department Administrative History
    Vice-Presidential Security File
John F. Kennedy Library, Boston, Mass.
  John F. Kennedy Papers
    National Security Files
    President's Office Files
Seely G. Mudd Library, Princeton University, Princeton, N.J.
  John Foster Dulles Papers
National Archives of the United States, Washington, D.C.
  Record Group 59, Records of the Department of State
  Record Group 218, Records of the Joint Chiefs of Staff

Record Group 273, Records of the National Security Council
Record Group 330, Records of the Office of the Secretary of Defense
Public Record Office, Kew, England
   Air Ministry (AIR)
   Cabinet Office (CAB)
   Ministry of Defence (DEFE)
   Foreign Office (FO)
   Prime Minister's Office (PREM)
Harry S. Truman Library, Independence, Mo.
   Harry S. Truman Papers
      President's Official Files

UNPUBLISHED U.S. GOVERNMENT DOCUMENTS

Congressional Budget Office

"Equipping the Total Army and POMCUS Sets 5 and 6." Staff Working Paper.
   July 1984.

Department of Defense

Draft Memorandum for the President. "NATO Strategy and Force Struc-
   ture." Sept. 21, 1966.
Draft Memorandum for the President. "NATO Strategy and Force Struc-
   ture." Jan. 16, 1968.
Draft Memorandum for the President. "NATO Strategy and Force Struc-
   ture." Jan. 7, 1969.
Draft Memorandum for the President. "Recommended FY 1964–68 General
   Purpose Forces," Dec. 3, 1962.
Draft Memorandum for the President. "The Role of Tactical Nuclear Forces
   in NATO Strategy." Jan. 15, 1965.
"Improving NATO's Conventional Capabilities." June 1984.
"NATO Center Region Military Balance Study, 1978–1984." 1979.
"Principal Remarks by Secretary of Defense-Designate James R. Schlesin-
   ger at NATO Defense Planning Committee Ministerial Meeting, Brussels,
   June 7, 1973."
"Remarks by Secretary McNamara, NATO Ministerial Meeting, May 5, 1962,
   Restricted Session."
"Statement of Secretary of Defense Robert S. McNamara Before the Senate
   Subcommittee on Department of Defense Appropriations on the Fiscal
   Year 1963–67 Defense Program and 1963 Defense Budget." Feb. 14, 1962.
"Statement of Secretary of Defense Robert S. McNamara Before the Sen-
   ate Subcommittee on Department of Defense Appropriations on the Fis-
   cal Year 1964–68 Defense Program and 1964 Defense Budget." Apr. 24,
   1963.
"Statement of Secretary of Defense Robert S. McNamara Before a Joint Session of the Senate Armed Services Committee and the Senate Subcommit-

tee on Department of Defense Appropriations on the Fiscal Year 1965–69 Defense Program and 1965 Defense Budget." Feb. 3, 1964.

"Statement of Secretary of Defense Robert S. McNamara Before the Senate Armed Services Committee on the Fiscal Year 1969–73 Defense Program and 1969 Defense Budget." Jan. 22, 1968.

PUBLISHED U.S. GOVERNMENT DOCUMENTS

Congress

CONGRESSIONAL BUDGET OFFICE

Strengthening NATO: POMCUS and Other Approaches. Feb. 1979.

GENERAL ACCOUNTING OFFICE

NATO's New Defense Program: Issues for Consideration. Report by the Comptroller General. July 1979.

The U.S.-German Wartime Host Nation Support Agreement. Nov. 1987.

U.S.-NATO Burden Sharing: Allies' Contributions to Common Defense During the 1980s. GAO/NSIAD-91-32. Oct. 1990.

HOUSE OF REPRESENTATIVES

Committee on Armed Services. *The American Commitment to NATO.* Report of the Special Subcommittee on North Atlantic Treaty Organization Commitments. 92d Cong., 2d sess., 1972.

———. *NATO Standardization, Interoperability, and Readiness.* Report of the Special Subcommittee on NATO Standardization, Interoperability, and Readiness. 95th Cong., 2d sess., 1978.

———. *Report of the Defense Burdensharing Panel.* 100th Cong., 2d sess., Aug. 1988.

———. Special Subcommittee on North Atlantic Treaty Organization Commitments. Hearings. 92d Cong., 1st and 2d sess., 1971–72.

Committee on Foreign Affairs. *Mutual Security Act of 1959.* Hearings. 86th Cong., 1st sess., 1959.

———. *NATO and Western Security in the 1980's: European Perceptions.* Report of a Staff Study Mission to Seven NATO Countries and Austria, January 2–18, 1980. Washington, D.C.: U.S. Government Printing Office, 1980.

———. *Selected Executive Session Hearings of the Committee, 1951–56. Historical Series.* Vol. 10, *Mutual Security Program, Part 2.* Washington, D.C.: U.S. Government Printing Office, 1980.

———. *Selected Executive Session Hearings of the Committee, 1951–56. Historical Series.* Vol. 15, *European Problems.* Washington, D.C.: U.S. Government Printing Office, 1980.

———. *Selected Executive Session Hearings of the Committee, 1957–60. Historical Series.* Vol. 19, *Mutual Security Program, Part 6.* Washington, D.C.: U.S. Government Printing Office, 1987.

———. *Selected Executive Session Hearings of the Committee, 1957–60. His-*

torical Series. Vol. 21. Washington, D.C.: U.S. Government Printing Office, 1987.

Committee on Government Operations. *The Implementation of the NATO Long-Term Defense Program (LTDP).* H. Rept. 97-37. 97th Cong., 1st sess., 1981.

OFFICE OF TECHNOLOGY ASSESSMENT

New Technology for NATO: Implementing Follow-on Forces Attack. OTA-ISC-309. Washington, D.C.: U.S. Government Printing Office, June 1987.

SENATE

Committee on Armed Services. *NATO and the New Soviet Threat.* Report of Sen. Sam Nunn and Sen. Dewey F. Bartlett. 95th Cong., 1st sess., Jan. 24, 1977.

———. *NATO: Can the Alliance Be Saved?* Report of Sen. Sam Nunn. 97th Cong., 2d sess., 1982. Committee Print.

———. *Policy Troops, and the NATO Alliance.* Report of Sen. Sam Nunn. 93rd Cong., 2d sess., Apr. 2, 1974. Committee Print.

Committee on Foreign Relations. *Crisis in the Atlantic Alliance: Origins and Implications.* 97th Cong., 2d sess., 1982.

———. *Executive Sessions of the Senate Foreign Relations Committee. Historical Series.* Vol. 10, *85th Cong., 2d Sess., 1958.* Washington, D.C.: U.S. Government Printing Office, 1980.

———. *Executive Sessions of the Senate Foreign Relations Committee. Historical Series.* Vol. 13, *87th Cong., 1st Sess., 1961.* Washington, D.C.: U.S. Government Printing Office, 1984.

———. *Mutual Security Act of 1958.* Hearings. 85th Cong., 2d sess., 1958.

———. *NATO and the Paris Accords Relating to West Germany.* Hearings. 84th Cong., 1st sess., 1955.

———. *Reviews of the World Situation, 1949–1950. Historical Series.* 81st Cong., 1st and 2d sess., 1974.

———. *United States Forces in Europe.* Hearings Before the Subcommittee on United States Security Agreements and Commitments Abroad, pt. 10. 91st Cong., 2d sess., 1970.

———. *U.S. Security Issues in Europe: Burden Sharing and Offset, MBFR and Nuclear Weapons.* Staff Report Prepared for the Use of the Subcommittee on U.S. Security Agreements and Commitments Abroad. 93d Cong., 1st sess., Dec. 2, 1973.

Committee on Foreign Relations and Committee on Armed Services. *Assignment of Ground Forces of the United States to Duty in the European Area.* Joint Hearings. 82d Cong., 1st sess., 1951.

———. *Mutual Defense Assistance Program, 1950.* Joint Hearings. 81st Cong., 2d sess., 1950.

———. Combined Subcommittee of Foreign Relations and Armed Services Committees. *United States Troops in Europe.* Report. 90th Cong., 2d sess., 1968. Joint Committee Print.

Executive Branch

Declassified Documents Reference System. Washington, D.C.: Carrollton Press, various years.

ARMS CONTROL AND DISARMAMENT AGENCY

Arms Control and Disarmament Agreements: Texts and Histories of Negotiations. Washington, D.C.: Arms Control and Disarmament Agency, 1980.

DEPARTMENT OF DEFENSE

Annual Report for Fiscal Year 1962. Washington, D.C.: U.S. Government Printing Office, 1963.

Annual Report for Fiscal Year 1963. Washington, D.C.: U.S. Government Printing Office, 1964.

Annual Report for Fiscal Year 1964. Washington, D.C.: U.S. Government Printing Office, 1964.

Annual Report for Fiscal Year 1965. Washington, D.C.: U.S. Government Printing Office, 1966.

Annual Report for Fiscal Year 1967. Washington, D.C.: U.S. Government Printing Office, 1968.

Annual Report for Fiscal Year 1968. Washington, D.C.: U.S. Government Printing Office, 1971.

Annual Report Fiscal Year 1970. "Statement of Secretary of Defense Clark M. Clifford: The Fiscal Year 1970–74 Defense Program and 1970 Defense Budget." Jan. 15, 1969.

Annual Report Fiscal Year 1972. "Statement of Secretary of Defense Melvin R. Laird Before the House Armed Services Committee on the FY 1972–1976 Defense Program and 1972 Defense Budget, March 9, 1971." Mar. 1, 1971.

Annual Report Fiscal Year 1973. "Statement of Secretary of Defense Melvin R. Laird Before the House Subcommittee on Department of Defense Appropriations on the FY 1973 Defense Budget and FY 1973–1977 Program, February 22, 1972." Feb. 8, 1972.

Annual Report Fiscal Year 1974. "Statement of Secretary of Defense Elliot L. Richardson Before the House Armed Services Committee on the FY 1974 Defense Budget and FY 1974–1978 Program, Tuesday, April 10, 1973." Mar. 29, 1973.

Annual Report Fiscal Year 1975. "Report of the Secretary of Defense James R. Schlesinger to the Congress on the FY 1975 Defense Budget and FY 1975–1979 Defense Program." Mar. 4, 1974.

Annual Report Fiscal Year 1976. "Report of Secretary of Defense James R. Schlesinger to the Congress on the FY 1976 and Transition Budgets, FY 1977 Authorization Request and FY 1976–1980 Defense Programs." Feb. 5, 1975.

Annual Report Fiscal Year 1977. "Report of Secretary of Defense Donald H. Rumsfeld to the Congress on the FY 1977 Budget and Its Implications for the FY 1978 Authorization Request and the FY 1977–1981 Defense Programs." Jan. 27, 1976.

Annual Report Fiscal Year 1978. "Report of Secretary of Defense Donald H. Rumsfeld to the Congress on the FY 1978 Budget, FY 1979 Authorization Request and FY 1978–1982 Defense Programs." Jan. 17, 1977.

Annual Report Fiscal Year 1979. Washington, D.C.: Department of Defense, Feb. 2, 1978.

Annual Report Fiscal Year 1980. Washington, D.C.: Department of Defense, Jan. 25, 1979.

Annual Report Fiscal Year 1982. Washington, D.C.: Department of Defense, Jan. 19, 1981.

Rationalization/Standardization Within NATO: Sixth Annual Report. Washington, D.C., Jan. 1980.

Rationalization/Standardization Within NATO: Seventh Annual Report. Washington, D.C., Jan. 1981.

Records of the Joint Chiefs of Staff, Part II: 1946–1953, Europe and NATO. Frederick, Md.: University Publications of America, 1980. 9 reels.

DEPARTMENT OF STATE

Foreign Relations of the United States, 1949. Vol. 4, *Western Europe.* Washington, D.C.: U.S. Government Printing Office, 1975.

Foreign Relations of the United States, 1950. Vol. 1, *National Security Affairs.* Washington, D.C.: U.S. Government Printing Office, 1977.

Foreign Relations of the United States, 1950. Vol. 3, *Western European Security.* Washington, D.C.: U.S. Government Printing Office, 1977.

Foreign Relations of the United States, 1950. Vol. 4, *Central and Eastern Europe; The Soviet Union.* Washington, D.C.: U.S. Government Printing Office, 1980.

Foreign Relations of the United States, 1951. Vol. 1, *National Security Affairs; Foreign Economic Policy.* Washington, D.C.: U.S. Government Printing Office, 1979.

Foreign Relations of the United States, 1951. Vol. 3, *European Security and the German Question.* Washington, D.C.: U.S. Government Printing Office, 1981.

Foreign Relations of the United States, 1952–1954. Vol. 2, *National Security Affairs.* Washington, D.C.: U.S. Government Printing Office, 1984.

Foreign Relations of the United States, 1952–1954. Vol. 5, *Western European Security.* Washington, D.C.: U.S. Government Printing Office, 1983.

Foreign Relations of the United States, 1952–1954. Vol. 6, *Western Europe and Canada.* Washington, D.C.: U.S. Government Printing Office, 1986.

Foreign Relations of the United States, 1952–1954. Vol. 7, *Germany and Austria.* Washington, D.C.: U.S. Government Printing Office, 1987.

Foreign Relations of the United States, 1955–1957. Vol. 4, *Western European Security and Integration.* Washington, D.C.: U.S. Government Printing Office, 1986.

WHITE HOUSE

Public Papers of the Presidents of the United States: John F. Kennedy, 1961. Washington, D.C.: U.S. Government Printing Office, 1962.

Public Papers of the Presidents of the United States: Harry S. Truman, 1950. Washington, D.C.: U.S. Government Printing Office, 1965.

OTHER CITED SOURCES, PRIMARY AND SECONDARY

Abshire, David M. "NATO's Conventional Defense Improvement Effort: An Ongoing Imperative." *Washington Quarterly* 10 (Spring 1987): 49–60.

———. *Preventing World War III: A Realistic Grand Strategy.* New York: Harper & Row, 1988.

Acheson, Dean. *Present at the Creation: My Years in the State Department.* New York: Norton, 1969.

Aggarwal, Vinod. *Liberal Protectionism: The International Politics of the Organized Textile Trade.* Berkeley: University of California Press, 1985.

Allison, Graham T. *The Essence of Decision: Explaining the Cuban Missile Crisis.* Boston: Little, Brown, 1971.

Aron, Raymond. *Peace and War: A Theory of International Relations.* New York: Praeger, 1966.

Aron, Raymond, and Daniel Lerner, eds. *France Defeats EDC.* New York: Praeger, 1957.

Aspin, Les. "A Surprise Attack on NATO—Refocusing the Debate." *NATO Review* 25 (Aug. 1977): 6–13.

Axelrod, Robert. *The Evolution of Cooperation.* New York: Basic Books, 1984.

———. "An Evolutionary Approach to Norms." *American Political Science Review* 80 (Dec. 1986): 1095–1111.

Bacevich, A. J. *The Pentomic Era: The U.S. Army Between Korea and Vietnam.* Washington, D.C.: National Defense University Press, 1986.

Baldauf, Joerg. "Implementing Flexible Response: The US, Germany, and NATO's Conventional Forces." Ph.D. diss., Massachusetts Institute of Technology, 1987.

Barnet, Richard. *The Alliance: America, Europe, Japan—Makers of the Post-War World.* New York: Simon & Schuster, 1983.

Bartlett, C. J. *The Long Retreat: A Short History of British Defence Policy, 1945–1970.* New York: St. Martin's, 1971.

Baylis, John, and Alan Macmillan. "The British Global Strategy Paper of 1952." *Journal of Strategic Studies* 16 (June 1993): 200–26.

Beach, Hugh. "British Forces in Germany, 1945–85." In *The Defence Equation: British Military Systems Policy, Planning and Performance,* ed. by Martin Edmonds. London: Brassey's, 1986.

Beer, Francis A. *Integration and Disintegration in NATO: Processes of Alliance Cohesion and Prospects for Atlantic Community.* Columbus: Ohio State University Press, 1969.

Bell, Coral. *Negotiation from Strength: A Study in the Politics of Power.* New York: Knopf, 1963.

Bell, Michael J. "Flexible Response—Is It Still Relevant?" *NATO Review* 36 (Apr. 1988): 12–17.

Berman, Robert P. *Soviet Air Power in Transition.* Washington, D.C.: Brookings Institution, 1978.

Betts, Richard K. *Surprise Attack: Lessons for Defense Planning.* Washington, D.C.: Brookings Institution, 1982.

Blechman, Barry M., Robert P. Berman, Martin Binkin, Stuart E. Johnson, Robert G. Weinland, and Frederick W. Young. *The Soviet Military Buildup and U.S. Defense Spending.* Washington, D.C.: Brookings Institution, 1977.

Bowie, Christopher J., Mark Lorell, and John Lund. *Trends in NATO Central Region Tactical Fighter Inventories, 1950–2005.* N-3053-AF. Santa Monica, Calif.: RAND Corporation, 1990.

Braband, Rolf. "The Long Term Defence Programme." *NATO's Fifteen Nations* 27 (Oct.–Nov. 1982): 46–52.

Brinkley, Douglas G. "Dean Acheson and European Unity." In *NATO: The Founding of the Atlantic Alliance and the Integration of Europe,* edited by Francis H. Heller and John R. Gillingham. New York: St. Martin's, 1992.

Brodie, Bernard. "Conventional Capabilities in Europe." *Reporter* 28 (23 May 1963): 25–33.

———. "Unlimited Weapons and Limited War." *Reporter* 11 (18 Nov. 1954): 16–21.

Buchan, Alistair Francis. "The United States and the Security of Europe." In *Western Europe: The Trials of Partnership,* edited by David S. Landes. Lexington, Mass.: D. C. Heath, 1977.

Bull, Hedley. *The Anarchical Society: A Study of Order in World Politics.* New York: Columbia University Press, 1977.

Bundy, McGeorge, Morton H. Halperin, William W. Kaufmann, George F. Kennan, Robert S. McNamara, Madalene O'Donnell, Leon V. Sigal, Gerard C. Smith, Richard H. Ullman, and Paul C. Warnke. "Back from the Brink." *Atlantic,* Aug. 1986, pp. 35–41.

Bundy, McGeorge, George F. Kennan, Robert S. McNamara, and Gerard Smith. "Nuclear Weapons and the Atlantic Alliance." *Foreign Affairs* 60 (Spring 1982): 753–68.

Buteux, Paul. *Strategy, Doctrine, and the Politics of Alliance: Theatre Nuclear Force Modernisation in NATO.* Boulder, Colo.: Westview, 1983.

Calleo, David P. *Beyond Hegemony: The Future of the Western Alliance.* New York: Basic Books, 1987.

Calvocoressi, Peter. *Survey of International Affairs, 1952.* London: Oxford University Press, 1954.

Canby, Steven L. "NATO Muscle: More Shadow Than Substance." *Foreign Policy,* no. 8 (Fall 1972): 38–49.

Carlyle, Margaret, ed. *Documents on International Affairs, 1949–1950.* London: Oxford University Press, 1953.

Carrington, Peter. "The Eurogroup." *NATO Review* 20 (Feb. 1972): 10–11.

Carter, Jimmy. "Carter Remarks to the North Atlantic Council." *NATO Review* 25 (June 1977): 21–24.

Carton, Alain. "French Policy Regarding the Stationing of Forces in Germany." In *Homeward Bound? Allied Forces in the New Germany*, edited by David G. Haglund and Olaf Mager. Boulder, Colo.: Westview, 1992.

Chayes, Abram. "An Inquiry into the Workings of Arms Control Agreements." *Harvard Law Review* 85 (Mar. 1972): 905–69.

Cioc, Mark. *Pax Atomica: The Nuclear Defense Debate in West Germany During the Adenauer Era.* New York: Columbia University Press, 1988.

Cleveland, Harlan. *NATO: The Transatlantic Bargain.* New York: Harper & Row, 1970.

Cochran, Thomas B., William M. Arkin, and Milton M. Hoenig. *Nuclear Weapons Databook.* Vol. 1, *U.S. Nuclear Forces and Capabilities.* Cambridge, Mass.: Ballinger, 1984.

Combs, Gerald A. "From MC-26 to the New Look: The Great Eisenhower Reversal on Defense Policy." Paper presented at the annual meeting of the Society for the History of American Foreign Relations, Charlottesville, Virginia, May 1993.

Condit, Doris M. *History of the Office of the Secretary of Defense.* Vol. 2, *The Test of War, 1950–1953.* Washington, D.C.: Office of the Secretary of Defense, 1988.

Condit, Kenneth W. *The History of the Joint Chiefs of Staff: The Joint Chiefs of Staff and National Policy.* Vol. 2, *1947–1949.* Wilmington, Del.: Michael Glazier, 1979.

Craig, Gordon A. "Germany and NATO: The Rearmament Debate, 1950–1958." In *NATO and American Security*, edited by Klaus Knorr. Princeton: Princeton University Press, 1959.

Cromwell, William C. *The Eurogroup and NATO.* Research Monograph Series, no. 18. Philadelphia: Foreign Policy Research Institute, 1974.

Daalder, Ivo H. *The CFE Treaty: An Overview and an Assessment.* Washington, D.C.: Johns Hopkins Foreign Policy Institute, 1991.

——. "NATO Nuclear Targetting and the INF Treaty." *Journal of Strategic Studies* 11 (Sept. 1988): 265–91.

——. *The Nature and Practice of Flexible Response: NATO Strategy and Theater Nuclear Forces Since 1967.* New York: Columbia University Press, 1991.

Dean, Jonathan. *Meeting Gorbachev's Challenge: How to Build Down the NATO–Warsaw Pact Confrontation.* New York: St. Martin's, 1989.

——. *Watershed in Europe: Dismantling the East-West Military Confrontation.* Lexington, Mass.: Lexington Books, 1987.

Dean, Vera Micheles, and Howard C. Gary. "Military and Economic Strength of Western Europe." *Foreign Policy Reports* 26 (15 Oct. 1950): 118–28.

De Vos, Luc. "The Scutum Belgarum: The 1 (BE) Corps in Germany, 1945–1991." In *Homeward Bound? Allied Forces in the New Germany*, edited by David G. Haglund and Olaf Mager. Boulder, Colo.: Westview, 1992.

Dingman, Roger V. "Theories of, and Approaches to, Alliance Politics." In *Diplomacy: New Approaches in History, Theory, and Policy*, edited by Paul Gordon Lauren. New York: Free Press, 1979.

Dockrill, Saki. *Britain's Policy for West German Rearmament, 1950–1955*. Cambridge, Eng.: Cambridge University Press, 1991.

———. "The Evolution of Britain's Policy Towards a European Army." *Journal of Strategic Studies* 12 (Mar. 1989): 38–62.

Duffield, John S. "International Regimes and Alliance Behavior: Explaining NATO Conventional Force Levels." *International Organization* 46 (Autumn 1992): 819–55.

———. "The Soviet Military Threat to Western Europe: U.S. Estimates in the 1950s and 1960s." *Journal of Strategic Studies* 15 (June 1992): 208–27.

Dulles, John Foster. "Challenge and Response in United States Policy." *Foreign Affairs* 36 (Oct. 1957): 25–43.

———. "The Evolution of Foreign Policy." *Department of State Bulletin* 30 (25 Jan. 1954): 107–10.

———. "Policy for Security and Peace." *Foreign Affairs* 32 (Apr. 1954): 353–64.

Eckstein, Harry. "Case Study and Theory in Political Science." In *Handbook of Political Science*, Vol. 7, edited by Fred I. Greenstein and Nelson W. Polsby. Reading, Mass.: Addison-Wesley, 1975.

Eisenhower, Dwight D. "Let's Be Honest with Ourselves." *Saturday Evening Post*, Oct. 26, 1963, pp. 26–27.

———. *The White House Years: Mandate for Change, 1953–1956*. Garden City, N.Y.: Doubleday, 1963.

Enthoven, Alain C. "U.S. Forces in Europe: How Many? Doing What?" *Foreign Affairs* 53 (Apr. 1975): 513–32.

Enthoven, Alain C., and K. Wayne Smith. *How Much Is Enough? Shaping the Defense Program, 1961–1969*. New York: Harper & Row, 1971.

———. "What Forces for NATO? And from Whom?" *Foreign Affairs* 48 (Oct. 1969): 80–96.

Epstein, Joshua M. *Conventional Force Reductions: A Dynamic Assessment*. Washington, D.C.: Brookings Institution, 1990.

Etzold, Thomas H. "The End of the Beginning . . . NATO's Adoption of Nuclear Strategy." In *Western Security: The Formative Years: European and Atlantic Defence, 1947–1953*, edited by Olav K. Riste. New York: Columbia University Press, 1985.

Facer, Roger L. L. *Conventional Forces and the NATO Strategy of Flexible Response: Issues and Approaches*. R-3209-FF. Santa Monica, Calif.: RAND Corporation, 1985.

Federal Minister of Defense. *White Paper 1970: The Security of the Federal Republic of Germany and the Development of the Federal Armed Forces*. Bonn: Federal Minister of Defense, 1970.

———. *White Paper 1975/1976: The Security of the Federal Republic of Germany and the Development of the Federal Armed Forces*. Bonn: Federal Minister of Defense, 1976.

——. *White Paper 1979: The Security of the Federal Republic of Germany and the Development of the Federal Armed Forces.* Bonn: Federal Minister of Defense, 1979.

——. *White Paper 1983: The Security of the Federal Republic of Germany.* Bonn: Federal Minister of Defense, 1983.

Feinstein, Lee. "Soviet Cutbacks After the Revolutions." *Arms Control Today* 22 (July–Aug. 1990): 8–12.

Flanagan, Stephen J. "NATO and Central and Eastern Europe: From Liaison to Security Partnership." *Washington Quarterly* 15 (Spring 1992): 141–52.

——. *NATO's Conventional Defense.* Cambridge, Mass.: Ballinger, 1988.

Foreign Policy Research Institute. *The Three Per Cent Solution.* Philadelphia: Foreign Policy Research Institute, 1981.

Fursdon, Edward. *The European Defence Community: A History.* New York: St. Martin's, 1980.

Gaddis, John Lewis. "Containment and the Logic of Strategy." *National Interest*, no. 10 (Winter 1987–88): 27–38.

——. *Strategies of Containment: A Critical Appraisal of Postwar American National Security Policy.* New York: Oxford University Press, 1982.

Galvin, General John R. "Allied Command Europe—Buttressing the Means." *NATO's Sixteen Nations* 32 (Aug. 1987): 14–16.

——. "The INF Treaty—No Relief from the Burden of Defense." *NATO Review* 36 (Feb. 1988): 1–7.

Garthoff, Raymond L. *Detente and Confrontation: American-Soviet Relations from Nixon to Reagan.* Washington, D.C.: Brookings Institution, 1985.

——. "New Thinking in Soviet Military Doctrine." *Washington Quarterly* 11 (Summer 1988): 131–58.

Gellner, Charles R. "The Mansfield Proposals to Reduce U.S. Troops in Western Europe, 1967–1974." Washington, D.C.: Congressional Research Service, 1 Dec. 1982.

George, Alexander L. "Case Studies and Theory Development: The Method of Structured, Focused Comparison." In *Diplomacy: New Approaches in History, Theory, and Policy*, edited by Paul Gordon Lauren. New York: Free Press, 1979.

——. "The Causal Nexus Between Cognitive Beliefs and Decision-Making Behavior: The 'Operational Code' Belief System." In *Psychological Models in International Politics*, edited by Lawrence S. Falkowski. Boulder, Colo.: Westview, 1979.

George, Alexander L., Philip J. Farley, and Alexander Dallin, eds. *U.S.-Soviet Security Cooperation.* New York: Oxford University Press, 1988.

Ghebali, Victor-Yves. "Putting Europe Back Together." *International Defense Review* 22 (June 1989): 743–45.

Gilpin, Robert. *War and Change in World Politics.* Cambridge, Eng.: Cambridge University Press, 1981.

Golden, James. *The Dynamics of Change in NATO: A Burden-Sharing Perspective.* New York: Praeger, 1983.

Goodpaster, Andrew J. "The Development of SHAPE, 1950–1953." *International Organization* 9 (May 1955): 257–60.

Gordon, Lincoln. "Economic Aspects of Coalition Diplomacy—The NATO Experience." *International Organization* 10 (Oct. 1956): 529–43.

Gowing, Margaret. *Independence and Deterrence: Britain and Atomic Energy, 1945–1952*. Vol. 1. New York: Macmillan, 1974.

Greiner, Christian. "The Defence of Western Europe and the Rearmament of West Germany, 1947–1950." In *Western Security: The Formative Years: European and Atlantic Defence, 1947–1953*, edited by Olav Riste. New York: Columbia University Press, 1985.

Grosser, Alfred. *The Western Alliance: European-American Relations Since 1945*. New York: Vintage, 1982.

Gundersen, H. F. Zeiner. "Military Perspective." *NATO Review* 26 (June 1978): 5–9.

Haas, Peter M. "Do Regimes Matter? Epistemic Communities and Mediterranean Pollution Control." *International Organization* 43 (Summer 1989): 377–404.

Haggard, Stephan, and Beth A. Simmons. "Theories of International Regimes." *International Organization* 41 (Summer 1987): 491–517.

Haig, Alexander M., Jr. "NATO and the Security of the West." *NATO Review* 26 (Aug. 1978): 8–11.

Halperin, Morton H. *Bureaucratic Politics and Foreign Policy*. Washington, D.C.: Brookings Institution, 1974.

Hammond, Paul Y. "NSC 68." In *Strategy, Politics, and Defense Budgets*, edited by Warner R. Schilling, Paul Y. Hammond, and Glenn H. Snyder. New York: Columbia University Press, 1962.

Hanrieder, Wolfram F. *Germany, America, Europe: Forty Years of German Foreign Policy*. New Haven: Yale University Press, 1989.

Harrison, Michael M. *The Reluctant Ally: France and Atlantic Security*. Baltimore: Johns Hopkins University Press, 1981.

Healey, Denis. "Britain and NATO." In *NATO and American Security*, edited by Klaus Knorr. Princeton: Princeton University Press, 1959.

Hilsman, Roger. "NATO: The Developing Strategic Context." In *NATO and American Security*, edited by Klaus Knorr. Princeton: Princeton University Press, 1959.

Hoag, Malcolm. "NATO: Deterrent or Shield?" *Foreign Affairs* 36 (Jan. 1958): 278–92.

———. "The Place of Limited War in NATO Strategy." In *NATO and American Security*, edited by Klaus Knorr. Princeton: Princeton University Press, 1959.

Hodgson, Godfrey. "The Establishment." *Foreign Policy*, no. 10 (Spring 1973): 3–40.

Holloway, David. "Gorbachev's New Thinking." *Foreign Affairs* 68 (America and the World 1988–89): 66–81.

Holsti, Ole R., P. Terrence Hopmann, and John D. Sullivan. *Unity and Dis-*

integration in International Alliances: Comparative Studies. New York: Wiley, 1973.

Honig, Jan Willem. *Defense Policy in the North Atlantic Alliance: The Case of the Netherlands.* Westport, Conn.: Praeger, 1993.

Hunt, Kenneth. "NATO Without France: The Military Implications." *Adelphi Papers*, no. 32. London: Institute for Strategic Studies, 1966.

Huntington, Samuel P. *The Common Defense: Strategic Programs in National Politics.* New York: Columbia University Press, 1961.

———. "Conventional Deterrence and Conventional Retaliation in Europe." *International Security* 8 (Winter 1983–84): 32–56.

Ikenberry, G. John. "Conclusion: An Institutional Approach to American Foreign Economic Policy." In *The State and American Foreign Economic Policy*, edited by G. John Ikenberry, David A. Lake, and Michael Mastanduno. Ithaca, N.Y.: Cornell University Press, 1988.

International Institute for Strategic Studies. *The Military Balance.* London: International Institute for Strategic Studies, various years.

———. *Strategic Survey.* London: International Institute for Strategic Studies, various years.

Ireland, Timothy P. *Creating the Entangling Alliance: The Origins of the North Atlantic Treaty Organization.* Westport, Conn.: Greenwood, 1981.

Ismay, Lord Hastings. *NATO: The First Five Years, 1949–1954.* Paris: North Atlantic Treaty Organization, 1954.

Jervis, Robert. "The Impact of the Korean War on the Cold War." *Journal of Conflict Resolution* 24 (Dec. 1980): 563–92.

———. *Perception and Misperception in International Politics.* Princeton: Princeton University Press, 1976.

———. "Realism, Game Theory, and Cooperation." *World Politics* 40 (Apr. 1988): 317–49.

———. "Security Regimes." In *International Regimes*, edited by Stephen D. Krasner. Ithaca, N.Y.: Cornell University Press, 1983.

Joffe, Josef. "Europe's American Pacifier." *Foreign Policy*, no. 54 (Spring 1984): 64–82.

———. *The Limited Partnership: Europe, the United States, and the Burdens of Alliance.* Cambridge, Mass.: Ballinger, 1987.

Kaiser, Karl. "Germany's Unification." *Foreign Affairs* 70 (America and the World 1990–91): 179–205.

Kaiser, Karl, Georg Leber, Alois Mertes, and Franz-Josef Schulze. "Nuclear Weapons and the Preservation of Peace." *Foreign Affairs* 60 (Summer 1982): 1155–70.

Kaplan, Fred. *The Wizards of Armageddon.* New York: Simon & Schuster, 1983.

Kaplan, Lawrence S. *A Community of Interests: NATO and the Military Assistance Program, 1948–1951.* Washington, D.C.: U.S. Government Printing Office, 1980.

———. "The NATO-Indochina Connection, 1952–1954." Paper presented at the conference on the United States and West European Security, 1950–1955, Cambridge, Mass., Dec. 1987.

Kaplan, Lawrence S., and Kathleen A. Kellner. "Lemnitzer: Surviving the French Military Withdrawal." In *Generals in International Politics*, edited by Robert S. Jordan. Lexington: University Press of Kentucky, 1987.

Karber, Philip A. "NATO Doctrine and National Operational Priorities: The Central Front and the Flanks, Part I." In *Power and Policy: Doctrine, the Alliance, and Arms Control*. Part 3. *Adelphi Papers*, no. 207. London: International Institute for Strategic Studies, 1986.

Kaufmann, William W. *The McNamara Strategy*. New York: Harper & Row, 1964.

———. "The Requirements of Deterrence." In *Military Policy and National Security*, edited by William W. Kaufmann. Princeton: Princeton University Press, 1956.

Kelleher, Catherine McArdle. *Germany and the Politics of Nuclear Weapons*. New York: Columbia University Press, 1975.

Keliher, John G. *The Negotiations on Mutual and Balanced Force Reductions: The Search for Arms Control in Europe*. New York: Pergamon, n.d.

Kennedy, John F. *The Strategy of Peace*. New York: Harper, 1960.

Kennedy, Robert. "Precision ATGMs and NATO Defense." *Orbis* 22 (Winter 1979): 897–927.

Keohane, Robert O. *After Hegemony: Cooperation and Discord in the World Political Economy*. Princeton: Princeton University Press, 1984.

———. "Multilateralism: An Agenda for Research." *International Journal* 45 (Autumn 1990): 731–64.

———, ed. *Neorealism and Its Critics*. New York: Columbia University Press, 1986.

Kinnard, Douglas. *President Eisenhower and Strategy Management: A Study in Defense Politics*. Lexington: University Press of Kentucky, 1977.

Kissinger, Henry. *The White House Years*. Boston: Little, Brown, 1979.

Knorr, Klaus. *The Power of Nations: The Political Economy of International Relations*. New York: Basic Books, 1975.

———. "Threat Perception." In *Historical Dimensions of National Security*, edited by Klaus Knorr. Lawrence: University Press of Kansas, 1976.

———, ed. *NATO and American Security*. Princeton: Princeton University Press, 1959.

Knowlton, Lt. Col. William A. "Early Stages in the Organization of 'SHAPE.'" *International Organization* 13 (Winter 1959): 1–18.

Kohl, Wilfrid L. *French Nuclear Diplomacy*. Princeton: Princeton University Press, 1971.

———. "The Nixon-Kissinger Foreign Policy System and U.S.-European Relations." *World Politics* 28 (Oct. 1975): 1–43.

Komer, Robert W. "Is Conventional Defense of Europe Feasible?" *Naval War College Review* 35 (Sept.–Oct. 1982): 80–91.

———. "The Origins and the Objectives." *NATO Review* 26 (June 1978): 9–12.

———. "Ten Suggestions for Rationalizing NATO." *Survival* 19 (Mar.–Apr. 1977): 67–72.

———. "Treating NATO's Self-Inflicted Wound." *Foreign Policy*, no. 13 (Winter 1973–74): 34–48.

Komer, R. W., E. W. Boyd, H. J. McChrystal, E. L. Schwab, and W. E. Simms. *Alliance Defense in the Eighties.* R-1980-ARPA/ISA/DP&E. Santa Monica, Calif.: RAND Corporation, 1976.

Krasner, Stephen D. "Approaches to the State: Conceptions and Historical Dynamics." *Comparative Politics* 16 (Jan. 1984): 223–46.

———. "Regimes and the Limits of Realism: Regimes as Autonomous Variables." In *International Regimes*, edited by Stephen D. Krasner. Ithaca, N.Y.: Cornell University Press, 1983.

———. "Sovereignty: An Institutional Perspective." *Comparative Political Studies* 21 (Apr. 1988): 66–94.

———. "Structural Causes and Regime Consequences: Regimes as Intervening Variables." In *International Regimes*, edited by Stephen D. Krasner. Ithaca, N.Y.: Cornell University Press, 1983.

———, ed. *International Regimes*. Ithaca, N.Y.: Cornell University Press, 1983.

Kratochwil, Friedrich, and John Gerard Ruggie. "International Organization: A State of the Art on an Art of the State." *International Organization* 40 (Autumn 1986): 753–75.

Kugler, Richard L. *Commitment to Purpose: How Alliance Partnership Won the Cold War.* Santa Monica, Calif.: RAND Corporation, 1993.

———. *The Great Strategy Debate: NATO's Evolution in the 1960s.* N-3252-FF/RC. Santa Monica, Calif.: RAND Corporation, 1991.

———. *Laying the Foundations: The Evolution of NATO in the 1950s.* N-3105-FF/RC. Santa Monica, Calif.: RAND Corporation, 1990.

Kupchan, Charles A. "NATO and the Persian Gulf: Examining Intra-alliance Behavior." *International Organization* 42 (Spring 1988): 317–46.

Lauren, Paul Gordon, ed. *Diplomacy: New Approaches in History, Theory, and Policy.* New York: Free Press, 1979.

Lawrence, Richard D., and Jeffrey Record. *U.S. Force Structure in NATO: An Alternative.* Washington, D.C.: Brookings Institution, 1974.

Leffler, Melvyn P. *A Preponderance of Power: National Security, the Truman Administration, and the Cold War.* Stanford: Stanford University Press, 1992.

Legge, J. Michael. *Theater Nuclear Weapons and the NATO Strategy of Flexible Response.* R-2964-FF. Santa Monica, Calif.: RAND Corporation, 1983.

Legvold, Robert. "The Revolution in Soviet Foreign Policy." *Foreign Affairs* 68 (America and the World 1988–89): 82–98.

Leitenberg, Milton. "Background Information on Tactical Nuclear Weapons (Primarily in a European Context)." In *Tactical Nuclear Weapons: European Perspectives*, Stockholm International Peace Research Institute (SIPRI). London: Taylor and Francis, 1978.

Lepgold, Joseph. *The Declining Hegemon: The United States and European Defense, 1960–1990.* New York: Greenwood Press, 1990.

Liddell Hart, B. H. "The State and Prospect of Europe's Defense." *Virginia Quarterly Review* 29 (Spring 1953): 161–74.

Lipson, Charles. "International Cooperation in Economic and Security Affairs." *World Politics* 37 (Oct. 1984): 1–23.

Lowe, Karl H. "US Mobilization for Reinforcing Western Europe." In *NATO-Warsaw Pact Force Mobilization*, edited by Jeffrey Simon. Washington, D.C.: National Defense University Press, 1989.

Lowe, Karl, and Thomas-Durell Young. "Multinational Corps in NATO." *Survival* 33 (Jan.–Feb. 1991): 66–77.

Mager, Olaf. "The Continental Commitment: Britain's Forces in Germany." In *Homeward Bound? Allied Forces in the New Germany*, edited by David G. Haglund and Olaf Mager. Boulder, Colo.: Westview, 1992.

Mako, William P. *U.S. Ground Forces and the Defense of Central Europe.* Washington, D.C.: Brookings Institution, 1983.

Martin, Laurence W. "The American Decision to Rearm Germany." In *American Civil-Military Decisions*, edited by Harold Stein. Birmingham: University of Alabama Press, 1963.

McGeehan, Robert. *The German Rearmament Question: American Diplomacy and European Defense After World War II.* Urbana: University of Illinois Press, 1971.

McLellan, David S. *Dean Acheson: The State Department Years.* New York: Dodd, Mead, 1976.

McNamara, Robert S. "The Military Roles of Nuclear Weapons: Perceptions and Misperceptions." *Foreign Affairs* 60 (Fall 1983): 59–80.

Mearsheimer, John J. *Conventional Deterrence.* Ithaca, N.Y.: Cornell University Press, 1983.

———. "Nuclear Weapons and Deterrence in Europe." *International Security* 9 (Winter 1984–85): 19–46.

———. "Precision-Guided Munitions and Conventional Deterrence." *Survival* 31 (Mar.–Apr. 1979): 68–76.

Melandri, Pierre. "France and the Atlantic Alliance, 1950–1953." In *Western Security: The Formative Years: European and Atlantic Defence, 1947–1953*, edited by Olav Riste. New York: Columbia University Press, 1985.

Mendl, Wolf. *Deterrence and Persuasion: French Nuclear Armament in the Context of National Policy, 1945–1969.* New York: Praeger, 1970.

Meyer, Stephen M. "The Sources and Prospects of Gorbachev's New Political Thinking on Security." *International Security* 13 (Fall 1988): 124–63.

Midgley, John J., Jr. *Deadly Illusions: Army Policy for the Nuclear Battlefield.* Boulder, Colo.: Westview, 1986.

Milner, Helen V. *Resisting Protection: Global Industries and the Politics of International Trade.* Princeton: Princeton University Press, 1988.

Moodie, Michael, and Brent Fischmann. "An Action Plan for NATO." *NATO Review* 36 (Feb. 1988): 19–24.

Moravcsik, Andrew M. "Disciplining Trade Finance: The OECD Export

Credit Arrangement." *International Organization* 43 (Winter 1989): 173–205.

Morgenthau, Hans J. "Alliances in Theory and Practice." In *Alliance Policy in the Cold War*, edited by Arnold Wolfers. Baltimore: Johns Hopkins Press, 1959.

———. *Politics Among Nations: The Struggle for Power and Peace.* 4th ed. New York: Knopf, 1967.

Mumford, Bill. "Foreword: NATO Defence at the Summit." *NATO Review* 26 (June 1978): 3–5.

Murphy, Charles J. V. "A New Strategy for NATO." *Fortune* 47 (Jan. 1953): 80–85, 166–70.

NATO Information Service. *The Eurogroup.* Brussels: NATO Information Service, 1975.

———. *NATO Final Communiques, 1949–1974.* Brussels: NATO Information Service, n.d.

———. *NATO Final Communiques, 1975–1980.* Brussels: NATO Information Service, n.d.

———. *The North Atlantic Treaty Organisation: Facts and Figures.* Brussels: NATO Information Service, various years.

Neustadt, Richard E. *Alliance Politics.* New York: Columbia University Press, 1970.

Newhouse, John. *De Gaulle and the Anglo-Saxons.* New York: Viking, 1970.

Newhouse, John, with Melvin Croan, Edward R. Fried, and Timothy W. Stanley. *U.S. Troops in Europe: Issues, Costs, and Choices.* Washington, D.C.: Brookings Institution, 1971.

Nicholls, David. "Long Term Defence Planning." *NATO's Fifteen Nations* 26 (Apr.–May 1981): 26–32.

Nitze, Paul. H. *From Hiroshima to Glasnost: At the Center of Decision.* New York: Grove Weidenfeld, 1989.

Nixon, Richard M. *U.S. Foreign Policy for the 1970s: A New Strategy for Peace.* Washington, D.C.: U.S. Government Printing Office, 1970.

Norstad, Lauris. "NATO: Deterrent and Shield." *NATO Letter* 5 (Feb. 1957): 27–30.

———. "NATO: Strength and Spirit." *NATO Letter* 8 (Jan. 1960): 7–11.

———. "Text of General Norstad's Cincinnati Speech." *NATO Letter* 5 (Dec. 1957): 26–28.

Nunn, Sam. "Address by Senator Sam Nunn, 11 September 1976." *Survival* 19 (Jan.–Feb. 1977): 31–32.

———. "Deterring War in Europe: Some Basic Assumptions Need Revising." *NATO Review* 25 (Feb. 1977): 4–7.

Nye, Joseph S., Jr. "Nuclear Learning and U.S.-Soviet Security Regimes." *International Organization* 41 (Summer 1987): 371–402.

Odell, John S. *U.S. Monetary Policy: Markets, Power, and Ideas as Sources of Change.* Princeton: Princeton University Press, 1982.

Olson, Mancur, Jr., and Richard Zeckhauser. "An Economic Theory of Alliances." *Review of Economics and Statistics* 48 (Aug. 1966): 266–79.

Oneal, John R. "Testing the Theory of Collective Action: NATO Defense Burdens, 1950–1984." *Journal of Conflict Resolution* 34 (Sept. 1990): 426–48.

Oppenheimer, Joe. "Collective Goods and Alliances: A Reassessment." *Journal of Conflict Resolution* 23 (Sept. 1979): 387–407.

Osgood, Robert E. *NATO: The Entangling Alliance.* Chicago: University of Chicago Press, 1962.

Ostrom, Elinor. "An Agenda for the Study of Institutions." *Public Choice* 48 (1986): 3–25.

Oye, Kenneth A., ed. *Cooperation Under Anarchy.* Princeton: Princeton University Press, 1986.

Pierre, Andrew J. *Nuclear Politics: The British Experience with an Independent Nuclear Force, 1939–1970.* New York: Oxford University Press, 1972.

———, ed. *The Conventional Defense of Europe: New Technologies and New Strategies.* New York: Council on Foreign Relations, 1985.

Pond, Elizabeth. *After the Wall: American Policy Toward Germany.* New York: Priority Press, 1991.

———. "A Wall Destroyed: The Dynamics of German Unification in the GDR." *International Security* 15 (Fall 1990): 35–66.

Poole, Walter S. *The History of the Joint Chiefs of Staff: The Joint Chiefs of Staff and National Policy.* Vol. 4, *1950–1952.* Wilmington, Del.: Michael Glazier, 1980.

Posen, Barry R. *The Sources of Military Doctrine: France, Britain, and Germany Between the World Wars.* Ithaca, N.Y.: Cornell University Press, 1984.

Quinlan, Michael. "The LTDP from a National Viewpoint." *NATO Review* 26 (June 1978): 13–15.

Raj, Christopher S. *American Military in Europe: Controversy over NATO Burden Sharing.* New Delhi: ABC, 1983.

Rearden, Steven L. *History of the Office of the Secretary of Defense.* Vol. 1, *The Formative Years, 1947–1950.* Washington, D.C.: U.S. Government Printing Office, 1984.

Record, Jeffrey. *U.S. Nuclear Weapons in Europe: Issues and Alternatives.* Washington, D.C.: Brookings Institution, 1974.

Rempel, Roy. "Canada's Troop Deployment in Germany: Twilight of a Forty-Year Presence?" In *Homeward Bound! Allied Forces in the New Germany,* edited by David G. Haglund and Olav Mager. Boulder, Colo.: Westview, 1992.

Richardson, James. L. *Germany and the Atlantic Alliance: The Interaction of Strategy and Politics.* Cambridge, Mass.: Harvard University Press, 1966.

Richardson, Robert C. III. "NATO Nuclear Strategy: A Look Back." *Strategic Review* 9 (Spring 1981): 35–43.

Riste, Olav K. *Western Security: The Formative Years: European and Atlantic Defence, 1947–1953.* New York: Columbia University Press, 1985.

Rogers, Bernard. "The Atlantic Alliance: Prescriptions for a Difficult Decade." *Foreign Affairs* 60 (Summer 1982): 1145–56.

——. "Follow-on Forces Attack (FOFA): Myths and Reality." *NATO Review* 32 (Dec. 1984): 1–9.

——. "NATO's Conventional Defense Improvement Initiative: A New Approach to an Old Challenge." *NATO's Sixteen Nations* 31 (July 1986).

Rosenau, James N. "Before Cooperation: Hegemons, Regimes, and Habit-Driven Actors in World Politics." *International Organization* 40 (Autumn 1986): 849–94.

Rowny, Edward Leon. *Decision-Making Process in NATO*. Ann Arbor: Xerox University Microfilms, 1977.

Ruiz Palmer, Diego A. "National Contributions." *Armed Forces Journal International* 121 (May 1984): 27, 52–77.

Sagan, Scott C. *Moving Targets*. Princeton: Princeton University Press, 1989.

Schelling, Thomas C. *Arms and Influence*. New Haven: Yale University Press, 1966.

——. *The Strategy of Conflict*. Cambridge, Mass.: Harvard University Press, 1960.

Schlesinger, Arthur M., Jr. *A Thousand Days: John F. Kennedy in the White House*. Boston: Houghton Mifflin, 1965.

Schmidt, Helmut. *A Grand Strategy for the West: The Anachronism of National Strategies in an Interdependent World*. New Haven: Yale University Press, 1985.

Schwartz, David N. *NATO's Nuclear Dilemmas*. Washington, D.C.: Brookings Institution, 1983.

Schwartz, Thomas Alan. *America's Germany: John J. McCloy and the Federal Republic of Germany*. Cambridge, Mass.: Harvard University Press, 1991.

——. "The 'Skeleton Key'—American Foreign Policy, European Unity, and German Rearmanent, 1949–54." *Central European History* 19 (Dec. 1986): 369–85.

Siccama, Jan Geert. "Enduring Interests: The Netherlands and the Stationing Issue." In *Homeward Bound? Allied Forces in the New Germany*, edited by David G. Haglund and Olaf Mager. Boulder, Colo.: Westview, 1992.

Simon, Herbert A. "Human Nature in Politics: The Dialogue of Psychology with Political Science." *American Political Science Review* 79 (June 1985): 293–304.

Sloan, Stanley R. *NATO's Future: Toward a New Transatlantic Bargain*. Washington, D.C.: National Defense University Press, 1985.

Snidal, Duncan. "The Limits of Hegemonic Stability Theory." *International Organization* 39 (Autumn 1985): 580–614.

Snyder, Glenn H. "Alliance Theory: A Neorealist First Cut." *Journal of International Affairs* 44 (Spring–Summer 1990): 103–23.

——. *Deterrence and Defense: Toward a Theory of National Security*. Princeton: Princeton University Press, 1962.

——. "The 'New Look' of 1953." In *Strategy, Politics, and Defense Budgets*, Warner R. Schilling, Paul Y. Hammond, and Glenn H. Snyder. New York: Columbia University Press, 1962.

——. "The Security Dilemma in Alliance Politics." *World Politics* 36 (July 1984): 461–95.

Snyder, Jack. "The Gorbachev Revolution: Limiting Offensive Conventional Forces in Europe." *International Security* 12 (Spring 1988): 48–77.

——. "The Gorbachev Revolution: A Waning of Soviet Expansionism?" *International Security* 12 (Winter 1987–88): 93–131.

——. *The Ideology of the Offensive: Military Decision Making and the Disasters of 1914.* Ithaca, N.Y.: Cornell University Press, 1984.

Sorensen, Theodore C. *Kennedy.* New York: Harper & Row, 1965.

Sorley, Lewis. "Goodpaster: Maintaining Deterrence During Detente." In *Generals in International Politics: NATO's Supreme Allied Commander, Europe,* edited by Robert S. Jordan. Lexington: University Press of Kentucky, 1987.

Speier, Hans. *German Rearmanent and Atomic War: The Views of German Military and Political Leaders.* Evanston, Ill.: Row, Peterson, 1957.

Stanley, Timothy. *NATO in Transition: The Future of the Atlantic Alliance.* New York: Praeger for the Council on Foreign Relations, 1965.

Stebbins, Richard P. *The United States in World Affairs, 1951.* New York: Council on Foreign Relations, 1952.

——. *The United States in World Affairs, 1956.* New York: Council on Foreign Relations, 1957.

——. *The United States in World Affairs, 1957.* New York: Council on Foreign Relations, 1958.

Steering Group of the European Security Study. *Strengthening Conventional Deterrence in Europe: Proposals for the 1980's.* Report of the European Security Study. New York: St. Martin's, 1983.

Steinberg, James B. "Rethinking the Debate on Burden-Sharing." *Survival* 39 (Jan.–Feb. 1987): 56–78.

Steinbruner, John D. *The Cybernetic Theory of Decision: New Dimensions of Political Analysis.* Princeton: Princeton University Press, 1974.

Steininger, Rolf. "John Foster Dulles, the European Defense Community, and the German Question." In *John Foster Dulles and the Diplomacy of the Cold War,* edited by Richard H. Immerman. Princeton: Princeton University Press, 1990.

Stewart, James Moray. "Conventional Defence Improvements: Where Is the Alliance Going?" *NATO Review* 33 (Apr. 1985): 1–7.

Stromseth, Jane E. *The Origins of Flexible Response: NATO's Debate over Strategy in the 1960's.* New York: St. Martin's, 1988.

Summers, Robert, and Alan Heston. "A New Set of International Comparisons of Real Product and Price Level Estimates for 130 Countries, 1950–1985." *Review of Income and Wealth* 34 (Mar. 1988): 1–26.

Sutton, Boyd D., et al. "Deep Attack Concepts and the Defence of Central Europe." *Survival* 36 (Mar.–Apr. 1984): 50–69.

Taylor, Maxwell D. *The Uncertain Trumpet.* New York: Harper & Brothers, 1959.

Taylor, Trevor. "NATO and Central Europe." *NATO Review* 39 (Oct. 1991): 17–22.

Thies, Wallace J. "Alliances and Collective Goods." *Journal of Conflict Resolution* 31 (June 1987): 298–332.

Trachtenberg, Marc. *History and Strategy*. Princeton: Princeton University Press, 1991.

———. "A 'Wasting Asset': American Strategy and the Shifting Nuclear Balance, 1949–1954." *International Security* 13 (Winter 1988–89): 5–49.

Treverton, Gregory F. *The "Dollar Drain" and American Forces in Germany: Managing the Political Economy of Alliance*. Athens: Ohio University Press, 1978.

Van Orden, Geoffrey. "The Bundeswehr in Transition." *Survival* 33 (July–Aug. 1991): 352–70.

Voigt, Karsten. *Conventional Defence in Europe: A Comprehensive Evaluation*. Brussels: North Atlantic Assembly, 1985.

von Hassel, Kai-Uwe. "Detente Through Firmness." *Foreign Affairs* 42 (Jan. 1964): 184–94.

———. "Organizing Western Defense." *Foreign Affairs* 43 (Jan. 1965): 209–16.

Walker, Paul F. "Precision-Guided Weapons." *Scientific American* 245 (Aug. 1981): 21–29.

Wall, Irwin M. *The United States and the Making of Postwar France, 1945–1954*. New York: Cambridge University Press, 1991.

Walt, Stephen M. *The Origins of Alliances*. Ithaca, N.Y.: Cornell University Press, 1987.

Waltz, Kenneth N. "Reflections on *Theory of International Politics*: A Response to My Critics." In *Neorealism and Its Critics*, edited by Robert O. Keohane. New York: Columbia University Press, 1986.

———. *Theory of International Politics*. Reading, Mass.: Addison-Wesley, 1979.

Wampler, Robert A. "From Lisbon to M.C. 48." Paper presented at the conference on the United States and West European Security, 1950–1955, Cambridge, Mass., Dec. 1987.

———. "NATO Strategic Planning and Nuclear Weapons, 1950–1957." *Nuclear History Program Occasional Paper 6*. College Park: Center for International Security Studies at Maryland, University of Maryland, 1990.

Ward, Michael Don. "Research Gaps in Alliance Dynamics." *Monograph Series in World Affairs* 19, no. 1 (Denver: Graduate School of International Studies, University of Denver, 1982).

Warner, Edward L. III. "New Thinking and Old Realities in Soviet Defence Policy." *Survival* 31 (Jan.–Feb. 1989): 13–33.

Warner, Geoffrey. "The British Labour Government and the Atlantic Alliance, 1949–1951." In *Western Security: The Formative Years: European and Atlantic Defence, 1947–1953*, edited by Olav K. Riste. New York: Columbia University Press, 1985.

Watson, Robert J. *The History of the Joint Chiefs of Staff: The Joint Chiefs of Staff and National Policy*. Vol. 5, *1953–1954*. Washington, D.C.: U.S. Government Printing Office, 1987.

Watt, Angus. "The Hand of Friendship—The Military Contacts Programme." *NATO Review* 40 (Feb. 1992): 19–22.

Weinberger, Caspar W. *Fighting for Peace: Seven Critical Years in the Pentagon.* New York: Warner, 1990.

Wells, Samuel F., Jr. "The First Cold War Buildup: Europe in United States Strategy and Policy, 1950–1953." In *Western Security: The Formative Years: European and Atlantic Defence, 1947–1953,* edited by Olav K. Riste. New York: Columbia University Press, 1985.

Wendt, James C., and Nanette Brown, *Improving the NATO Force Planning Process: Lessons from Past Efforts.* R-3383-USDP. Santa Monica, Calif.: RAND Corporation, 1986.

Whitt, Darnell M. II. "Quarrel on the Rhine: The Trilateral Talks of 1966–1967." Ph.D. diss., Johns Hopkins University, 1977.

Wiggershaus, Norbert. "The Decision for a West German Defence Contribution." In *Western Security: The Formative Years: European and Atlantic Defence, 1947–1953,* edited by Olav K. Riste. New York: Columbia University Press, 1985.

Williams, Phil. "The Nunn Amendment, Burden-Sharing and US Troops in Europe." *Survival* 27 (Jan.–Feb. 1985): 2–10.

———. *The Senate and U.S. Troops in Europe.* New York: St. Martin's, 1985.

———. "Whatever Happened to the Mansfield Amendment?" *Survival* 18 (July–Aug. 1976): 146–53.

Williamson, Samuel R., Jr. *The Politics of Grand Strategy: Britain and France Prepare for War, 1904–1914.* Cambridge, Mass.: Harvard University Press, 1969.

Wilmot, Chester. "If NATO Had to Fight." *Foreign Affairs* 31 (Jan. 1953): 200–214.

Wolfe, Thomas W. *Soviet Power and Europe, 1945–1970.* Baltimore: Johns Hopkins University Press, 1970.

Yochelson, John. "The American Military Presence in Europe: Current Debate in the United States." *Orbis* 15 (Fall 1971): 784–807.

Yost, David S. *France and Conventional Defense in Central Europe.* Boulder, Colo.: Westview, 1985.

Young, Oran R. *International Cooperation: Building Regimes for Natural Resources and the Environment.* Ithaca, N.Y.: Cornell University Press, 1989.

Young, Robert J. *In Command of France: French Foreign Policy and Military Planning, 1933–1940.* Cambridge, Mass.: Harvard University Press, 1978.

Index

In this index an "f" after a number indicates a separate reference on the next page, and an "ff" indicates separate references on the next two pages. A continuous discussion over two or more pages is indicated by a span of page numbers, e.g., "57–59." *Passim* is used for a cluster of references in close but not consecutive sequence.

Library of Congress Cataloging-in-Publication Data

Duffield, John S.

Power rules : the evolution of NATO's conventional force posture / John S. Duffield

p. cm.

Includes bibliographical references and index.

ISBN 0-8047-2396-6 (cloth)

1. North Atlantic Treaty Organization—Armed Forces—History. 2. Warfare, Conventional. I. Title.

UA646.3.D817   1995

355'.031'091821—dc20

94-25006
CIP

♾ This book is printed on acid-free paper.

Printed in the USA
CPSIA information can be obtained
at www.ICGtesting.com
JSHW021322221024
72173JS00012B/1636/J

9 780804 723961